# CAMBRIDGE LIBRARY COLLECTION

*Books of enduring scholarly value*

## Physical Sciences

From ancient times, humans have tried to understand the workings of the world around them. The roots of modern physical science go back to the very earliest mechanical devices such as levers and rollers, the mixing of paints and dyes, and the importance of the heavenly bodies in early religious observance and navigation. The physical sciences as we know them today began to emerge as independent academic subjects during the early modern period, in the work of Newton and other 'natural philosophers', and numerous sub-disciplines developed during the centuries that followed. This part of the Cambridge Library Collection is devoted to landmark publications in this area which will be of interest to historians of science concerned with individual scientists, particular discoveries, and advances in scientific method, or with the establishment and development of scientific institutions around the world.

## The Foundations of Science

A member of the Académie française, Henri Poincaré (1854–1912) was one of the greatest mathematicians and theoretical physicists of the late nineteenth and early twentieth centuries. His discovery of chaotic motion laid the foundations of modern chaos theory, and he was acknowledged by Einstein as a key contributor in the field of special relativity. He earned his enduring reputation as a philosopher of mathematics and science with this elegantly written work, which was first published in French as three separate essays: *Science and Hypothesis* (1902), *The Value of Science* (1905), and *Science and Method* (1908). Poincaré asserts that much scientific work is a matter of convention, and that intuition and prediction play key roles. George Halsted's authorised 1913 English translation retains Poincaré's lucid prose style, presenting complex ideas for both professional scientists and those readers interested in the history of mathematics and the philosophy of science.

Cambridge University Press has long been a pioneer in the reissuing of out-of-print titles from its own backlist, producing digital reprints of books that are still sought after by scholars and students but could not be reprinted economically using traditional technology. The Cambridge Library Collection extends this activity to a wider range of books which are still of importance to researchers and professionals, either for the source material they contain, or as landmarks in the history of their academic discipline.

Drawing from the world-renowned collections in the Cambridge University Library and other partner libraries, and guided by the advice of experts in each subject area, Cambridge University Press is using state-of-the-art scanning machines in its own Printing House to capture the content of each book selected for inclusion. The files are processed to give a consistently clear, crisp image, and the books finished to the high quality standard for which the Press is recognised around the world. The latest print-on-demand technology ensures that the books will remain available indefinitely, and that orders for single or multiple copies can quickly be supplied.

The Cambridge Library Collection brings back to life books of enduring scholarly value (including out-of-copyright works originally issued by other publishers) across a wide range of disciplines in the humanities and social sciences and in science and technology.

# The Foundations of Science

*Science and Hypothesis,*
*The Value of Science,*
*Science and Method*

HENRI POINCARÉ

TRANSLATED BY GEORGE BRUCE HALSTED

CAMBRIDGE
UNIVERSITY PRESS

# CAMBRIDGE
## UNIVERSITY PRESS

University Printing House, Cambridge, CB2 8BS, United Kingdom

Cambridge University Press is part of the University of Cambridge.
It furthers the University's mission by disseminating knowledge in the pursuit of
education, learning and research at the highest international levels of excellence.

www.cambridge.org
Information on this title: www.cambridge.org/9781108069496

© in this compilation Cambridge University Press 2015

This edition first published 1913
This digitally printed version 2015

ISBN 978-1-108-06949-6 Paperback

# SCIENCE AND EDUCATION

A SERIES OF VOLUMES FOR THE PROMOTION OF
SCIENTIFIC RESEARCH AND EDUCATIONAL PROGRESS

EDITED BY J. McKEEN CATTELL

VOLUME I—THE FOUNDATIONS OF SCIENCE

# THE FOUNDATIONS OF SCIENCE

SCIENCE AND HYPOTHESIS
THE VALUE OF SCIENCE
SCIENCE AND METHOD

BY

## H. POINCARÉ

AUTHORIZED TRANSLATION BY

## GEORGE BRUCE HALSTED

WITH A SPECIAL PREFACE BY POINCARÉ, AND AN INTRODUCTION
BY JOSIAH ROYCE, HARVARD UNIVERSITY

THE SCIENCE PRESS
NEW YORK AND GARRISON, N. Y.
1913

PRESS OF
THE NEW ERA PRINTING COMPANY
LANCASTER, PA.

# CONTENTS

CONTENTS

## THE VALUE OF SCIENCE

### PART I. *The Mathematical Sciences*

### PART II. *The Physical Sciences*

# HENRI POINCARÉ

Sir George Darwin, worthy son of an immortal father, said, referring to what Poincaré was to him and to his work: "He must be regarded as the presiding genius—or, shall I say, my patron saint?"

Henri Poincaré was born April 29, 1854, at Nancy, where his father was a physician highly respected. His schooling was broken into by the war of 1870–71, to get news of which he learned to read the German newspapers. He outclassed the other boys of his age in all subjects and in 1873 passed highest into the École Polytechnique, where, like John Bolyai at Maros Vásárhely, he followed the courses in mathematics without taking a note and without the syllabus. He proceeded in 1875 to the School of Mines, and was *Nommé*, March 26, 1879. But he won his doctorate in the University of Paris, August 1, 1879, and was appointed to teach in the Faculté des Sciences de Caen, December 1, 1879, whence he was quickly called to the University of Paris, teaching there from October 21, 1881, until his death, July 17, 1912. So it is an error to say he started as an engineer. At the early age of thirty-two he became a member of l'Académie des Sciences, and, March 5, 1908, was chosen Membre de l'Académie Française. July 1, 1909, the number of his writings was 436.

His earliest publication was in 1878, and was not important. Afterward came an essay submitted in competition for the Grand Prix offered in 1880, but it did not win. Suddenly there came a change, a striking fire, a bursting forth, in February, 1881, and Poincaré tells us the very minute it happened. Mounting an omnibus, "at the moment when I put my foot upon the step, the idea came to me, without anything in my previous thoughts seeming to foreshadow it, that the transformations I had used to define the Fuchsian functions were identical with those of non-Euclidean geometry." Thereby was opened a perspective new and immense. Moreover, the magic wand of his whole

ix

life-work had been grasped, the Aladdin's lamp had been rubbed, non-Euclidean geometry, whose necromancy wâs to open up a new theory of our universe, whose brilliant exposition was commenced in his book *Science and Hypothesis,* which has been translated into six languages and has already had a circulation of over 20,000. The non-Euclidean notion is that of the possibility of alternative laws of nature, which in the Introduction to the *Électricité et Optique,* 1901, is thus put: "If therefore a phenomenon admits of a complete mechanical explanation, it will admit of an infinity of others which will account equally well for all the peculiarities disclosed by experiment."

The scheme of laws of nature so largely due to Newton is merely one of an infinite number of conceivable rational schemes for helping us master and make experience; it is *commode,* convenient; but perhaps another may be vastly more advantageous. The old conception of *true* has been revised. The first expression of the new idea occurs on the title page of John Bolyai's marvelous *Science Absolute of Space,* in the phrase "haud unquam a priori decidenda."

With bearing on the history of the earth and moon system and the origin of double stars, in formulating the geometric criterion of stability, Poincaré proved the existence of a previously unknown pear-shaped figure, with the possibility that the progressive deformation of this figure with increasing angular velocity might result in the breaking up of the rotating body into two detached masses. Of his treatise *Les Méthodes nouvelles de la Méchanique céleste,* Sir George Darwin says: "It is probable that for half a century to come it will be the mine from which humbler investigators will excavate their materials." Brilliant was his appreciation of Poincaré in presenting the gold medal of the Royal Astronomical Society. The three others most akin in genius are linked with him by the Sylvester medal of the Royal Society, the Lobachevski medal of the Physico-Mathematical Society of Kazan, and the Bolyai prize of the Hungarian Academy of Sciences. His work must be reckoned with the greatest mathematical achievements of mankind.

The kernel of Poincaré's power lies in an oracle Sylvester often quoted to me as from Hesiod: The whole is less than its part.

He penetrates at once the divine simplicity of the perfectly general case, and thence descends, as from Olympus, to the special concrete earthly particulars.

A combination of seemingly extremely simple analytic and geometric concepts gave necessary general conclusions of immense scope from which sprang a disconcerting wilderness of possible deductions. And so he leaves a noble, fruitful heritage.

Says Love: "His right is recognized now, and it is not likely that future generations will revise the judgment, to rank among the greatest mathematicians of all time."

GEORGE BRUCE HALSTED.

# SCIENCE AND HYPOTHESIS

# AUTHOR'S PREFACE TO THE TRANSLATION

I AM exceedingly grateful to Dr. Halsted, who has been so good as to present my book to American readers in a translation, clear and faithful.

Every one knows that this savant has already taken the trouble to translate many European treatises and thus has powerfully contributed to make the new continent understand the thought of the old.

Some people love to repeat that Anglo-Saxons have not the same way of thinking as the Latins or as the Germans; that they have quite another way of understanding mathematics or of understanding physics; that this way seems to them superior to all others; that they feel no need of changing it, nor even of knowing the ways of other peoples.

In that they would beyond question be wrong, but I do not believe that is true, or, at least, that is true no longer. For some time the English and Americans have been devoting themselves much more than formerly to the better understanding of what is thought and said on the continent of Europe.

To be sure, each people will preserve its characteristic genius, and it would be a pity if it were otherwise, supposing such a thing possible. If the Anglo-Saxons wished to become Latins, they would never be more than bad Latins; just as the French, in seeking to imitate them, could turn out only pretty poor Anglo-Saxons.

And then the English and Americans have made scientific conquests they alone could have made; they will make still more of which others would be incapable. It would therefore be deplorable if there were no longer Anglo-Saxons.

But continentals have on their part done things an Englishman could not have done, so that there is no need either for wishing all the world Anglo-Saxon.

Each has his characteristic aptitudes, and these aptitudes

should be diverse, else would the scientific concert resemble a quartet where every one wanted to play the violin.

And yet it is not bad for the violin to know what the violoncello is playing, and *vice versa.*

This it is that the English and Americans are comprehending more and more; and from this point of view the translations undertaken by Dr. Halsted are most opportune and timely.

Consider first what concerns the mathematical sciences. It is frequently said the English cultivate them only in view of their applications and even that they despise those who have other aims; that speculations too abstract repel them as savoring of metaphysic.

The English, even in mathematics, are to proceed always from the particular to the general, so that they would never have an idea of entering mathematics, as do many Germans, by the gate of the theory of aggregates. They are always to hold, so to speak, one foot in the world of the senses, and never burn the bridges keeping them in communication with reality. They thus are to be incapable of comprehending or at least of appreciating certain theories more interesting than utilitarian, such as the non-Euclidean geometries. According to that, the first two parts of this book, on number and space, should seem to them void of all substance and would only baffle them.

But that is not true. And first of all, are they such uncompromising realists as has been said? Are they absolutely refractory, I do not say to metaphysic, but at least to everything metaphysical?

Recall the name of Berkeley, born in Ireland doubtless, but immediately adopted by the English, who marked a natural and necessary stage in the development of English philosophy.

Is this not enough to show they are capable of making ascensions otherwise than in a captive balloon?

And to return to America, is not the *Monist* published at Chicago, that review which even to us seems bold and yet which finds readers?

And in mathematics? Do you think American geometers are concerned only about applications? Far from it. The part of the science they cultivate most devotedly is the theory of groups of substitutions, and under its most abstract form, the farthest removed from the practical.

Moreover, Dr. Halsted gives regularly each year a review of all productions relative to the non-Euclidean geometry, and he has about him a public deeply interested in his work. He has initiated this public into the ideas of Hilbert, and he has even written an elementary treatise on 'Rational Geometry,' based on the principles of the renowned German savant.

To introduce this principle into teaching is surely this time to burn all bridges of reliance upon sensory intuition, and this is, I confess, a boldness which seems to me almost rashness.

The American public is therefore much better prepared than has been thought for investigating the origin of the notion of space.

Moreover, to analyze this concept is not to sacrifice reality to I know not what phantom. The geometric language is after all only a language. Space is only a word that we have believed a thing. What is the origin of this word and of other words also? What things do they hide? To ask this is permissible; to forbid it would be, on the contrary, to be a dupe of words; it would be to adore a metaphysical idol, like savage peoples who prostrate themselves before a statue of wood without daring to take a look at what is within.

In the study of nature, the contrast between the Anglo-Saxon spirit and the Latin spirit is still greater.

The Latins seek in general to put their thought in mathematical form; the English prefer to express it by a material representation.

Both doubtless rely only on experience for knowing the world; when they happen to go beyond this, they consider their fore-knowledge as only provisional, and they hasten to ask its definitive confirmation from nature herself.

But experience is not all, and the savant is not passive; he does not wait for the truth to come and find him, or for a chance meeting to bring him face to face with it. He must go to meet it, and it is for his thinking to reveal to him the way leading thither. For that there is need of an instrument; well, just there begins the difference—the instrument the Latins ordinarily choose is not that preferred by the Anglo-Saxons.

For a Latin, truth can be expressed only by equations; it must obey laws simple, logical, symmetric and fitted to satisfy minds in love with mathematical elegance.

The Anglo-Saxon to depict a phenomenon will first be engrossed in making a *model,* and he will make it with common materials, such as our crude, unaided senses show us them. He also makes a hypothesis, he assumes implicitly that nature, in her finest elements, is the same as in the complicated aggregates which alone are within the reach of our senses. He concludes from the body to the atom.

Both therefore make hypotheses, and this indeed is necessary, since no scientist has ever been able to get on without them. The essential thing is never to make them unconsciously.

From this point of view again, it would be well for these two sorts of physicists to know something of each other; in studying the work of minds so unlike their own, they will immediately recognize that in this work there has been an accumulation of hypotheses.

Doubtless this will not suffice to make them comprehend that they on their part have made just as many; each sees the mote without seeing the beam; but by their criticisms they will warn their rivals, and it may be supposed these will not fail to render them the same service.

The English procedure often seems to us crude, the analogies they think they discover to us seem at times superficial; they are not sufficiently interlocked, not precise enough; they sometimes permit incoherences, contradictions in terms, which shock a geometric spirit and which the employment of the mathematical method would immediately have put in evidence. But most often it is, on the other hand, very fortunate that they have not perceived these contradictions; else would they have rejected their model and could not have deduced from it the brilliant results they have often made to come out of it.

And then these very contradictions, when they end by perceiving them, have the advantage of showing them the hypothetical character of their conceptions, whereas the mathematical method, by its apparent rigor and inflexible course, often inspires in us a confidence nothing warrants, and prevents our looking about us.

From another point of view, however, the two conceptions are very unlike, and if all must be said, they are very unlike because of a common fault.

The English wish to make the world out of what we see. I mean what we see with the unaided eye, not the microscope, nor that still more subtile microscope, the human head guided by scientific induction.

The Latin wants to make it out of formulas, but these formulas are still the quintessenced expression of what we see. In a word, both would make the unknown out of the known, and their excuse is that there is no way of doing otherwise.

And yet is this legitimate, if the unknown be the simple and the known the complex?

Shall we not get of the simple a false idea, if we think it like the complex, or worse yet if we strive to make it out of elements which are themselves compounds?

Is not each great advance accomplished precisely the day some one has discovered under the complex aggregate shown by our senses something far more simple, not even resembling it—as when Newton replaced Kepler's three laws by the single law of gravitation, which was something simpler, equivalent, yet unlike?

One is justified in asking if we are not on the eve of just such a revolution or one even more important. Matter seems on the point of losing its mass, its solidest attribute, and resolving itself into electrons. Mechanics must then give place to a broader conception which will explain it, but which it will not explain.

So it was in vain the attempt was made in England to construct the ether by material models, or in France to apply to it the laws of dynamic.

The ether it is, the unknown, which explains matter, the known; matter is incapable of explaining the ether.

POINCARÉ.

# INTRODUCTION

BY PROFESSOR JOSIAH ROYCE

HARVARD UNIVERSITY

THE treatise of a master needs no commendation through the words of a mere learner. But, since my friend and former fellow student, the translator of this volume, has joined with another of my colleagues, Professor Cattell, in asking me to undertake the task of calling the attention of my fellow students to the importance and to the scope of M. Poincaré's volume, I accept the office, not as one competent to pass judgment upon the book, but simply as a learner, desirous to increase the number of those amongst us who are already interested in the type of researches to which M. Poincaré has so notably contributed.

## I

The branches of inquiry collectively known as the Philosophy of Science have undergone great changes since the appearance of Herbert Spencer's *First Principles,* that volume which a large part of the general public in this country used to regard as the representative compend of all modern wisdom relating to the foundations of scientific knowledge. The summary which M. Poincaré gives, at the outset of his own introduction to the present work, where he states the view which the 'superficial observer' takes of scientific truth, suggests, not indeed Spencer's own most characteristic theories, but something of the spirit in which many disciples of Spencer interpreting their master's formulas used to conceive the position which science occupies in dealing with experience. It was well known to them, indeed, that experience is a constant guide, and an inexhaustible source both of novel scientific results and of unsolved problems; but the fundamental Spencerian principles of science, such as 'the persistence of force,' the 'rhythm of motion' and the rest, were treated by Spencer himself as demonstrably objective, although

9

indeed 'relative' truths, capable of being tested once for all by the 'inconceivability of the opposite,' and certain to hold true for the whole 'knowable' universe. Thus, whether one dwelt upon the results of such a mathematical procedure as that to which M. Poincaré refers in his opening paragraphs, or whether, like Spencer himself, one applied the 'first principles' to regions of less exact science, this confidence that a certain orthodoxy regarding the principles of science was established forever was characteristic of the followers of the movement in question. Experience, lighted up by reason, seemed to them to have predetermined for all future time certain great theoretical results regarding the real constitution of the 'knowable' cosmos. Whoever doubted this doubted 'the verdict of science.'

Some of us well remember how, when Stallo's 'Principles and Theories of Modern Physics' first appeared, this sense of scientific orthodoxy was shocked amongst many of our American readers and teachers of science. I myself can recall to mind some highly authoritative reviews of that work in which the author was more or less sharply taken to task for his ignorant presumption in speaking with the freedom that he there used regarding such sacred possessions of humanity as the fundamental concepts of physics. That very book, however, has quite lately been translated into German as a valuable contribution to some of the most recent efforts to reconstitute a modern 'philosophy of nature.' And whatever may be otherwise thought of Stallo's critical methods, or of his results, there can be no doubt that, at the present moment, if his book were to appear for the first time, nobody would attempt to discredit the work merely on account of its disposition to be agnostic regarding the objective reality of the concepts of the kinetic theory of gases, or on account of its call for a logical rearrangement of the fundamental concepts of the theory of energy. We are no longer able so easily to know heretics at first sight.

For we now appear to stand in this position: The control of natural phenomena, which through the sciences men have attained, grows daily vaster and more detailed, and in its details more assured. Phenomena men know and predict better than ever. But regarding the most general theories, and the most fundamental, of science, there is no longer any notable scientific orthodoxy. Thus, as knowledge grows firmer and wider, conceptual construction becomes less rigid. The field of the theoretical philosophy of nature—yes, the field of the logic of science—this whole region is to-day an open one. Whoever will work there must indeed accept the verdict of experience regarding what happens in the natural world. So far he is indeed bound. But he may undertake without hindrance from mere tradition the task of trying afresh to reduce what happens to conceptual unity. The circle-squares and the inventors of devices for perpetual motion are indeed still as unwelcome in scientific company as they were in the days when scientific orthodoxy was more rigidly defined; but that is not because the foundations of geometry are now viewed as completely settled, beyond controversy, nor yet because the 'persistence of force' has been finally so defined as to make the 'opposite inconceivable' and the doctrine of energy beyond the reach of novel formulations. No, the circle-squarers and the inventors of devices for perpetual motion are to-day discredited, not because of any unorthodoxy of their general philosophy of nature, but because their views regarding special facts and processes stand in conflict with certain equally special results of science which themselves admit of very various general theoretical interpretations. Certain properties of the irrational number $\pi$ are known, in sufficient multitude to justify the mathematician in declining to listen to the arguments of the circle-squarer; but, despite great advances, and despite the assured results of Dedekind, of Cantor, of Weierstrass and of various others, the general theory of the logic of the numbers, rational and irrational, still presents several important features of great obscurity; and the philosophy of the concepts of geometry yet remains, in several very notable respects, unconquered territory, despite the work of Hilbert and of Pieri, and of our author himself. The ordinary inventors of the perpetual motion machines still stand in conflict with accepted generalizations; but nobody knows as yet what the final form of the theory of energy will be, nor can any one say precisely what place the phenomena of the radioactive bodies will occupy in that theory. The alchemists would not be welcome workers in modern laboratories; yet some sorts of transformation and of evolution of the elements are to-day matters which theory can find it convenient, upon occasion, to treat as more or less exactly definable possibilities; while some newly observed phenomena tend to indicate, not indeed that the ancient hopes of the alchemists were well founded, but that the ultimate constitution of matter is something more fluent, less invariant, than the theoretical orthodoxy of a recent period supposed.    Again, regarding the foundations of biology, a theoretical orthodoxy grows less possible, less definable, less conceivable (even as a hope) the more knowledge advançes.    Once 'mechanism' and 'vitalism' were mutually contradictory theories regarding the ultimate constitution of living bodies.    Now they are obviously becoming more and more 'points of view,' diverse but not necessarily conflicting.    So far as you find it convenient to limit your study of vital processes to those phenomena which distinguish living matter from all other natural obects, you may assume, in the modern 'pragmatic' sense, the attitude of a 'neo-vitalist.'    So far, however, as you are able to lay stress, with good results, upon the many ways in which the life processes can be assimilated to those studied in physics and in chemistry, you work as if you were a partisan of 'mechanics.'    In any case, your special science prospers by reason of the empirical discoveries that you make.    And your theories, whatever they are, must not run counter to any positive empirical results.    But otherwise, scientific orthodoxy no longer predetermines what alone it is respectable for you to think about the nature of living substance.

This gain in the freedom of theory, coming, as it does, side by side with a constant increase of a positive knowledge of nature, lends itself to various interpretations, and raises various obvious questions.

## II

One of the most natural of these interpretations, one of the most obvious of these questions, may be readily stated.    Is not the lesson of all these recent discussions simply this, that general theories are simply vain, that a philosophy of nature is an idle dream, and that the results of science are coextensive with the range of actual empirical observation and of successful prediction? If this is indeed the lesson, then the decline of theoretical orthodoxy in science is—like the eclipse of dogma in religion—merely a further lesson in pure positivism, another proof that man does best when he limits himself to thinking about what can be found in human experience, and in trying to plan what can be done to make human life more controllable and more reasonable. What we are free to do as we please—is it any longer a serious business? What we are free to think as we please—is it of any further interest to one who is in search of truth? If certain general theories are mere conceptual constructions, which to-day are, and to-morrow are cast into the oven, why dignify them by the name of philosophy? Has science any place for such theories? Why be a 'neo-vitalist,' or an 'evolutionist,' or an 'atomist,' or an 'Energetiker'? Why not say, plainly: ''Such and such phenomena, thus and thus described, have been observed; such and such experiences are to be expected, since the hypotheses by the terms of which we are required to expect them have been verified too often to let us regard the agreement with experience as due merely to chance; so much then with reasonable assurance we know; all else is silence—or else is some matter to be tested by another experiment?'' Why not limit our philosophy of science strictly to such a counsel of resignation? Why not substitute, for the old scientific orthodoxy, simply a confession of ignorance, and a resolution to devote ourselves to the business of enlarging the bounds of actual empirical knowledge?

Such comments upon the situation just characterized are frequently made. Unfortunately, they seem not to content the very age whose revolt from the orthodoxy of traditional theory, whose uncertainty about all theoretical formulations, and whose vast wealth of empirical discoveries and of rapidly advancing special researches, would seem most to justify these very comments. Never has there been better reason than there is to-day to be content, if rational man could be content, with a pure positivism. The splendid triumphs of special research in the most various fields, the constant increase in our practical control over nature—these, our positive and growing possessions, stand in glaring contrast to the failure of the scientific orthodoxy of a former period to fix the outlines of an ultimate creed about the nature of the knowable universe. Why not 'take the cash and let the credit go'? Why pursue the elusive theoretical 'unification' any further, when what we daily get from our sciences is an increasing wealth of detailed information and of practical guidance?

As a fact, however, the known answer of our own age to these very obvious comments is a constant multiplication of new efforts towards large and unifying theories. If theoretical orthodoxy is no longer clearly definable, theoretical construction was never more rife. The history of the doctrine of evolution, even in its most recent phases, when the theoretical uncertainties regarding the 'factors of evolution' are most insisted upon, is full of illustrations of this remarkable union of scepticism in critical work with courage regarding the use of the scientific imagination. The history of those controversies regarding theoretical physics, some of whose principal phases M. Poincaré, in his book, sketches with the hand of the master, is another illustration of the consciousness of the time. Men have their freedom of thought in these regions; and they feel the need of making constant and constructive use of this freedom. And the men who most feel this need are by no means in the majority of cases professional metaphysicians—or students who, like myself, have to view all these controversies amongst the scientific theoreticians from without as learners. These large theoretical constructions are due, on the contrary, in a great many cases to special workers, who have been driven to the freedom of philosophy by the oppression of experience, and who have learned in the conflict with special problems the lesson that they now teach in the form of general ideas regarding the philosophical aspects of science.

Why, then, does science actually need general theories, despite the fact that these theories inevitably alter and pass away? What is the service of a philosophy of science, when it is certain that the philosophy of science which is best suited to the needs of one generation must be superseded by the advancing insight of the next generation? Why must that which endlessly grows, namely, man's knowledge of the phenomenal order of nature, be constantly united in men's minds with that which is certain to decay, namely, the theoretical formulation of special knowledge in more or less completely unified systems of doctrine?

I understand our author's volume to be in the main an answer to this question. To be sure, the compact and manifold teachings which this text contains relate to a great many different special issues. A student interested in the problems of the philosophy of mathematics, or in the theory of probabilities, or in the nature and office of mathematical physics, or in still other problems belonging to the wide field here discussed, may find what he wants here and there in the text, even in case the general issues which give the volume its unity mean little to him, or even if he differs from the author's views regarding the principal issues of the book. But in the main, this volume must be regarded as what its title indicates—a critique of the nature and place of hypothesis in the work of science and a study of the logical relations of theory and fact. The result of the book is a substantial justification of the scientific utility of theoretical construction—an abandonment of dogma, but a vindication of the rights of the constructive reason.

### III

The most notable of the results of our author's investigation of the logic of scientific theories relates, as I understand his work, to a topic which the present state of logical investigation, just summarized, makes especially important, but which has thus far been very inadequately treated in the text-books of inductive logic. The useful hypotheses of science are of two kinds:

1. The hypotheses which are valuable *precisely* because they are either verifiable or else refutable through a definite appeal to the tests furnished by experience; and

2. The hypotheses which, despite the fact that experience suggests them, are valuable *despite*, or even *because*, of the fact that experience can *neither* confirm nor refute them. The contrast between these two kinds of hypotheses is a prominent topic of our author's discussion.

Hypotheses of the general type which I have here placed first in order are the ones which the text-books of inductive logic and those summaries of scientific method which are customary in the course of the elementary treatises upon physical science are already accustomed to recognize and to characterize. The value of such hypotheses is indeed undoubted. But hypotheses of the type which I have here named in the second place are far less frequently recognized in a perfectly explicit way as useful aids in the work of special science. One usually either fails to admit their presence in scientific work, or else remains silent as to the reasons of their usefulness. Our author's treatment of the work of science is therefore especially marked by the fact that he explicitly makes prominent both the existence and the scientific importance of hypotheses of this second type. They occupy in his discussion a place somewhat analogous to each of the two distinct positions occupied by the 'categories' and the 'forms of sensibility,' on the one hand, and by the 'regulative principles of the reason,' on the other hand, in the Kantian theory of our knowledge of nature. That is, these hypotheses which can neither be confirmed nor refuted by experience appear, in M. Poincaré's account, partly (like the conception of ' continuous quantity') as devices of the understanding whereby we give conceptual unity and an invisible connectedness to certain types of phenomenal facts which come to us in a discrete form and in a confused variety; and partly (like the larger organizing concepts of science) as principles regarding the structure of the world in its wholeness; i. e., as principles in the light of which we try to interpret our experience, so as to give to it a totality and an inclusive unity such as Euclidean space, or such as the world of the theory of energy is conceived to possess. Thus viewed, M. Poincaré's logical theory of this second class of hypotheses undertakes to accomplish, with modern means and in the light of to-day's issues, a part of what Kant endeavored to accomplish in his theory of scientific knowledge with the limited means which were at his disposal. Those aspects of science which are determined by the use of the hypotheses of this second kind appear in our author's account as constituting an essential human way of viewing nature, an interpretation rather than a portrayal or a prediction of the objective facts of nature, an adjustment of our conceptions of things to the internal needs of our intelligence, rather than a grasping of things as they are in themselves.

To be sure, M. Poincaré's view, in this portion of his work, obviously differs, meanwhile, from that of Kant, as well as this agrees, in a measure, with the spirit of the Kantian epistemology. I do not mean therefore to class our author as a Kantian. For Kant, the interpretations imposed by the 'forms of sensibility,' and by the 'categories of the understanding,' upon our doctrine of nature are rigidly predetermined by the unalterable 'form' of our intellectual powers. We 'must' thus view facts, whatever the data of sense must be. This, of course, is not M. Poincaré's view. A similarly rigid predetermination also limits the Kantian 'ideas of the reason' to a certain set of principles whose guidance of the course of our theoretical investigations is indeed only 'regulative,' but is 'a priori,' and so unchangeable. For M. Poincaré, on the contrary, all this adjustment of our interpretations of experience to the needs of our intellect is something far less rigid and unalterable, and is constantly subject to the suggestions of experience. We must indeed interpret in our own way; but our way is itself only relatively determinate; it is essentially more or less plastic; other interpretations of experience are conceivable. Those that we use are merely the ones found to be most convenient. But this convenience is not absolute necessity. Unverifiable and irrefutable hypotheses in science are indeed, in general, indispensable aids to the organization and to the guidance of our interpretation of experience. But it is experience itself which points out to us what lines of interpretation will prove most convenient. Instead of Kant's rigid list of *a priori* 'forms,' we consequently have in M. Poincaré's account a set of conventions, neither wholly subjective and arbitrary, nor yet imposed upon us unambiguously by the external compulsion of experience. The organization of science, so far as this organization is due to hypotheses of the kind here in question, thus resembles that of a constitutional government—neither absolutely necessary, nor yet determined apart from the will of the subjects, nor yet accidental—a free, yet not a capricious establishment of good order, in conformity with empirical needs.

3

Characteristic remains, however, for our author, as, in his decidedly contrasting way, for Kant, the thought that *without principles which at every stage transcend precise confirmation through such experience as is then accessible the organization of experience is impossible.* Whether one views these principles as conventions or as *a priori* 'forms,' they may therefore be described as hypotheses, but as hypotheses that, while lying at the basis of our actual physical sciences, at once refer to experience and help us in dealing with experience, and are yet neither confirmed nor refuted by the experiences which we possess or which we can hope to attain.

Three special instances or classes of instances, according to our author's account, may be used as illustrations of this general type of hypotheses. They are: (1) The hypothesis of the existence of continuous extensive *quanta* in nature; (2) The principles of geometry; (3) The principles of mechanics and of the general theory of energy. In case of each of these special types of hypotheses we are at first disposed, apart from reflection, to say that we *find* the world to be thus or thus, so that, for instance, we can confirm the thesis according to which nature contains continuous magnitudes; or can prove or disprove the physical truth of the postulates of Euclidean geometry; or can confirm by definite experience the objective validity of the principles of mechanics. A closer examination reveals, according to our author, the incorrectness of all such opinions. Hypotheses of these various special types are needed; and their usefulness can be empirically shown. They are in touch with experience; and that they are not merely arbitrary conventions is also verifiable. They are not *a priori* necessities; and we can easily conceive intelligent beings whose experience could be best interpreted without using these hypotheses. Yet these hypotheses are *not* subject to direct confirmation or refutation by experience. They stand then in sharp contrast to the scientific hypotheses of the other, and more frequently recognized, type, *i. e.,* to the hypotheses which *can* be tested by a definite appeal to experience. To these other hypotheses our author attaches, of course, great importance. His treatment of them is full of a living appreciation of the significance of empirical investigation. But the central problem of the logic of science thus becomes the problem of the relation between the two fundamentally distinct types of hypotheses, *i. e.*, between those which can not be verified or refuted through experience, and those which can be empirically tested.

IV

The detailed treatment which M. Poincaré gives to the problem thus defined must be learned from his text. It is no part of my purpose to expound, to defend or to traverse any of his special conclusions regarding this matter. Yet I can not avoid observing that, while M. Poincaré strictly confines his illustrations and his expressions of opinion to those regions of science wherein, as special investigator, he is himself most at home, the issues which he thus raises regarding the logic of science are of even more critical importance and of more impressive interest when one applies M. Poincaré's methods to the study of the concepts and presuppositions of the organic and of the historical and social sciences, than when one confines one's attention, as our author here does, to the physical sciences. It belongs to the province of an introduction like the present to point out, however briefly and inadequately, that the significance of our author's ideas extends far beyond the scope to which he chooses to confine their discussion.

The historical sciences, and in fact all those sciences such as geology, and such as the evolutionary sciences in general, undertake theoretical constructions which relate to past time. Hypotheses relating to the more or less remote past stand, however, in a position which is very interesting from the point of view of the logic of science. Directly speaking, no such hypothesis is capable of confirmation or of refutation, because we can not return into the past to verify by our own experience what then happened. Yet indirectly, such hypotheses may lead to predictions of coming experience. These latter will be subject to control. Thus, Schliemann's confidence that the legend of Troy had a definite historical foundation led to predictions regarding what certain excavations would reveal. In a sense somewhat different from that which filled Schliemann's enthusiastic mind, these predictions proved verifiable. The result has been a considerable change in the attitude of historians toward the legend of Troy. Geological investigation leads to predictions regarding the order of the strata or the course of mineral veins in a district, regarding the fossils which may be discovered in given formations, and so on. These hypotheses are subject to the control of experience. The various theories of evolutionary doctrine include many hypotheses capable of confirmation and of refutation by empirical tests. Yet, despite all such empirical control, it still remains true that whenever a science is mainly concerned with the remote past, whether this science be archeology, or geology, or anthropology, or Old Testament history, the principal theoretical constructions always include features which no appeal to present or to accessible future experience can ever definitely test. Hence the suspicion with which students of experimental science often regard the theoretical constructions of their confrères of the sciences that deal with the past. The origin of the races of men, of man himself, of life, of species, of the planet; the hypotheses of anthropologists, of archeologists, of students of 'higher criticism'—all these are matters which the men of the laboratory often regard with a general incredulity as belonging not at all to the domain of true science. Yet no one can doubt the importance and the inevitableness of endeavoring to apply scientific method to these regions also. Science needs theories regarding the past history of the world. And no one who looks closer into the methods of these sciences of past time can doubt that verifiable and unverifiable hypotheses are in all these regions inevitably interwoven; so that, while experience is always the guide, the attitude of the investigator towards experience is determined by interests which have to be partially due to what I should call that 'internal meaning,' that human interest in rational theoretical construction which inspires the scientific inquiry; and the theoretical constructions which prevail in such sciences are neither unbiased reports of the actual constitution of an external reality, nor yet arbitrary constructions of fancy. These constructions in fact resemble in a measure those which M. Poincaré in this book has analyzed in the case of geometry. They are constructions molded, but *not* predetermined in their details, by experience. We report facts; we let the facts speak; but we, as we investigate, in the popular phrase, 'talk back' to the facts. We interpret as well as report. Man is not merely made for science, but science is made for man. It expresses his deepest intellectual needs, as well as his careful observations. It is an effort to bring internal meanings into harmony with external verifications. It attempts therefore to control, as well as to submit, to conceive with rational unity, as well as to accept data. Its arts are those directed towards self-possession as well as towards an imitation of the outer reality which we find. It seeks therefore a disciplined freedom of thought. The discipline is as essential as the freedom; but the latter has also its place. The theories of science are human, as well as objective, internally rational, as well as (when that is possible) subject to external tests.

In a field very different from that of the historical sciences, namely, in a science of observation and of experiment, which is at the same time an organic science, I have been led in the course of some study of the history of certain researches to notice the existence of a theoretical conception which has proved extremely fruitful in guiding research, but which apparently resembles in a measure the type of hypotheses of which M. Poincaré speaks when he characterizes the principles of mechanics and of the theory of energy. I venture to call attention here to this conception, which seems to me to illustrate M. Poincaré's view of the functions of hypothesis in scientific work.

The modern science of pathology is usually regarded as dating from the earlier researches of Virchow, whose 'Cellular Pathology' was the outcome of a very careful and elaborate induction. Virchow, himself, felt a strong aversion to mere speculation. He endeavored to keep close to observation, and to relieve medical science from the control of fantastic theories, such as those of the *Naturphilosophen* had been. Yet Virchow's researches were, as early as 1847, or still earlier, already under the guidance of a theoretical presupposition which he himself states as follows: "We have learned to recognize," he says, "that diseases are not autonomous organisms, that they are no entities that have entered into the body, that they are no parasites which take root in the body, but *that they merely show us the course of*

*the vital processes under altered conditions''* ('dass sie nur
Ablauf der Lebenserscheinungen unter veränderten Bedingungen darstellen').

The enormous importance of this theoretical presupposition
for all the early successes of modern pathological investigation
is generally recognized by the experts. I do not doubt this
opinion. It appears to be a commonplace of the history of this
science. But in Virchow's later years this very presupposition
seemed to some of his contemporaries to be called in question by
the successes of recent bacteriology. The question arose whether
the theoretical foundations of Virchow's pathology had not been
set aside. And in fact the theory of the parasitical origin of
a vast number of diseased conditions has indeed come upon an
empirical basis to be generally recognized. Yet to the end of his
own career Virchow stoutly maintained that in all its essential
significance his own fundamental principle remained quite un-
touched by the newer discoveries. And, as a fact, this view
could indeed be maintained. For if diseases proved to be the
consequences of the presence of parasites, the diseases themselves, so far as they belonged to the diseased organism, were
still not the parasites, but were, as before, the reaction of the
organism to the *veränderte Bedingungen* which the presence of
the parasites entailed. So Virchow could well insist. And if
the famous principle in question is only stated with sufficient
generality, it amounts simply to saying that if a disease involves a change in an organism, and if this change is subject to
law at all, then the nature of the organism and the reaction of
the organism to whatever it is which causes the disease must be
understood in case the disease is to be understood.

For this very reason, however, Virchow's theoretical principle
in its most general form *could be neither confirmed nor refuted
by experience.* It would remain empirically irrefutable, so far
as I can see, even if we should learn that the devil was the
true cause of all diseases. For the devil himself would then
simply predetermine the *veränderte Bedingungen* to which the
diseased organism would be reacting. Let bullets or bacteria,
poisons or compressed air, or the devil be the *Bedingungen* to
which a diseased organism reacts, the postulate that Virchow

states in the passage just quoted will remain irrefutable, if only this postulate be interpreted to meet the case. For the principle in question merely says that whatever entity it may be, bullet, or poison, or devil, that affects the organism, the disease is not that entity, but is the resulting alteration in the process of the organism.

I insist, then, that this principle of Virchow's is no trial supposition, no scientific hypothesis in the narrower sense—capable of being submitted to precise empirical tests. It is, on the contrary, a very precious *leading idea,* a theoretical interpretation of phenomena, in the light of which observations are to be made—'a regulative principle' of research. It is equivalent to a resolution to search for those detailed connections which link the processes of disease to the normal process of the organism. Such a search undertakes to find the true unity, whatever that may prove to be, wherein the pathological and the normal processes are linked. Now without some such leading idea, the cellular pathology itself could never have been reached; because the empirical facts in question would never have been observed. Hence this principle of Virchow's was indispensable to the growth of his science. Yet it was not a verifiable and not a refutable hypothesis. One value of unverifiable and irrefutable hypotheses of this type lies, then, in the sort of empirical inquiries which they initiate, inspire, organize and guide. In these inquiries hypotheses in the narrower sense, that is, trial propositions which are to be submitted to definite empirical control, are indeed everywhere present. And the use of the other sort of principles lies wholly in their application to experience. Yet without what I have just proposed to call the 'leading ideas' of a science, that is, its principles of an unverifiable and irrefutable character, suggested, but not to be finally tested, by experience, the hypotheses in the narrower sense would lack that guidance which, as M. Poincaré has shown, the larger ideas of science give to empirical investigation.

## V

I have dwelt, no doubt, at too great length upon one aspect only of our author's varied and well-balanced discussion of the problems and concepts of scientific theory. Of the hypotheses in the narrower sense and of the value of direct empirical control, he has also spoken with the authority and the originality which belong to his position. And in dealing with the foundations of mathematics he has raised one or two questions of great philosophical import into which I have no time, even if I had the right, to enter here. In particular, in speaking of the essence of mathematical reasoning, and of the difficult problem of what makes possible novel results in the field of pure mathematics, M. Poincaré defends a thesis regarding the office of 'demonstration by recurrence'—a thesis which is indeed disputable, which has been disputed and which I myself should be disposed, so far as I at present understand the matter, to modify in some respects, even in accepting the spirit of our author's assertion. Yet there can be no doubt of the importance of this thesis, and of the fact that it defines a characteristic that is indeed fundamental in a wide range of mathematical research. The philosophical problems that lie at the basis of recurrent proofs and processes are, as I have elsewhere argued, of the most fundamental importance.

These, then, are a few hints relating to the significance of our author's discussion, and a few reasons for hoping that our own students will profit by the reading of the book as those of other nations have already done.

Of the person and of the life-work of our author a few words are here, in conclusion, still in place, addressed, not to the students of his own science, to whom his position is well known, but to the general reader who may seek guidance in these pages.

Jules Henri Poincaré was born at Nancy, in 1854, the son of a professor in the Faculty of Medicine at Nancy. He studied at the École Polytechnique and at the École des Mines, and later received his doctorate in mathematics in 1879. In 1883 he began courses of instruction in mathematics at the École Polytechnique; in 1886 received a professorship of mathematical physics in the Faculty of Sciences at Paris; then became member of the Academy of Sciences at Paris, in 1887, and devoted his life to instruction and investigation in the regions of pure mathematics, of mathematical physics and of celestial mechanics. His list of published treatises relating to various branches of his chosen sciences is long; and his original memoirs have included several momentous investigations, which have gone far to transform more than one branch of research. His presence at the International Congress of Arts and Science in St. Louis was one of the most noticeable features of that remarkable gathering of distinguished foreign guests. In Poincaré the reader meets, then, not one who is primarily a speculative student of general problems for their own sake, but an original investigator of the highest rank in several distinct, although interrelated, branches of modern research. The theory of functions—a highly recondite region of pure mathematics— owes to him advances of the first importance, for instance, the definition of a new type of functions. The 'problem of the three bodies,' a famous and fundamental problem of celestial mechanics, has received from his studies a treatment whose significance has been recognized by the highest authorities. His international reputation has been confirmed by the conferring of more than one important prize for his researches. His membership in the most eminent learned societies of various nations is widely extended; his volumes bearing upon various branches of mathematics and of mathematical physics are used by special students in all parts of the learned world; in brief, he is, as geometer, as analyst and as a theoretical physicist, a leader of his age.

Meanwhile, as contributor to the philosophical discussion of the bases and methods of science, M. Poincaré has long been active. When, in 1893, the admirable *Revue de Métaphysique et de Morale* began to appear, M. Poincaré was soon found amongst the most satisfactory of the contributors to the work of that journal, whose office it has especially been to bring philosophy and the various special sciences (both natural and moral) into a closer mutual understanding. The discussions brought together in the present volume are in large part the outcome of M. Poincaré's contributions to the *Revue de Métaphysique et de Morale*. The reader of M. Poincaré's book is in presence, then, of a great special investigator who is also a philosopher.

# SCIENCE AND HYPOTHESIS

## INTRODUCTION

For a superficial observer, scientific truth is beyond the possibility of doubt; the logic of science is infallible, and if the scientists are sometimes mistaken, this is only from their mistaking its rules.

"The mathematical verities flow from a small number of self-evident propositions by a chain of impeccable reasonings; they impose themselves not only on us, but on nature itself. They fetter, so to speak, the Creator and only permit him to choose between some relatively few solutions. A few experiments then will suffice to let us know what choice he has made. From each experiment a crowd of consequences will follow by a series of mathematical deductions, and thus each experiment will make known to us a corner of the universe."

Behold what is for many people in the world, for scholars getting their first notions of physics, the origin of scientific certitude. This is what they suppose to be the rôle of experimentation and mathematics. This same conception, a hundred years ago, was held by many savants who dreamed of constructing the world with as little as possible taken from experiment.

On a little more reflection it was perceived how great a place hypothesis occupies; that the mathematician can not do without it, still less the experimenter. And then it was doubted if all these constructions were really solid, and believed that a breath would overthrow them. To be skeptical in this fashion is still to be superficial. To doubt everything and to believe everything are two equally convenient solutions; each saves us from thinking.

Instead of pronouncing a summary condemnation, we ought therefore to examine with care the rôle of hypothesis; we shall then recognize, not only that it is necessary, but that usually it is legitimate. We shall also see that there are several sorts of hypotheses; that some are verifiable, and once confirmed by experiment become fruitful truths; that others, powerless to lead us astray, may be useful to us in fixing our ideas; that others, finally, are hypotheses only in appearance and are reducible to disguised definitions or conventions.

These last are met with above all in mathematics and the related sciences. Thence precisely it is that these sciences get their rigor; these conventions are the work of the free activity of our mind, which, in this domain, recognizes no obstacle. Here our mind can affirm, since it decrees; but let us understand that while these decrees are imposed upon *our* science, which, without them, would be impossible, they are not imposed upon nature. Are they then arbitrary? No, else were they sterile. Experiment leaves us our freedom of choice, but it guides us by aiding us to discern the easiest way. Our decrees are therefore like those of a prince, absolute but wise, who consults his council of state.

Some people have been struck by this character of free convention recognizable in certain fundamental principles of the sciences. They have wished to generalize beyond measure, and, at the same time, they have forgotten that liberty is not license. Thus they have reached what is called *nominalism,* and have asked themselves if the savant is not the dupe of his own definitions and if the world he thinks he discovers is not simply created by his own caprice.[1] Under these conditions science would be certain, but deprived of significance.

If this were so, science would be powerless. Now every day we see it work under our very eyes. That could not be if it taught us nothing of reality. Still, the things themselves are not what it can reach, as the naïve dogmatists think, but only the relations between things. Outside of these relations there is no knowable reality.

Such is the conclusion to which we shall come, but for that we must review the series of sciences from arithmetic and geometry to mechanics and experimental physics.

---

[1] See Le Roy, 'Science et Philosophie,' *Revue de Métaphysique et de Morale,* 1901.

What is the nature of mathematical reasoning? Is is really deductive, as is commonly supposed? A deeper analysis shows us that it is not, that it partakes in a certain measure of the nature of inductive reasoning, and just because of this is it so fruitful. None the less does it retain its character of rigor absolute; this is the first thing that had to be shown.

Knowing better now one of the instruments which mathematics puts into the hands of the investigator, we had to analyze another fundamental notion, that of mathematical magnitude. Do we find it in nature, or do we ourselves introduce it there? And, in this latter case, do we not risk marring everything? Comparing the rough data of our senses with that extremely complex and subtle concept which mathematicians call magnitude, we are forced to recognize a difference; this frame into which we wish to force everything is of our own construction; but we have not made it at random. We have made it, so to speak, by measure and therefore we can make the facts fit into it without changing what is essential in them.

Another frame which we impose on the world is space. Whence come the first principles of geometry? Are they imposed on us by logic? Lobachevski has proved not, by creating non-Euclidean geometry. Is space revealed to us by our senses? Still no, for the space our senses could show us differs absolutely from that of the geometer. Is experience the source of geometry? A deeper discussion will show us it is not. We therefore conclude that the first principles of geometry are only conventions; but these conventions are not arbitrary and if transported into another world (that I call the non-Euclidean world and seek to imagine), then we should have been led to adopt others.

In mechanics we should be led to analogous conclusions, and should see that the principles of this science, though more directly based on experiment, still partake of the conventional character of the geometric postulates. Thus far nominalism triumphs; but now we arrive at the physical sciences, properly so called. Here the scene changes; we meet another sort of hypotheses and we see their fertility. Without doubt, at first blush, the theories seem to us fragile, and the history of science proves to us how ephemeral they are; yet they do not entirely perish, and of each of them something remains. It is this something we must seek to disentangle, since there and there alone is the veritable reality.

The method of the physical sciences rests on the induction which makes us expect the repetition of a phenomenon when the circumstances under which it first happened are reproduced. If *all* these circumstances could be reproduced at once, this principle could be applied without fear; but that will never happen; some of these circumstances will always be lacking. Are we absolutely sure they are unimportant? Evidently not. That may be probable, it can not be rigorously certain. Hence the important rôle the notion of probability plays in the physical sciences. The calculus of probabilities is therefore not merely a recreation or a guide to players of baccarat, and we must seek to go deeper with its foundations. Under this head I have been able to give only very incomplete results, so strongly does this vague instinct which lets us discern probability defy analysis.

After a study of the conditions under which the physicist works, I have thought proper to show him at work. For that I have taken instances from the history of optics and of electricity. We shall see whence have sprung the ideas of Fresnel, of Maxwell, and what unconscious hypotheses were made by Ampère and the other founders of electrodynamics.

# PART I

## NUMBER AND MAGNITUDE

### CHAPTER I

#### ON THE NATURE OF MATHEMATICAL REASONING

#### I

THE very possibility of the science of mathematics seems an insoluble contradiction. If this science is deductive only in appearance, whence does it derive that perfect rigor no one dreams of doubting? If, on the contrary, all the propositions it enunciates can be deduced one from another by the rules of formal logic, why is not mathematics reduced to an immense tautology? The syllogism can teach us nothing essentially new, and, if everything is to spring from the principle of identity, everything should be capable of being reduced to it. Shall we then admit that the enunciations of all those theorems which fill so many volumes are nothing but devious ways of saying $A$ is $A$?

Without doubt, we can go back to the axioms, which are at the source of all these reasonings. If we decide that these can not be reduced to the principle of contradiction, if still less we see in them experimental facts which could not partake of mathematical necessity, we have yet the resource of classing them among synthetic *a priori* judgments. This is not to solve the difficulty, but only to baptize it; and even if the nature of synthetic judgments were for us no mystery, the contradiction would not have disappeared, it would only have moved back; syllogistic reasoning remains incapable of adding anything to the data given it; these data reduce themselves to a few axioms, and we should find nothing else in the conclusions.

No theorem could be new if no new axiom intervened in its demonstration; reasoning could give us only the immediately

31

evident verities borrowed from direct intuition; it would be only an intermediary parasite, and therefore should we not have good reason to ask whether the whole syllogistic apparatus did not serve solely to disguise our borrowing?

The contradiction will strike us the more if we open any book on mathematics; on every page the author will announce his intention of generalizing some proposition already known. Does the mathematical method proceed from the particular to the general, and, if so, how then can it be called deductive?

If finally the science of number were purely analytic, or could be analytically derived from a small number of synthetic judgments, it seems that a mind sufficiently powerful could at a glance perceive all its truths; nay more, we might even hope that some day one would invent to express them a language sufficiently simple to have them appear self-evident to an ordinary intelligence.

If we refuse to admit these consequences, it must be conceded that mathematical reasoning has of itself a sort of creative virtue and consequently differs from the syllogism.

The difference must even be profound. We shall not, for example, find the key to the mystery in the frequent use of that rule according to which one and the same uniform operation applied to two equal numbers will give identical results.

All these modes of reasoning, whether or not they be reducible to the syllogism properly so called, retain the analytic character, and just because of that are powerless.

## II

The discussion is old; Leibnitz tried to prove 2 and 2 make 4; let us look a moment at his demonstration.

I will suppose the number 1 defined and also the operation $x + 1$ which consists in adding unity to a given number $x$.

These definitions, whatever they be, do not enter into the course of the reasoning.

I define then the numbers 2, 3 and 4 by the equalities

$$(1) \quad 1 + 1 = 2; \quad (2) \quad 2 + 1 = 3; \quad (3) \quad 3 + 1 = 4.$$

In the same way, I define the operation $x + 2$ by the relation:

$$(4) \quad x + 2 = (x + 1) + 1.$$

That presupposed, we have

$$2 + 1 + 1 = 3 + 1 \qquad \text{(Definition 2)},$$
$$3 + 1 = 4 \qquad \text{(Definition 3)},$$
$$2 + 2 = (2 + 1) + 1 \qquad \text{(Definition 4)},$$

whence

$$2 + 2 = 4 \quad \text{Q. E. D.}$$

It can not be denied that this reasoning is purely analytic. But ask any mathematician: 'That is not a demonstration properly so called,' he will say to you: 'that is a verification.' We have confined ourselves to comparing two purely conventional definitions and have ascertained their identity; we have learned nothing new. *Verification* differs from true demonstration precisely because it is purely analytic and because it is sterile. It is sterile because the conclusion is nothing but the premises translated into another language. On the contrary, true demonstration is fruitful because the conclusion here is in a sense more general than the premises.

The equality $2 + 2 = 4$ is thus susceptible of a verification only because it is particular. Every particular enunciation in mathematics can always be verified in this same way. But if mathematics could be reduced to a series of such verifications, it would not be a science. So a chess-player, for example, does not create a science in winning a game. There is no science apart from the general.

It may even be said the very object of the exact sciences is to spare us these direct verifications.

## III

Let us, therefore, see the geometer at work and seek to catch his process.

The task is not without difficulty; it does not suffice to open a work at random and analyze any demonstration in it.

We must first exclude geometry, where the question is complicated by arduous problems relative to the rôle of the postulates, to the nature and the origin of the notion of space. For analogous reasons we can not turn to the infinitesimal analysis.

4

We must seek mathematical thought where it has remained pure, that is, in arithmetic.

A choice still is necessary; in the higher parts of the theory of numbers, the primitive mathematical notions have already undergone an elaboration so profound that it becomes difficult to analyze them.

It is, therefore, at the beginning of arithmetic that we must expect to find the explanation we seek, but it happens that precisely in the demonstration of the most elementary theorems the authors of the classic treatises have shown the least precision and rigor. We must not impute this to them as a crime; they have yielded to a necessity; beginners are not prepared for real mathematical rigor; they would see in it only useless and irksome subtleties; it would be a waste of time to try prematurely to make them more exacting; they must pass over rapidly, but without skipping stations, the road traversed slowly by the founders of the science.

Why is so long a preparation necessary to become habituated to this perfect rigor, which, it would seem, should naturally impress itself upon all good minds? This is a logical and psychological problem well worthy of study.

But we shall not take it up; it is foreign to our purpose; all I wish to insist on is that, not to fail of our purpose, we must recast the demonstrations of the most elementary theorems and give them, not the crude form in which they are left, so as not to harass beginners, but the form that will satisfy a skilled geometer.

DEFINITION OF ADDITION.—I suppose already defined the operation $x + 1$, which consists in adding the number 1 to a given number $x$.

This definition, whatever it be, does not enter into our subsequent reasoning.

We now have to define the operation $x + a$, which consists in adding the number $a$ to a given number $x$.

Supposing we have defined the operation

$$x + (a-1),$$

the operation $x + a$ will be defined by the equality

(1)  $$x + a = [x + (a-1)] + 1.$$

We shall know then what $x + a$ is when we know what $x + (a - 1)$ is, and as I have supposed that to start with we knew what $x + 1$ is, we can define successively and 'by recurrence' the operations $x + 2$, $x + 3$, etc.

This definition deserves a moment's attention; it is of a particular nature which already distinguishes it from the purely logical definition; the equality (1) contains an infinity of distinct definitions, each having a meaning only when one knows the preceding.

PROPERTIES OF ADDITION.—*Associativity.*—I say that

$$a + (b + c) = (a + b) + c.$$

In fact the theorem is true for $c = 1$; it is then written

$$a + (b + 1) = (a + b) + 1,$$

which, apart from the difference of notation, is nothing but the equality (1), by which I have just defined addition.

Supposing the theorem true for $c = \gamma$, I say it will be true for $c = \gamma + 1$.

In fact, supposing

$$(a + b) + \gamma = a + (b + \gamma),$$

it follows that

$$[(a + b) + \gamma] + 1 = [a + (b + \gamma)] + 1$$

or by definition (1)

$$(a + b) + (\gamma + 1) = a + (b + \gamma + 1) = a + [b + (\gamma + 1)],$$

which shows, by a series of purely analytic deductions, that the theorem is true for $\gamma + 1$.

Being true for $c = 1$, we thus see successively that so it is for $c = 2$, for $c = 3$, etc.

*Commutativity.*—1° I say that

$$a + 1 = 1 + a.$$

The theorem is evidently true for $a = 1$; we can *verify* by purely analytic reasoning that if it is true for $a = \gamma$ it will be true for $a = \gamma + 1$; for then

$$(\gamma + 1) + 1 = (1 + \gamma) + 1 = 1 + (\gamma + 1);$$

now it is true for $a = 1$, therefore it will be true for $a = 2$, for $a = 3$, etc., which is expressed by saying that the enunciated proposition is demonstrated by recurrence.

2° I say that

$$a + b = b + a.$$

The theorem has just been demonstrated for $b = 1$; it can be *verified* analytically that if it is true for $b = \beta$, it will be true for $b = \beta + 1$.

The proposition is therefore established by recurrence.

DEFINITION OF MULTIPLICATION.—We shall define multiplication by the equalities.

(1)　　　　　　　　$a \times 1 = a.$

(2)　　　　　　　　$a \times b = [a \times (b - 1)] + a.$

Like equality (1), equality (2) contains an infinity of definitions; having defined $a \times 1$, it enables us to define successively: $a \times 2$, $a \times 3$, etc.

PROPERTIES OF MULTIPLICATION.—*Distributivity.*—I say that

$$(a + b) \times c = (a \times c) + (b \times c).$$

We verify analytically that the equality is true for $c = 1$; then that if the theorem is true for $c = \gamma$, it will be true for $c = \gamma + 1$.

The proposition is, therefore, demonstrated by recurrence.

*Commutativity.*—1° I say that

$$a \times 1 = 1 \times a.$$

The theorem is evident for $a = 1$.

We verify analytically that if it is true for $a = a$, it will be true for $a = a + 1$.

2° I say that

$$a \times b = b \times a.$$

The theorem has just been proven for $b = 1$. We could verify analytically that if it is true for $b = \beta$, it will be true for $b = \beta + 1$.

## IV

Here I stop this monotonous series of reasonings. But this very monotony has the better brought out the procedure which is uniform and is met again at each step.

This procedure is the demonstration by recurrence. We first establish a theorem for $n = 1$; then we show that if it is true of $n - 1$, it is true of $n$, and thence conclude that it is true for all the whole numbers.

We have just seen how it may be used to demonstrate the rules of addition and multiplication, that is to say, the rules of the algebraic calculus; this calculus is an instrument of transformation, which lends itself to many more differing combinations than does the simple syllogism; but it is still an instrument purely analytic, and incapable of teaching us anything new. If mathematics had no other instrument, it would therefore be forthwith arrested in its development; but it has recourse anew to the same procedure, that is, to reasoning by recurrence, and it is able to continue its forward march.

If we look closely, at every step we meet again this mode of reasoning, either in the simple form we have just given it, or under a form more or less modified.

Here then we have the mathematical reasoning *par excellence,* and we must examine it more closely.

## V

The essential characteristic of reasoning by recurrence is that it contains, condensed, so to speak, in a single formula, an infinity of syllogisms.

That this may the better be seen, I will state one after another these syllogisms which are, if you will allow me the expression, arranged in 'cascade.'

These are of course hypothetical syllogisms.

The theorem is true of the number 1.

Now, if it is true of 1, it is true of 2.

Therefore it is true of 2.

Now, if it is true of 2, it is true of 3.

Therefore it is true of 3, and so on.

We see that the conclusion of each syllogism serves as minor to the following.

Furthermore the majors of all our syllogisms can be reduced to a single formula.

If the theorem is true of $n-1$, so it is of $n$.

We see, then, that in reasoning by recurrence we confine ourselves to stating the minor of the first syllogism, and the general formula which contains as particular cases all the majors.

This never-ending series of syllogisms is thus reduced to a phrase of a few lines.

It is now easy to comprehend why every particular consequence of a theorem can, as I have explained above, be verified by purely analytic procedures.

If instead of showing that our theorem is true of all numbers, we only wish to show it true of the number 6, for example, it will suffice for us to establish the first 5 syllogisms of our cascade; 9 would be necessary if we wished to prove the theorem for the number 10; more would be needed for a larger number; but, however great this number might be, we should always end by reaching it, and the analytic verification would be possible.

And yet, however far we thus might go, we could never rise to the general theorem, applicable to all numbers, which alone can be the object of science. To reach this, an infinity of syllogisms would be necessary; it would be necessary to overleap an abyss that the patience of the analyst, restricted to the resources of formal logic alone, never could fill up.

I asked at the outset why one could not conceive of a mind sufficiently powerful to perceive at a glance the whole body of mathematical truths.

The answer is now easy; a chess-player is able to combine four moves, five moves, in advance, but, however extraordinary he may be, he will never prepare more than a finite number of them; if he applies his faculties to arithmetic, he will not be able to perceive its general truths by a single direct intuition; to arrive at the smallest theorem he can not dispense with the aid of reasoning by recurrence, for this is an instrument which enables us to pass from the finite to the infinite.

This instrument is always useful, for, allowing us to overleap at a bound as many stages as we wish, it spares us verifications, long, irksome and monotonous, which would quickly become impracticable. But it becomes indispensable as soon as we aim at the general theorem, to which analytic verification would bring us continually nearer without ever enabling us to reach it.

In this domain of arithmetic, we may think ourselves very far from the infinitesimal analysis, and yet, as we have just seen, the idea of the mathematical infinite already plays a preponderant rôle, and without it there would be no science, because there would be nothing general.

## VI

The judgment on which reasoning by recurrence rests can be put under other forms; we may say, for example, that in an infinite collection of different whole numbers there is always one which is less than all the others.

We can easily pass from one enunciation to the other and thus get the illusion of having demonstrated the legitimacy of reasoning by recurrence. But we shall always be arrested, we shall always arrive at an undemonstrable axiom which will be in reality only the proposition to be proved translated into another language.

We can not therefore escape the conclusion that the rule of reasoning by recurrence is irreducible to the principle of contradiction.

Neither can this rule come to us from experience; experience could teach us that the rule is true for the first ten or hundred numbers; for example, it can not attain to the indefinite series of numbers, but only to a portion of this series, more or less long but always limited.

Now if it were only a question of that, the principle of contradiction would suffice; it would always allow of our developing as many syllogisms as we wished; it is only when it is a question of including an infinity of them in a single formula, it is only before the infinite that this principle fails, and there too, experience becomes powerless. This rule, inaccessible to analytic demonstration and to experience, is the veritable type of the synthetic *a priori* judgment. On the other hand, we can not think of seeing in it a convention, as in some of the postulates of geometry.

Why then does this judgment force itself upon us with an irresistible evidence? It is because it is only the affirmation of the power of the mind which knows itself capable of conceiving the indefinite repetition of the same act when once this act is possible. The mind has a direct intuition of this power, and experience can only give occasion for using it and thereby becoming conscious of it.

But, one will say, if raw experience can not legitimatize reasoning by recurrence, is it so of experiment aided by induction? We see successively that a theorem is true of the number 1, of the number 2, of the number 3 and so on; the law is evident, we say, and it has the same warranty as every physical law based on observations, whose number is very great but limited.

Here is, it must be admitted, a striking analogy with the usual procedures of induction. But there is an essential difference. Induction applied to the physical sciences is always uncertain, because it rests on the belief in a general order of the universe, an order outside of us. Mathematical induction, that is, demonstration by recurrence, on the contrary, imposes itself necessarily because it is only the affirmation of a property of the mind itself.

## VII

Mathematicians, as I have said before, always endeavor to *generalize* the propositions they have obtained. and, to seek no other example, we have just proved the equality:

$$a + 1 = 1 + a$$

and afterwards used it to establish the equality

$$a + b = b + a$$

which is manifestly more general.

Mathematics can, therefore, like the other sciences, proceed from the particular to the general.

This is a fact which would have appeared incomprehensible to us at the outset of this study, but which is no longer mysterious to us, since we have ascertained the analogies between demonstration by recurrence and ordinary induction.

Without doubt recurrent reasoning in mathematics and inductive reasoning in physics rest on different foundations, but their march is parallel, they advance in the same sense, that is to say, from the particular to the general.

Let us examine the case a little more closely.

To demonstrate the equality

$$a + 2 = 2 + a$$

it suffices to twice apply the rule

(1) $$a + 1 = 1 + a$$

and write

(2) $$a + 2 = a + 1 + 1 = 1 + a + 1 = 1 + 1 + a = 2 + a.$$

The equality (2) thus deduced in purely analytic way from the equality (1) is, however, not simply a particular case of it; it is something quite different.

We can not therefore even say that in the really analytic and deductive part of mathematical reasoning we proceed from the general to the particular in the ordinary sense of the word.

The two members of the equality (2) are simply combinations more complicated than the two members of the equality (1), and analysis only serves to separate the elements which enter into these combinations and to study their relations.

Mathematicians proceed therefore 'by construction,' they 'construct' combinations more and more complicated. Coming back then by the analysis of these combinations, of these aggregates, so to speak, to their primitive elements, they perceive the relations of these elements and from them deduce the relations of the aggregates themselves.

This is a purely analytical proceeding, but it is not, however, a proceeding from the general to the particular, because evidently the aggregates can not be regarded as more particular than their elements.

Great importance, and justly, has been attached to this procedure of 'construction,' and some have tried to see in it the necessary and sufficient condition for the progress of the exact sciences.

Necessary, without doubt; but sufficient, no.

For a construction to be useful and not a vain toil for the mind, that it may serve as stepping-stone to one wishing to mount, it must first of all possess a sort of unity enabling us to see in it something besides the juxtaposition of its elements.

Or, more exactly, there must be some advantage in considering the construction rather than its elements themselves.

What can this advantage be?

Why reason on a polygon, for instance, which is always decomposable into triangles, and not on the elementary triangles?

It is because there are properties appertaining to polygons of any number of sides and that may be immediately applied to any particular polygon.

Usually, on the contrary, it is only at the cost of the most prolonged exertions that they could be found by studying directly the relations of the elementary triangles. The knowledge of the general theorem spares us these efforts.

A construction, therefore, becomes interesting only when it can be ranged beside other analogous constructions, forming species of the same genus.

If the quadrilateral is something besides the juxtaposition of two triangles, this is because it belongs to the genus polygon.

Moreover, one must be able to demonstrate the properties of the genus without being forced to establish them successively for each of the species.

To attain that, we must necessarily mount from the particular to the general, ascending one or more steps.

The analytic procedure 'by construction' does not oblige us to descend, but it leaves us at the same level.

We can ascend only by mathematical induction, which alone can teach us something new. Without the aid of this induction, different in certain respects from physical induction, but quite as fertile, construction would be powerless to create science.

Observe finally that this induction is possible only if the same operation can be repeated indefinitely. That is why the theory of chess can never become a science, for the different moves of the same game do not resemble one another.

# CHAPTER II

## MATHEMATICAL MAGNITUDE AND EXPERIENCE

To learn what mathematicians understand by a continuum, one should not inquire of geometry. The geometer always seeks to represent to himself more or less the figures he studies, but his representations are for him only instruments; in making geometry he uses space just as he does chalk; so too much weight should not be attached to non-essentials, often of no more importance than the whiteness of the chalk.

The pure analyst has not this rock to fear. He has disengaged the science of mathematics from all foreign elements, and can answer our question: 'What exactly is this continuum about which mathematicians reason?' Many analysts who reflect on their art have answered already; Monsieur Tannery, for example, in his *Introduction à la théorie des fonctions d'une variable*.

Let us start from the scale of whole numbers; between two consecutive steps, intercalate one or more intermediary steps, then between these new steps still others, and so on indefinitely. Thus we shall have an unlimited number of terms; these will be the numbers called fractional, rational or commensurable. But this is not yet enough; between these terms, which, however, are already infinite in number, it is still necessary to intercalate others called irrational or incommensurable. A remark before going further. The continuum so conceived is only a collection of individuals ranged in a certain order, infinite in number, it is true, but *exterior* to one another. This is not the ordinary conception, wherein is supposed between the elements of the continuum a sort of intimate bond which makes of them a whole, where the point does not exist before the line, but the line before the point. Of the celebrated formula, 'the continuum is unity in multiplicity,' only the multiplicity remains, the unity has disappeared. The analysts are none the less right in defining their continuum as they do, for they always reason on just this as soon as they pique themselves on their rigor. But this is enough to apprise us that the veritable mathematical continuum is a very different thing from that of the physicists and that of the metaphysicians.

It may also be said perhaps that the mathematicians who are content with this definition are dupes of words, that it is necessary to say precisely what each of these intermediary steps is, to explain how they are to be intercalated and to demonstrate that it is possible to do it. But that would be wrong; the only property of these steps which is used in their reasonings[1] is that of being before or after such and such steps; therefore also this alone should occur in the definition.

So how the intermediary terms should be intercalated need not concern us; on the other hand, no one will doubt the possibility of this operation, unless from forgetting that possible, in the language of geometers, simply means free from contradiction.

Our definition, however, is not yet complete, and I return to it after this over-long digression.

DEFINITION OF INCOMMENSURABLES.—The mathematicians of the Berlin school, Kronecker in particular, have devoted themselves to constructing this continuous scale of fractional and irrational numbers without using any material other than the whole number. The mathematical continuum would be, in this view, a pure creation of the mind, where experience would have no part.

The notion of the rational number seeming to them to present no difficulty, they have chiefly striven to define the incommensurable number. But before producing here their definition, I must make a remark to forestall the astonishment it is sure to arouse in readers unfamiliar with the customs of geometers.

Mathematicians study not objects, but relations between objects; the replacement of these objects by others is therefore indifferent to them, provided the relations do not change. The matter is for them unimportant, the form alone interests them.

Without recalling this, it would scarcely be comprehensible that Dedekind should designate by the name *incommensurable number* a mere symbol, that is to say, something very different

[1] With those contained in the special conventions which serve to define addition and of which we shall speak later.

from the ordinary idea of a quantity, which should be measurable and almost tangible.

Let us see now what Dedekind's definition is:

The commensurable numbers can in an infinity of ways be partitioned into two classes, such that any number of the first class is greater than any number of the second class.

It may happen that among the numbers of the first class there is one smaller than all the others; if, for example, we range in the first class all numbers greater than 2, and 2 itself, and in the second class all numbers less than 2, it is clear that 2 will be the least of all numbers of the first class. The number 2 may be chosen as symbol of this partition.

It may happen, on the contrary, that among the numbers of the second class is one greater than all the others; this is the case, for example, if the first class comprehends all numbers greater than 2, and the second all numbers less than 2, and 2 itself. Here again the number 2 may be chosen as symbol of this partition.

But it may equally well happen that neither is there in the first class a number less than all the others, nor in the second class a number greater than all the others. Suppose, for example, we put in the first class all commensurable numbers whose squares are greater than 2 and in the second all whose squares are less than 2. There is none whose square is precisely 2. Evidently there is not in the first class a number less than all the others, for, however near the square of a number may be to 2, we can always find a commensurable number whose square is still closer to 2.

In Dedekind's view, the incommensurable number

$$\sqrt{2} \text{ or } (2)^{\frac{1}{2}}$$

is nothing but the symbol of this particular mode of partition of commensurable numbers; and to each mode of partition corresponds thus a number, commensurable or not, which serves as its symbol.

But to be content with this would be to forget too far the origin of these symbols; it remains to explain how we have been led to attribute to them a sort of concrete existence, and, besides, does not the difficulty begin even for the fractional numbers themselves? Should we have the notion of these numbers if we had not previously known a matter that we conceive as infinitely divisible, that is to say, a continuum?

THE PHYSICAL CONTINUUM.—We ask ourselves then if the notion of the mathematical continuum is not simply drawn from experience. If it were, the raw data of experience, which are our sensations, would be susceptible of measurement. We might be tempted to believe they really are so, since in these latter days the attempt has been made to measure them and a law has even been formulated, known as Fechner's law, according to which sensation is proportional to the logarithm of the stimulus.

But if we examine more closely the experiments by which it has been sought to establish this law, we shall be led to a diametrically opposite conclusion. It has been observed, for example, that a weight $A$ of 10 grams and a weight $B$ of 11 grams produce identical sensations, that the weight $B$ is just as indistinguishable from a weight $C$ of 12 grams, but that the weight $A$ is easily distinguished from the weight $C$. Thus the raw results of experience may be expressed by the following relations:

$$A = B, \quad B = C, \quad A < C,$$

which may be regarded as the formula of the physical continuum.

But here is an intolerable discord with the principle of contradiction, and the need of stopping this has compelled us to invent the mathematical continuum.

We are, therefore, forced to conclude that this notion has been created entirely by the mind, but that experience has given the occasion.

We can not believe that two quantities equal to a third are not equal to one another, and so we are led to suppose that $A$ is different from $B$ and $B$ from $C$, but that the imperfection of our senses has not permitted of our distinguishing them.

CREATION OF THE MATHEMATICAL CONTINUUM.—*First Stage.* So far it would suffice, in accounting for the facts, to intercalate between $A$ and $B$ a few terms, which would remain discrete. What happens now if we have recourse to some instrument to supplement the feebleness of our senses, if, for example, we make use of a microscope? Terms such as $A$ and $B$, before indistinguishable, appear now distinct; but between $A$ and $B$, now become distinct, will be intercalated a new term, $D$, that we can distinguish neither from $A$ nor from $B$. Despite the employ ment of the most highly perfected methods, the raw results of our experience will always present the characteristics of the physical continuum with the contradiction which is inherent in it.

We shall escape it only by incessantly intercalating new terms between the terms already distinguished, and this operation must be continued indefinitely. We might conceive the stopping of this operation if we could imagine some instrument sufficiently powerful to decompose the physical continuum into discrete elements, as the telescope resolves the milky way into stars. But this we can not imagine; in fact, it is with the eye we observe the image magnified by the microscope, and consequently this image must always retain the characteristics of visual sensation and consequently those of the physical continuum.

Nothing distinguishes a length observed directly from the half of this length doubled by the microscope. The whole is homogeneous with the part; this is a new contradiction, or rather it would be if the number of terms were supposed finite; in fact, it is clear that the part containing fewer terms than the whole could not be similar to the whole.

The contradiction ceases when the number of terms is regarded as infinite; nothing hinders, for example, considering the aggregate of whole numbers as similar to the aggregate of even numbers, which, however, is only a part of it; and, in fact, to each whole number corresponds an even number, its double.

But it is not only to escape this contradiction contained in the empirical data that the mind is led to create the concept of a continuum, formed of an indefinite number of terms.

All happens as in the sequence of whole numbers. We have the faculty of conceiving that a unit can be added to a collection of units; thanks to experience, we have occasion to exercise this faculty and we become conscious of it; but from this moment we feel that our power has no limit and that we can count indefinitely, though we have never had to count more than a finite number of objects.

Just so, as soon as we have been led to intercalate means between two consecutive terms of a series, we feel that this operation can be continued beyond all limit, and that there is, so to speak, no intrinsic reason for stopping.

As an abbreviation, let me call a mathematical continuum of the first order every aggregate of terms formed according to the same law as the scale of commensurable numbers. If we afterwards intercalate new steps according to the law of formation of incommensurable numbers, we shall obtain what we will call a continuum of the second order.

*Second Stage.*—We have made hitherto only the first stride; we have explained the origin of continua of the first order; but it is necessary to see why even they are not sufficient and why the incommensurable numbers had to be invented.

If we try to imagine a line, it must have the characteristics of the physical continuum, that is to say, we shall not be able to represent it except with a certain breadth. Two lines then will appear to us under the form of two narrow bands, and, if we are content with this rough image, it is evident that if the two lines cross they will have a common part.

But the pure geometer makes a further effort; without entirely renouncing the aid of the senses, he tries to reach the concept of the line without breadth, of the point without extension. This he can only attain to by regarding the line as the limit toward which tends an ever narrowing band, and the point as the limit toward which tends an ever lessening area. And then, our two bands, however narrow they may be, will always have a common area, the smaller as they are the narrower, and whose limit will be what the pure geometer calls a point.

This is why it is said two lines which cross have a point in common, and this truth seems intuitive.

But it would imply contradiction if lines were conceived as continua of the first order, that is to say, if on the lines traced by the geometer should be found only points having for coordinates rational numbers. The contradiction would be manifest as soon as one affirmed, for example, the existence of straights and circles.

It is clear, in fact, that if the points whose coordinates are commensurable were alone regarded as real, the circle inscribed in a square and the diagonal of this square would not intersect, since the coordinates of the point of intersection are incommensurable.

That would not yet be sufficient, because we should get in this way only certain incommensurable numbers and not all those numbers.

But conceive of a straight line divided into two rays. Each of these rays will appear to our imagination as a band of a certain breadth; these bands moreover will encroach one on the other, since there must be no interval between them. The common part will appear to us as a point which will always remain when we try to imagine our bands narrower and narrower, so that we admit as an intuitive truth that if a straight is cut into two rays their common frontier is a point; we recognize here the conception of Dedekind, in which an incommensurable number was regarded as the common frontier of two classes of rational numbers.

Such is the origin of the continuum of the second order, which is the mathematical continuum properly so called.

*Résumé.*—In recapitulation, the mind has the faculty of creating symbols, and it is thus that it has constructed the mathematical continuum, which is only a particular system of symbols. Its power is limited only by the necessity of avoiding all contradiction; but the mind only makes use of this faculty if experience furnishes it a stimulus thereto.

In the case considered, this stimulus was the notion of the physical continuum, drawn from the rough data of the senses. But this notion leads to a series of contradictions from which it is necessary successively to free ourselves. So we are forced to imagine a more and more complicated system of symbols. That at which we stop is not only exempt from internal contradiction (it was so already at all the stages we have traversed), but neither is it in contradiction with various propositions called intuitive, which are derived from empirical notions more or less elaborated.

MEASURABLE MAGNITUDE.—The magnitudes we have studied hitherto are not *measurable;* we can indeed say whether a given one of these magnitudes is greater than another, but not whether it is twice or thrice as great.

So far, I have only considered the order in which our terms are ranged. But for most applications that does not suffice. We must learn to compare the interval which separates any two terms. Only on this condition does the continuum become a measurable magnitude and the operations of arithmetic applicable.

This can only be done by the aid of a new and special *convention*. We will *agree* that in such and such a case the interval comprised between the terms *A* and *B* is equal to the interval which separates *C* and *D*. For example, at the beginning of our work we have set out from the scale of the whole numbers and we have supposed intercalated between two consecutive steps *n* intermediary steps; well, these new steps will be by convention regarded as equidistant.

This is a way of defining the addition of two magnitudes, because if the interval *AB* is by definition equal to the interval *CD*, the interval *AD* will be by definition the sum of the intervals *AB* and *AC*.

This definition is arbitrary in a very large measure. It is not completely so, however. It is subjected to certain conditions and, for example, to the rules of commutativity and associativity of addition. But provided the definition chosen satisfies these rules, the choice is indifferent, and it is useless to particularize it.

VARIOUS REMARKS.—We can now discuss several important questions:

1° Is the creative power of the mind exhausted by the creation of the mathematical continuum?

No: the works of Du Bois-Reymond demonstrate it in a striking way.

We know that mathematicians distinguish between infinitesimals of different orders and that those of the second order are infinitesimal, not only in an absolute way, but also in relation to those of the first order. It is not difficult to imagine infinitesimals of fractional or even of irrational order, and thus we find again that scale of the mathematical continuum which has been dealt with in the preceding pages.

Further, there are infinitesimals which are infinitely small in relation to those of the first order, and, on the contrary, infinitely great in relation to those of order $1 + \epsilon$, and that however small $\epsilon$ may be. Here, then, are new terms intercalated in our series, and if I may be permitted to revert to the phraseology lately employed which is very convenient though not consecrated by usage, I shall say that thus has been created a sort of continuum of the third order.

It would be easy to go further, but that would be idle; one would only be imagining symbols without possible application, and no one will think of doing that. The continuum of the third order, to which the consideration of the different orders of infinitesimals leads, is itself not useful enough to have won citizenship, and geometers regard it only as a mere curiosity. The mind uses its creative faculty only when experience requires it.

2° Once in possession of the concept of the mathematical continuum, is one safe from contradictions analogous to those which gave birth to it?

No, and I will give an example.

One must be very wise not to regard it as evident that every curve has a tangent; and in fact if we picture this curve and a straight as two narrow bands we can always so dispose them that they have a part in common without crossing. If we imagine then the breadth of these two bands to diminish indefinitely, this common part will always subsist and, at the limit, so to speak, the two lines will have a point in common without crossing, that is to say, they will be tangent.

The geometer who reasons in this way, consciously or not, is only doing what we have done above to prove two lines which cut have a point in common, and his intuition might seem just as legitimate.

It would deceive him however. We can demonstrate that there are curves which have no tangent, if such a curve is defined as an analytic continuum of the second order.

Without doubt some artifice analogous to those we have discussed above would have sufficed to remove the contradiction; but, as this is met with only in very exceptional cases, it has received no further attention.

Instead of seeking to reconcile intuition with analysis, we have been content to sacrifice one of the two, and as analysis must remain impeccable, we have decided against intuition.

THE PHYSICAL CONTINUUM OF SEVERAL DIMENSIONS.—We have discussed above the physical continuum as derived from the immediate data of our senses, or, if you wish, from the rough results of Fechner's experiments; I have shown that these results are summed up in the contradictory formulas

$$A = B, \qquad B = C, \qquad A < C.$$

Let us now see how this notion has been generalized and how from it has come the concept of many-dimensional continua.

Consider any two aggregates of sensations. Either we can discriminate them one from another, or we can not, just as in Fechner's experiments a weight of 10 grams can be distinguished from a weight of 12 grams, but not from a weight of 11 grams. This is all that is required to construct the continuum of several dimensions.

Let us call one of these aggregates of sensations an *element*. That will be something analogous to the *point* of the mathematicians; it will not be altogether the same thing however. We can not say our element is without extension, since we can not distinguish it from neighboring elements and it is thus surrounded by a sort of haze. If the astronomical comparison may be allowed, our 'elements' would be like nebulae, whereas the mathematical points would be like stars.

That being granted, a system of elements will form a *continuum* if we can pass from any one of them to any other, by a series of consecutive elements such that each is indistinguishable from the preceding. This *linear series* is to the *line* of the mathematician what an isolated *element* was to the point.

Before going farther, I must explain what is meant by a *cut*. Consider a continuum $C$ and remove from it certain of its elements which for an instant we shall regard as no longer belonging to this continuum. The aggregate of the elements so removed will be called a cut. It may happen that, thanks to this cut, $C$ may be *subdivided* into several distinct continua, the aggregate of the remaining elements ceasing to form a unique continuum.

There will then be on $C$ two elements, $A$ and $B$, that must be regarded as belonging to two distinct continua, and this will be recognized because it will be impossible to find a linear series of consecutive elements of $C$, each of these elements indistinguishable from the preceding, the first being $A$ and the last $B$, *without one of the elements of this series being indistinguishable from one of the elements of the cut.*

On the contrary, it may happen that the cut made is insufficient to subdivide the continuum $C$. To classify the physical continua, we will examine precisely what are the cuts which must be made to subdivide them.

If a physical continuum $C$ can be subdivided by a cut reducing to a finite number of elements all distinguishable from one another (and consequently forming neither a continuum, nor several continua), we shall say $C$ is a *one-dimensional* continuum.

If, on the contrary, $C$ can be subdivided only by cuts which are themselves continua, we shall say $C$ has several dimensions. If cuts which are continua of one dimension suffice, we shall say $C$ has two dimensions; if cuts of two dimensions suffice, we shall say $C$ has three dimensions, and so on.

Thus is defined the notion of the physical continuum of several dimensions, thanks to this very simple fact that two aggregates of sensations are distinguishable or indistinguishable.

THE MATHEMATICAL CONTINUUM OF SEVERAL DIMENSIONS.— Thence the notion of the mathematical continuum of $n$ dimensions has sprung quite naturally by a process very like that we discussed at the beginning of this chapter. A point of such a continuum, you know, appears to us as defined by a system of $n$ distinct magnitudes called its coordinates.

These magnitudes need not always be measurable; there is, for instance, a branch of geometry independent of the measurement of these magnitudes, in which it is only a question of knowing, for example, whether on a curve $ABC$, the point $B$ is between the points $A$ and $C$, and not of knowing whether the arc $AB$ is equal to the arc $BC$ or twice as great. This is what is called *Analysis Situs*.

This is a whole body of doctrine which has attracted the

attention of the greatest geometers and where we see flow one from another a series of remarkable theorems. What distinguishes these theorems from those of ordinary geometry is that they are purely qualitative and that they would remain true if the figures were copied by a draughtsman so awkward as to grossly distort the proportions and replace straights by strokes more or less curved.

Through the wish to introduce measure next into the continuum just defined this continuum becomes space, and geometry is born. But the discussion of this is reserved for Part Second.

# PART II
## SPACE

---

### CHAPTER III

#### THE NON-EUCLIDEAN GEOMETRIES

EVERY conclusion supposes premises; these premises themselves either are self-evident and need no demonstration, or can be established only by relying upon other propositions, and since we can not go back thus to infinity, every deductive science, and in particular geometry, must rest on a certain number of undemonstrable axioms. All treatises on geometry begin, therefore, by the enunciation of these axioms. But among these there is a distinction to be made: Some, for example, 'Things which are equal to the same thing are equal to one another,' are not propositions of geometry, but propositions of analysis. I regard them as analytic judgments *a priori*, and shall not concern myself with them.

But I must lay stress upon other axioms which are peculiar to geometry. Most treatises enunciate three of these explicitly:

1° Through two points can pass only one straight;

2° The straight line is the shortest path from one point to another;

3° Through a given point there is not more than one parallel to a given straight.

Although generally a proof of the second of these axioms is omitted, it would be possible to deduce it from the other two and from those, much more numerous, which are implicitly admitted without enunciating them, as I shall explain further on.

It was long sought in vain to demonstrate likewise the third axiom, known as *Euclid's Postulate*. What vast effort has been wasted in this chimeric hope is truly unimaginable. Finally, in the first quarter of the nineteenth century, and almost at the same time, a Hungarian and a Russian, Bolyai and Lobachevski, established irrefutably that this demonstration is impossible; they have almost rid us of inventors of geometries 'sans postulatum'; since then the Académie des Sciences receives only about one or two new demonstrations a year.

The question was not exhausted; it soon made a great stride by the publication of Riemann's celebrated memoir entitled: *Ueber die Hypothesen welche der Geometrie zu Grunde liegen.* This paper has inspired most of the recent works of which I shall speak further on, and among which it is proper to cite those of Beltrami and of Helmholtz.

THE BOLYAI-LOBACHEVSKI GEOMETRY.—If it were possible to deduce Euclid's postulate from the other axioms, it is evident that in denying the postulate and admitting the other axioms, we should be led to contradictory consequences; it would therefore be impossible to base on such premises a coherent geometry.

Now this is precisely what Lobachevski did.

He assumes at the start that: *Through a given point can be drawn two parallels to a given straight.*

And he retains besides all Euclid's other axioms. From these hypotheses he deduces a series of theorems among which it is impossible to find any contradiction, and he constructs a geometry whose faultless logic is inferior in nothing to that of the Euclidean geometry.

The theorems are, of course, very different from those to which we are accustomed, and they can not fail to be at first a little disconcerting.

Thus the sum of the angles of a triangle is always less than two right angles, and the difference between this sum and two right angles is proportional to the surface of the triangle.

It is impossible to construct a figure similar to a given figure but of different dimensions.

If we divide a circumference into $n$ equal parts, and draw tangents at the points of division, these $n$ tangents will form a polygon if the radius of the circle is small enough; but if this radius is sufficiently great they will not meet.

It is useless to multiply these examples; Lobachevski's propositions have no relation to those of Euclid, but they are not less logically bound one to another.

RIEMANN'S GEOMETRY.—Imagine a world uniquely peopled by beings of no thickness (height) ; and suppose these 'infinitely flat' animals are all in the same plane and can not get out. Admit besides that this world is sufficiently far from others to be free from their influence. While we are making hypotheses, it costs us no more to endow these beings with reason and believe them capable of creating a geometry. In that case, they will certainly attribute to space only two dimensions.

But suppose now that these imaginary animals, while remaining without thickness, have the form of a spherical, and not of a plane, figure, and are all on the same sphere without power to get off. What geometry will they construct? First it is clear they will attribute to space only two dimensions; what will play for them the rôle of the straight line will be the shortest path from one point to another on the sphere, that is to say, an arc of a great circle; in a word, their geometry will be the spherical geometry.

What they will call space will be this sphere on which they must stay, and on which happen all the phenomena they can know. Their space will therefore be *unbounded* since on a sphere one can always go forward without ever being stopped, and yet it will be *finite;* one can never find the end of it, but one can make a tour of it.

Well, Riemann's geometry is spherical geometry extended to three dimensions. To construct it, the German mathematician had to throw overboard, not only Euclid's postulate, but also the first axiom: *Only one straight can pass through two points.*

On a sphere, through two given points we can draw *in general* only one great circle (which, as we have just seen, would play the rôle of the straight for our imaginary beings) ; but there is an exception: if the two given points are diametrically opposite, an infinity of great circles can be drawn through them.

In the same way, in Riemann's geometry (at least in one of its forms), through two points will pass in general only a single straight; but there are exceptional cases where through two points an infinity of straights can pass.

There is a sort of opposition between Riemann's geometry and that of Lobachevski.

Thus the sum of the angles of a triangle is:

Equal to two right angles in Euclid's geometry;

Less than two right angles in that of Lobachevski;

Greater than two right angles in that of Riemann.

The number of straights through a given point that can be drawn coplanar to a given straight, but nowhere meeting it, is equal:

To one in Euclid's geometry;

To zero in that of Riemann;

To infinity in that of Lobachevski.

Add that Riemann's space is finite, although unbounded, in the sense given above to these two words.

THE SURFACES OF CONSTANT CURVATURE.—One objection still remained possible. The theorems of Lobachevski and of Riemann present no contradiction; but however numerous the consequences these two geometers have drawn from their hypotheses, they must have stopped before exhausting them, since their number would be infinite; who can say then that if they had pushed their deductions farther they would not have eventually reached some contradiction?

This difficulty does not exist for Riemann's geometry, provided it is limited to two dimensions; in fact, as we have seen, two-dimensional Riemannian geometry does not differ from spherical geometry, which is only a branch of ordinary geometry, and consequently is beyond all discussion.

Beltrami, in correlating likewise Lobachevski's two-dimensional geometry with a branch of ordinary geometry, has equally refuted the objection so far as it is concerned.

Here is how he accomplished it. Consider any figure on a surface. Imagine this figure traced on a flexible and inextensible canvas applied over this surface in such a way that when the canvas is displaced and deformed, the various lines of this figure can change their form without changing their length. In general, this flexible and inextensible figure can not be displaced without leaving the surface; but there are certain particular surfaces for which such a movement would be possible; these are the surfaces of constant curvature.

If we resume the comparison made above and imagine beings without thickness living on one of these surfaces, they will regard as possible the motion of a figure all of whose lines remain constant in length. On the contrary, such a movement would appear absurd to animals without thickness living on a surface of variable curvature.

These surfaces of constant curvature are of two sorts: Some are *of positive curvature,* and can be deformed so as to be applied over a sphere. The geometry of these surfaces reduces itself therefore to the spherical geometry, which is that of Riemann.

The others are *of negative curvature.* Beltrami has shown that the geometry of these surfaces is none other than that of Lobachevski. The two-dimensional geometries of Riemann and Lobachevski are thus correlated to the Euclidean geometry.

INTERPRETATION OF NON-EUCLIDEAN GEOMETRIES.—So vanishes the objection so far as two-dimensional geometries are concerned.

It would be easy to extend Beltrami's reasoning to three-dimensional geometries. The minds that space of four dimensions does not repel will see no difficulty in it, but they are few. I prefer therefore to proceed otherwise.

Consider a certain plane, which I shall call the fundamental plane, and construct a sort of dictionary, by making correspond each to each a double series of terms written in two columns, just as correspond in the ordinary dictionaries the words of two languages whose significance is the same:

*Space:* Portion of space situated above the fundamental plane.

*Plane:* Sphere cutting the fundamental plane orthogonally.

*Straight:* Circle cutting the fundamental plane orthogonally.

*Sphere:* Sphere.

*Circle:* Circle.

*Angle:* Angle.

*Distance between two points:* Logarithm of the cross ratio of these two points and the intersections of the fundamental plane with a circle passing through these two points and cutting it orthogonally. Etc., Etc.

Now take Lobachevski's theorems and translate them with the aid of this dictionary as we transate a German text with the aid of a German-English dictionary. *We shall thus obtain theorems of the ordinary geometry.* For example, that theorem of Lobachevski: 'the sum of the angles of a triangle is less than two right angles' is translated thus: "If a curvilinear triangle has for sides circle-arcs which prolonged would cut orthogonally the fundamental plane, the sum of the angles of this curvilinear triangle will be less than two right angles." Thus, however far the consequences of Lobachevski's hypotheses are pushed, they will never lead to a contradiction. In fact, if two of Lobachevski's theorems were contradictory, it would be the same with the translations of these two theorems, made by the aid of our dictionary, but these translations are theorems of ordinary geometry and no one doubts that the ordinary geometry is free from contradiction. Whence comes this certainty and is it justified? That is a question I can not treat here because it would require to be enlarged upon, but which is very interesting and I think not insoluble.

Nothing remains then of the objection above formulated. This is not all. Lobachevski's geometry, susceptible of a concrete interpretation, ceases to be a vain logical exercise and is capable of applications; I have not the time to speak here of these applications, nor of the aid that Klein and I have gotten from them for the integration of linear differential equations.

This interpretation moreover is not unique, and several dictionaries analogous to the preceding could be constructed, which would enable us by a simple 'translation' to transform Lobachevski's theorems into theorems of ordinary geometry.

THE IMPLICIT AXIOMS.—Are the axioms explicitly enunciated in our treatises the sole foundations of geometry? We may be assured of the contrary by noticing that after they are successively abandoned there are still left over some propositions common to the theories of Euclid, Lobachevski and Riemann. These propositions must rest on premises the geometers admit without enunciation. It is interesting to try to disentangle them from the classic demonstrations.

Stuart Mill has claimed that every definition contains an

axiom, because in defining one affirms implicitly the existence of the object defined. This is going much too far; it is rare that in mathematics a definition is given without its being followed by the demonstration of the existence of the object defined, and when this is dispensed with it is generally because the reader can easily supply it. It must not be forgotten that the word existence has not the same sense when it refers to a mathematical entity and when it is a question of a material object. A mathematical entity exists, provided its definition implies no contradiction, either in itself, or with the propositions already admitted.

But if Stuart Mill's observation can not be applied to all definitions, it is none the less just for some of them. The plane is sometimes defined as follows:

The plane is a surface such that the straight which joins any two of its points is wholly on this surface.

This definition manifestly hides a new axiom; it is true we might change it, and that would be preferable, but then we should have to enunciate the axiom explicitly.

Other definitions would suggest reflections not less important.

Such, for example, is that of the equality of two figures; two figures are equal when they can be superposed; to superpose them one must be displaced until it coincides with the other; but how shall it be displaced? If we should ask this, no doubt we should be told that it must be done without altering the shape and as a rigid solid. The vicious circle would then be evident.

In fact this definition defines nothing; it would have no meaning for a being living in a world where there were only fluids. If it seems clear to us, that is because we are used to the properties of natural solids which do not differ much from those of the ideal solids, all of whose dimensions are invariable.

Yet, imperfect as it may be, this definition implies an axiom.

The possibility of the motion of a rigid figure is not a self-evident truth, or at least it is so only in the fashion of Euclid's postulate and not as an analytic judgment *a priori* would be.

Moreover, in studying the definitions and the demonstrations of geometry, we see that one is obliged to admit without proof not only the possibility of this motion, but some of its properties besides.

This is at once seen from the definition of the straight line. Many defective definitions have been given, but the true one is that which is implied in all the demonstrations where the straight line enters:

"It may happen that the motion of a rigid figure is such that all the points of a line belonging to this figure remain motionless while all the points situated outside of this line move. Such a line will be called a straight line." We have designedly, in this enunciation, separated the definition from the axiom it implies.

Many demonstrations, such as those of the cases of the equality of triangles, of the possibility of dropping a perpendicular from a point to a straight, presume propositions which are not enunciated, for they require the admission that it is possible to transport a figure in a certain way in space.

THE FOURTH GEOMETRY.—Among these implicit axioms, there is one which seems to me to merit some attention, because when it is abandoned a fourth geometry can be constructed as coherent as those of Euclid, Lobachevski and Riemann.

To prove that a perpendicular may always be erected at a point $A$ to a straight $AB$, we consider a straight $AC$ movable around the point $A$ and initially coincident with the fixed straight $AB$; and we make it turn about the point $A$ until it comes into the prolongation of $AB$.

Thus two propositions are presupposed: First, that such a rotation is possible, and next that it may be continued until the two straights come into the prolongation one of the other.

If the first point is admitted and the second rejected, we are led to a series of theorems even stranger than those of Lobachevski and Riemann, but equally exempt from contradiction.

I shall cite only one of these theorems and that not the most singular: *A real straight may be perpendicular to itself.*

LIE'S THEOREM.—The number of axioms implicitly introduced in the classic demonstrations is greater than necessary, and it would be interesting to reduce it to a minimum. It may first be asked whether this reduction is possible, whether the number of necessary axioms and that of imaginable geometries are not infinite.

A theorem of Sophus Lie dominates this whole discussion. It may be thus enunciated:

Suppose the following premises are admitted:

1° Space has $n$ dimensions;

2° The motion of a rigid figure is possible;

3° It requires $p$ conditions to determine the position of this figure in space.

*The number of geometries compatible with these premises will be limited.*

I may even add that if $n$ is given, a superior limit can be assigned to $p$.

If therefore the possibility of motion is admitted, there can be invented only a finite (and even a rather small) number of three-dimensional geometries.

RIEMANN'S GEOMETRIES.—Yet this result seems contradicted by Riemann, for this savant constructs an infinity of different geometries, and that to which his name is ordinarily given is only a particular case.

All depends, he says, on how the length of a curve is defined. Now, there is an infinity of ways of defining this length, and each of them may be the starting point of a new geometry.

That is perfectly true, but most of these definitions are incompatible with the motion of a rigid figure, which in the theorem of Lie is supposed possible. These geometries of Riemann, in many ways so interesting, could never therefore be other than purely analytic and would not lend themselves to demonstrations analogous to those of Euclid.

ON THE NATURE OF AXIOMS.—Most mathematicians regard Lobachevski's geometry only as a mere logical curiosity; some of them, however, have gone farther. Since several geometries are possible, is it certain ours is the true one? Experience no doubt teaches us that the sum of the angles of a triangle is equal to two right angles; but this is because the triangles we deal with are too little; the difference, according to Lobachevski, is proportional to the surface of the triangle; will it not perhaps become sensible when we shall operate on larger triangles or when our measurements shall become more precise? The Euclidean geometry would thus be only a provisional geometry.

To discuss this opinion, we should first ask ourselves what is the nature of the geometric axioms.

Are they synthetic *a priori* judgments, as Kant said?

They would then impose themselves upon us with such force that we could not conceive the contrary proposition, nor build upon it a theoretic edifice. There would be no non-Euclidean geometry.

To be convinced of it take a veritable synthetic *a priori* judgment, the following, for instance, of which we have seen the preponderant rôle in the first chapter:

*If a theorem is true for the number 1, and if it has been proved that it is true of $n + 1$ provided it is true of $n$, it will be true of all the positive whole numbers.*

Then try to escape from that and, denying this proposition, try to found a false arithmetic analogous to non-Euclidean geometry—it can not be done; one would even be tempted at first blush to regard these judgments as analytic.

Moreover, resuming our fiction of animals without thickness, we can hardly admit that these beings, if their minds are like ours, would adopt the Euclidean geometry which would be contradicted by all their experience.

Should we therefore conclude that the axioms of geometry are experimental verities? But we do not experiment on ideal straights or circles; it can only be done on material objects. On what then could be based experiments which should serve as foundation for geometry? The answer is easy.

We have seen above that we constantly reason as if the geometric figures behaved like solids. What geometry would borrow from experience would therefore be the properties of these bodies. The properties of light and its rectilinear propagation have also given rise to some of the propositions of geometry, and in particular those of projective geometry, so that from this point of view one would be tempted to say that metric geometry is the study of solids, and projective, that of light.

But a difficulty remains, and it is insurmountable. If geometry were an experimental science, it would not be an exact science, it would be subject to a continual revision. Nay, it would from this very day be convicted of error, since we know that there is no rigorously rigid solid.

The *axioms of geometry therefore are neither synthetic* a priori *judgments nor experimental facts.*

They are *conventions;* our choice among all possible conventions is *guided* by experimental facts; but it remains *free* and is limited only by the necessity of avoiding all contradiction. Thus it is that the postulates can remain *rigorously* true even though the experimental laws which have determined their adoption are only approximative.

In other words, *the axioms of geometry* (I do not speak of those of arithmetic) *are merely disguised definitions.*

Then what are we to think of that question: Is the Euclidean geometry true?

It has no meaning.

As well ask whether the metric system is true and the old measures false; whether Cartesian coordinates are true and polar coordinates false. One geometry can not be more true than another; it can only be *more convenient.*

Now, Euclidean geometry is, and will remain, the most convenient:

1° Because it is the simplest; and it is so not only in consequence of our mental habits, or of I know not what direct intuition that we may have of Euclidean space; it is the simplest in itself, just as a polynomial of the first degree is simpler than one of the second; the formulas of spherical trigonometry are more complicated than those of plane trigonometry, and they would still appear so to an analyst ignorant of their geometric signification.

2° Because it accords sufficiently well with the properties of natural solids, those bodies which our hands and our eyes compare and with which we make our instruments of measure.

# CHAPTER IV

## SPACE AND GEOMETRY

LET us begin by a little paradox.

Beings with minds like ours, and having the same senses as we, but without previous education, would receive from a suitably chosen external world impressions such that they would be led to construct a geometry other than that of Euclid and to localize the phenomena of that external world in a non-Euclidean space, or even in a space of four dimensions.

As for us, whose education has been accomplished by our actual world, if we were suddenly transported into this new world, we should have no difficulty in referring its phenomena to our Euclidean space. Conversely, if these beings were transported into our environment, they would be led to relate our phenomena to non-Euclidean space.

Nay more; with a little effort we likewise could do it. A person who should devote his existence to it might perhaps attain to a realization of the fourth dimension.

GEOMETRIC SPACE AND PERCEPTUAL SPACE.—It is often said the images of external objects are localized in space, even that they can not be formed except on this condition. It is also said that this space, which serves thus as a ready prepared *frame* for our sensations and our representations, is identical with that of the geometers, of which it possesses all the properties.

To all the good minds who think thus, the preceding statement must have appeared quite extraordinary. But let us see whether they are not subject to an illusion that a more profound analysis would dissipate.

What, first of all, are the properties of space, properly so called? I mean of that space which is the object of geometry and which I shall call *geometric space*.

The following are some of the most essential:

1° It is continuous;

2° It is infinite;

3° It has three dimensions;

4° It is homogeneous, that is to say, all its points are identical one with another;

5° It is isotropic, that is to say, all the straights which pass through the same point are identical one with another.

Compare it now to the frame of our representations and our sensations, which I may call *perceptual space*.

VISUAL SPACE.—Consider first a purely visual impression, due to an image formed on the bottom of the retina.

A cursory analysis shows us this image as continuous, but as possessing only two dimensions; this already distinguishes from geometric space what we may call *pure visual space*.

Besides, this image is enclosed in a limited frame.

Finally, there is another difference not less important: *this pure visual space is. not homogeneous.* All the points of the retina, aside from the images which may there be formed, do not play the same rôle. The yellow spot can in no way be regarded as identical with a point on the border of the retina. In fact, not only does the same object produce there much more vivid impressions, but in every *limited* frame the point occupying the center of the frame will never appear as equivalent to a point near one of the borders.

No doubt a more profound analysis would show us that this continuity of visual space and its two dimensions are only an illusion; it would separate it therefore still more from geometric space, but we shall not dwell on this remark.

Sight, however, enables us to judge of distances and consequently to perceive a third dimension. But every one knows that this perception of the third dimension reduces itself to the sensation of the effort at accommodation it is necessary to make, and to that of the convergence which must be given to the two eyes, to perceive an object distinctly.

These are muscular sensations altogether different from the visual sensations which have given us the notion of the first two dimensions. The third dimension therefore will not appear to us as playing the same rôle as the other two. What may be called *complete visual space* is therefore not an isotropic space.

It has, it is true, precisely three dimensions, which means that the elements of our visual sensations (those at least which combine to form the notion of extension) will be completely defined when three of them are known; to use the language of mathematics, they will be functions of three independent variables.

But examine the matter a little more closely. The third dimension is revealed to us in two different ways: by the effort of accommodation and by the convergence of the eyes.

No doubt these two indications are always concordant, there is a constant relation between them, or, in mathematical terms, the two variables which measure these two muscular sensations do not appear to us as independent; or again, to avoid an appeal to mathematical notions already rather refined, we may go back to the language of the preceding chapter and enunciate the same fact as follows: If two sensations of convergence, $A$ and $B$, are indistinguishable, the two sensations of accommodation, $A'$ and $B'$, which respectively accompany them, will be equally indistinguishable.

But here we have, so to speak, an experimental fact; *a priori* nothing prevents our supposing the contrary, and if the contrary takes place, if these two muscular sensations vary independently of one another, we shall have to take account of one more independent variable, and 'complete visual space' will appear to us as a physical continuum of four dimensions.

We have here even, I will add, a fact of *external* experience. Nothing prevents our supposing that a being with a mind like ours, having the same sense organs that we have, may be placed in a world where light would only reach him after having traversed reflecting media of complicated form. The two indications which serve us in judging distances would cease to be connected by a constant relation. A being who should achieve in such a world the education of his senses would no doubt attribute four dimensions to complete visual space.

TACTILE SPACE AND MOTOR SPACE.—'Tactile space' is still more complicated than visual space and farther removed from geometric space. It is superfluous to repeat for touch the discussion I have given for sight.

But apart from the data of sight and touch, there are other sensations which contribute as much and more than they to the genesis of the notion of space. These are known to every one; they accompany all our movements, and are usually called muscular sensations.

The corresponding frame constitutes what may be called *motor space.*

Each muscle gives rise to a special sensation capable of augmenting or of diminishing, so that the totality of our muscular sensations will depend upon as many variables as we have muscles. From this point of view, *motor space would have as many dimensions as we have muscles.*

I know it will be said that if the muscular sensations contribute to form the notion of space, it is because we have the sense of the *direction* of each movement and that it makes an integrant part of the sensation. If this were so, if a muscular sensation could not arise except accompanied by this geometric sense of direction, geometric space would indeed be a form imposed upon our sensibility.

But I perceive nothing at all of this when I analyze my sensations.

What I do see is that the sensations which correspond to movements in the same direction are connected in my mind by a mere *association of ideas.* It is to this association that what we call 'the sense of direction' is reducible. This feeling therefore can not be found in a single sensation.

This association is extremely complex, for the contraction of the same muscle may correspond, according to the position of the limbs, to movements of very different direction.

Besides, it is evidently acquired; it is, like all associations of ideas, the result of a *habit;* this habit itself results from very numerous *experiences;* without any doubt, if the education of our senses had been accomplished in a different environment, where we should have been subjected to different impressions, contrary habits would have arisen and our muscular sensations would have been associated according to other laws.

CHARACTERISTICS OF PERCEPTUAL SPACE.—Thus perceptual space, under its triple form, visual, tactile and motor, is essentially different from geometric space.

It is neither homogeneous, nor isotropic; one can not even say that it has three dimensions.

It is often said that we 'project' into geometric space the objects of our external perception; that we 'localize' them.

Has this a meaning, and if so what?

Does it mean that we *represent* to ourselves external objects in geometric space?

Our representations are only the reproduction of our sensations; they can therefore be ranged only in the same frame as these, that is to say, in perceptual space.

It is as impossible for us to represent to ourselves external bodies in geometric space, as it is for a painter to paint on a plane canvas objects with their three dimensions.

Perceptual space is only an image of geometric space, an image altered in shape by a sort of perspective, and we can represent to ourselves objects only by bringing them under the laws of this perspective.

Therefore we do not *represent* to ourselves external bodies in geometric space, but we *reason* on these bodies as if they were situated in geometric space.

When it is said then that we 'localize' such and such an object at such and such a point of space, what does it mean?

*It simply means that we represent to ourselves the movements it would be necessary to make to reach that object;* and one may not say that to represent to oneself these movements, it is necessary to project the movements themselves in space and that the notion of space must, consequently, pre-exist.

When I say that we represent to ourselves these movements, I mean only that we represent to ourselves the muscular sensations which accompany them and which have no geometric character whatever, which consequently do not at all imply the pre-existence of the notion of space.

CHANGE OF STATE AND CHANGE OF POSITION.—But, it will be said, if the idea of geometric space is not imposed upon our mind, and if, on the other hand, none of our sensations can furnish it, how could it have come into existence?

This is what we have now to examine, and it will take some time, but I can summarize in a few words the attempt at explanation that I am about to develop.

*None of our sensations, isolated, could have conducted us to the idea of space; we are led to it only in studying the laws, according to which these sensations succeed each other.*

We see first that our impressions are subject to change; but among the changes we ascertain we are soon led to make a distinction.

At one time we say that the objects which cause these impressions have changed state, at another time that they have changed position, that they have only been displaced.

Whether an object changes its state or merely its position, this is always translated for us in the same manner: *by a modification in an aggregate of impressions.*

How then could we have been led to distinguish between the two? It is easy to account for. If there has only been a change of position, we can restore the primitive aggregate of impressions by making movements which replace us opposite the mobile object in the same *relative* situation. We thus *correct* the modification that happened and we reestablish the initial state by an inverse modification.

If it is a question of sight, for example, and if an object changes its place before our eye, we can 'follow it with the eye' and maintain its image on the same point of the retina by appropriate movements of the eyeball.

These movements we are conscious of because they are voluntary and because they are accompanied by muscular sensations, but that does not mean that we represent them to ourselves in geometric space.

So what characterizes change of position, what distinguishes it from change of state, is that it can always be corrected in this way.

It may therefore happen that we pass from the totality of impressions $A$ to the totality $B$ in two different ways:

1° Involuntarily and without experiencing muscular sensations; this happens when it is the object which changes place;

2° Voluntarily and with muscular sensations; this happens when the object is motionless, but we move so that the object has relative motion with reference to us.

If this be so, the passage from the totality $A$ to the totality $B$ is only a change of position.

It follows from this that sight and touch could not have given us the notion of space without the aid of the 'muscular sense.'

Not only could this notion not be derived from a single sensation or even *from a series of sensations,* but what is more, an *immobile* being could never have acquired it, since, not being able to *correct* by his movements the effects of the changes of position of exterior objects, he would have had no reason whatever to distinguish them from changes of state. Just as little could he have acquired it if his motions had not been voluntary or were unaccompanied by any sensations.

CONDITIONS OF COMPENSATION.—How is a like compensation possible, of such sort that two changes, otherwise independent of each other, reciprocally correct each other?

A mind already familiar with geometry would reason as follows: Evidently, if there is to be compensation, the various parts of the external object, on the one hand, and the various sense organs, on the other hand, must be in the same *relative* position after the double change. And, for that to be the case, the various parts of the external object must likewise have retained in reference to each other the same relative position, and the same must be true of the various parts of our body in regard to each other.

In other words, the external object, in the first change, must be displaced as is a rigid solid, and so must it be with the whole of our body in the second change which corrects the first.

Under these conditions, compensation may take place.

But we who as yet know nothing of geometry, since for us the notion of space is not yet formed, we can not reason thus, we can not foresee *a priori* whether compensation is possible. But experience teaches us that it sometimes happens, and it is from this experimental fact that we start to distinguish changes of state from changes of position.

SOLID BODIES AND GEOMETRY.—Among surrounding objects there are some which frequently undergo displacements susceptible of being thus corrected by a correlative movement of our own body; these are the *solid bodies.* The other objects, whose form is variable, only exceptionally undergo like displacements (change of position without change of form). When a body changes its place *and its shape,* we can no longer, by appropriate movements, bring back our sense-organs into the same *relative* situation with regard to this body; consequently we can no longer reestablish the primitive totality of impressions.

It is only later, and as a consequence of new experiences, that we learn how to decompose the bodies of variable form into smaller elements, such that each is displaced almost in accordance with the same laws as solid bodies. Thus we distinguish 'deformations' from other changes of state; in these deformations, each element undergoes a mere change of position, which can be corrected, but the modification undergone by the aggregate is more profound and is no longer susceptible of correction by a correlative movement.

Such a notion is already very complex and must have been relatively late in appearing; moreover it could not have arisen if the observation of solid bodies had not already taught us to distinguish changes of position.

*Therefore, if there were no solid bodies in nature, there would be no geometry.*

Another remark also deserves a moment's attention. Suppose a solid body to occupy successively the positions $a$ and $\beta$; in its first position, it will produce on us the totality of impressions $A$, and in its second position the totality of impressions $B$. Let there be now a second solid body, having qualities entirely different from the first, for example, a different color. Suppose it to pass from the position $a$, where it gives us the totality of impressions $A'$, to the position $\beta$, where it gives the totality of impressions $B'$.

In general, the totality $A$ will have nothing in common with the totality $A'$, nor the totality $B$ with the totality $B'$. The transition from the totality $A$ to the totality $B$ and that from the totality $A'$ to the totality $B'$ are therefore two changes which *in themselves* have in general nothing in common.

And yet we regard these two changes both as displacements and, furthermore, we consider them as the *same* displacement. How can that be?

It is simply because they can both be corrected by the *same* correlative movement of our body.

'Correlative movement' therefore constitutes the *sole connection* between two phenomena which otherwise we never should have dreamt of likening.

On the other hand, our body, thanks to the number of its articulations and muscles, may make a multitude of different movements; but all are not capable of 'correcting' a modification of external objects; only those will be capable of it in which our whole body, or at least all those of our sense-organs which come into play, are displaced as a whole, that is, without their relative positions varying, or in the fashion of a solid body.

To summarize:

1° We are led at first to distinguish two categories of phenomena:

Some, involuntary, unaccompanied by muscular sensations, are attributed by us to external objects; these are external changes;

Others, opposite in character and attributed by us to the movements of our own body, are internal changes;

2° We notice that certain changes of each of these categories may be corrected by a correlative change of the other category;

3° We distinguish among external changes those which have thus a correlative in the other category; these we call displacements; and just so among the internal changes, we distinguish those which have a correlative in the first category.

Thus are defined, thanks to this reciprocity, a particular class of phenomena which we call displacements.

*The laws of these phenomena constitute the object of geometry.*

LAW OF HOMOGENEITY.—The first of these laws is the law of homogeneity.

Suppose that, by an external change $a$, we pass from the totality of impressions $A$ to the totality $B$, then that this change $a$ is corrected by a correlative voluntary movement $\beta$, so that we are brought back to the totality $A$.

Suppose now that another external change $a'$ makes us pass anew from the totality $A$ to the totality $B$.

Experience teaches us that this change $a'$ is, like $a$, susceptible of being corrected by a correlative voluntary movement

$\beta'$ and that this movement $\beta'$ corresponds to the same muscular sensations as the movement $\beta$ which corrected $a$.

This fact is usually enunciated by saying that *space is homogeneous and isotropic.*

It may also be said that a movement which has once been produced may be repeated a second and a third time, and so on, without its properties varying.

In the first chapter, where we discussed the nature of mathematical reasoning, we saw the importance which must be attributed to the possibility of repeating indefinitely the same operation.

It is from this repetition that mathematical reasoning gets its power; it is, therefore, thanks to the law of homogeneity, that it has a hold on the geometric facts.

For completeness, to the law of homogeneity should be added a multitude of other analogous laws, into the details of which I do not wish to enter, but which mathematicians sum up in a word by saying that displacements form 'a group.'

THE NON-EUCLIDEAN WORLD.—If geometric space were a frame imposed on *each* of our representations, considered individually, it would be impossible to represent to ourselves an image stripped of this frame, and we could change nothing of our geometry.

But this is not the case; geometry is only the résumé of the laws according to which these images succeed each other. Nothing then prevents us from imagining a series of representations, similar in all points to our ordinary representations, but succeeding one another according to laws different from those to which we are accustomed.

We can conceive then that beings who received their education in an environment where these laws were thus upset might have a geometry very different from ours.

Suppose, for example, a world enclosed in a great sphere and subject to the following laws:

The temperature is not uniform; it is greatest at the center, and diminishes in proportion to the distance from the center, to sink to absolute zero when the sphere is reached in which this world is enclosed.

To specify still more precisely the law in accordance with which this temperature varies: Let $R$ be the radius of the limiting sphere; let $r$ be the distance of the point considered from the center of this sphere. The absolute temperature shall be proportional to $R^2 - r^2$.

I shall further suppose that, in this world, all bodies have the same coefficient of dilatation, so that the length of any rule is proportional to its absolute temperature.

Finally, I shall suppose that a body transported from one point to another of different temperature is put immediately into thermal equilibrium with its new environment.

Nothing in these hypotheses is contradictory or unimaginable.

A movable object will then become smaller and smaller in proportion as it approaches the limit-sphere.

Note first that, though this world is limited from the point of view of our ordinary geometry, it will appear infinite to its inhabitants.

In fact, when these try to approach the limit-sphere, they cool off and become smaller and smaller. Therefore the steps they take are also smaller and smaller, so that they can never reach the limiting sphere.

If, for us, geometry is only the study of the laws according to which rigid solids move, for these imaginary beings it will be the study of the laws of motion of solids *distorted by the differences of temperature* just spoken of.

No doubt, in our world, natural solids likewise undergo variations of form and volume due to warming or cooling. But we neglect these variations in laying the foundations of geometry, because, besides their being very slight, they are irregular and consequently seem to us accidental.

In our hypothetical world, this would no longer be the case, and these variations would follow regular and very simple laws.

Moreover, the various solid pieces of which the bodies of its inhabitants would be composed would undergo the same variations of form and volume.

I will make still another hypothesis; I will suppose light traverses media diversely refractive and such that the index of refraction is inversely proportional to $R^2 - r^2$. It is easy to see that, under these conditions, the rays of light would not be rectilinear, but circular.

To justify what precedes, it remains for me to show that certain changes in the position of external objects can be *corrected* by correlative movements of the sentient beings inhabiting this imaginary world, and that in such a way as to restore the primitive aggregate of impressions experienced by these sentient beings.

Suppose in fact that an object is displaced, undergoing deformation, not as a rigid solid, but as a solid subjected to unequal dilatations in exact conformity to the law of temperature above supposed. Permit me for brevity to call such a movement a *non-Euclidean displacement.*

If a sentient being happens to be in the neighborhood, his impressions will be modified by the displacement of the object, but he can reestablish them by moving in a suitable manner. It suffices if finally the aggregate of the object and the sentient being, considered as forming a single body, has undergone one of those particular displacements I have just called non-Euclidean. This is possible if it be supposed that the limbs of these beings dilate according to the same law as the other bodies of the world they inhabit.

Although from the point of view of our ordinary geometry there is a deformation of the bodies in this displacement and their various parts are no longer in the same relative position, nevertheless we shall see that the impressions of the sentient being have once more become the same.

In fact, though the mutual distances of the various parts may have varied, yet the parts originally in contact are again in contact. Therefore the tactile impressions have not changed.

On the other hand, taking into account the hypothesis made above in regard to the refraction and the curvature of the rays of light, the visual impressions will also have remained the same.

These imaginary beings will therefore like ourselves be led to classify the phenomena they witness and to distinguish among them the 'changes of position' susceptible of correction by a correlative voluntary movement.

If they construct a geometry, it will not be, as ours is, the study of the movements of our rigid solids; it will be the study of the changes of position which they will thus have distinguished and which are none other than the 'non-Euclidean displacements'; *it will be non-Euclidean geometry.*

Thus beings like ourselves, educated in such a world, would not have the same geometry as ours.

THE WORLD OF FOUR DIMENSIONS.—We can represent to ourselves a four-dimensional world just as well as a non-Euclidean.

The sense of sight, even with a single eye, together with the muscular sensations relative to the movements of the eyeball, would suffice to teach us space of three dimensions.

The images of external objects are painted on the retina, which is a two-dimensional canvas; they are *perspectives.*

But, as eye and objects are movable, we see in succession various perspectives of the same body, taken from different points of view.

At the same time, we find that the transition from one perspective to another is often accompanied by muscular sensations.

If the transition from the perspective *A* to the perspective *B*, and that from the perspective *A'* to the perspective *B'* are accompanied by the same muscular sensations, we liken them one to the other as operations of the same nature.

Studying then the laws according to which these operations combine, we recognize that they form a group, which has the same structure as that of the movements of rigid solids.

Now, we have seen that it is from the properties of this group we have derived the notion of geometric space and that of three dimensions.

We understand thus how the idea of a space of three dimensions could take birth from the pageant of these perspectives, though each of them is of only two dimensions, since *they follow one another according to certain laws.*

Well, just as the perspective of a three-dimensional figure can be made on a plane, we can make that of a four-dimensional figure on a picture of three (or of two) dimensions. To a geometer this is only child's play.

We can even take of the same figure several perspectives from several different points of view.

We can easily represent to ourselves these perspectives, since they are of only three dimensions.

Imagine that the various perspectives of the same object succeed one another, and that the transition from one to the other is accompanied by muscular sensations.

We shall of course consider two of these transitions as two operations of the same nature when they are associated with the same muscular sensations.

Nothing then prevents us from imagining that these operations combine according to any law we choose, for example, so as to form a group with the same structure as that of the movements of a rigid solid of four dimensions.

Here there is nothing unpicturable, and yet these sensations are precisely those which would be felt by a being possessed of a two-dimensional retina who could move in space of four dimensions. In this sense we may say the fourth dimension is imaginable.

CONCLUSIONS.—We see that experience plays an indispensable rôle in the genesis of geometry; but it would be an error thence to conclude that geometry is, even in part, an experimental science.

If it were experimental, it would be only approximative and provisional. And what rough approximation!

Geometry would be only the study of the movements of solids; but in reality it is not occupied with natural solids, it has for object certain ideal solids, absolutely rigid, which are only a simplified and very remote image of natural solids.

The notion of these ideal solids is drawn from all parts of our mind, and experience is only an occasion which induces us to bring it forth from them.

The object of geometry is the study of a particular 'group'; but the general group concept pre-exists, at least potentially, in our minds. It is imposed on us, not as form of our sense, but as form of our understanding.

Only, from among all the possible groups, that must be chosen which will be, so to speak, the *standard* to which we shall refer natural phenomena.

Experience guides us in this choice without forcing it upon

us; it tells us not which is the truest geometry, but which is the most *convenient*.

Notice that I have been able to describe the fantastic worlds above imagined *without ceasing to employ the language of ordinary geometry.*

And, in fact, we should not have to change it if transported thither.

Beings educated there would doubtless find it more convenient to create a geometry different from ours, and better adapted to their impressions. As for us, in face of the *same* impressions, it is certain we should find it more convenient not to change our habits.

# CHAPTER V

## EXPERIENCE AND GEOMETRY

1. ALREADY in the preceding pages I have several times tried to show that the principles of geometry are not experimental facts and that in particular Euclid's postulate can not be proven experimentally.

However decisive appear to me the reasons already given, I believe I should emphasize this point because here a false idea is profoundly rooted in many minds.

2. If we construct a material circle, measure its radius and circumference, and see if the ratio of these two lengths is equal to $\pi$, what shall we have done? We shall have made an experiment on the properties of the matter with which we constructed this *round thing,* and of that of which the measure used was made.

3. GEOMETRY AND ASTRONOMY.—The question has also been put in another way. If Lobachevski's geometry is true, the parallax of a very distant star will be finite; if Riemann's is true, it will be negative. These are results which seem within the reach of experiment, and there have been hopes that astronomical observations might enable us to decide between the three geometries.

But in astronomy 'straight line' means simply 'path of a ray of light.'

If therefore negative parallaxes were found, or if it were demonstrated that all parallaxes are superior to a certain limit, two courses would be open to us; we might either renounce Euclidean geometry, or else modify the laws of optics and suppose that light does not travel rigorously in a straight line.

It is needless to add that all the world would regard the latter solution as the more advantageous.

The Euclidean geometry has, therefore, nothing to fear from fresh experiments.

4. Is the position tenable, that certain phenomena, possible in Euclidean space, would be impossible in non-Euclidean space,

so that experience, in establishing these phenomena, would directly contradict the non-Euclidean hypothesis? For my part I think no such question can be put. To my mind it is precisely equivalent to the following, whose absurdity is patent to all eyes: are there lengths expressible in meters and centimeters, but which can not be measured in fathoms, feet and inches, so that experience, in ascertaining the existence of these lengths, would directly contradict the hypothesis that there are fathoms divided into six feet?

Examine the question more closely. I suppose that the straight line possesses in Euclidean space any two properties which I shall call $A$ and $B$; that in non-Euclidean space it still possesses the property $A$, but no longer has the property $B$; finally I suppose that in both Euclidean and non-Euclidean space the straight line is the only line having the property $A$.

If this were so, experience would be capable of deciding between the hypothesis of Euclid and that of Lobachevski. It would be ascertained that a definite concrete object, accessible to experiment, for example, a pencil of rays of light, possesses the property $A$; we should conclude that it is rectilinear, and then investigate whether or not it has the property $B$.

But *this is not so;* no property exists which, like this property $A$, can be an absolute criterion enabling us to recognize the straight line and to distinguish it from every other line.

Shall we say, for instance: "the following is such a property: the straight line is a line such that a figure of which this line forms a part can be moved without the mutual distances of its points varying and so that all points of this line remain fixed"?

This, in fact, is a property which, in Euclidean or non-Euclidean space, belongs to the straight and belongs only to it. But how shall we ascertain experimentally whether it belongs to this or that concrete object? It will be necessary to measure distances, and how shall one know that any concrete magnitude which I have measured with my material instrument really represents the abstract distance?

We have only pushed back the difficulty.

In reality the property just enunciated is not a property of the straight line alone, it is a property of the straight line and distance. For it to serve as absolute criterion, we should have to be able to establish not only that it does not also belong to a line other than the straight and to distance, but in addition that it does not belong to a line other than the straight and to a magnitude other than distance. Now this is not true.

It is therefore impossible to imagine a concrete experiment which can be interpreted in the Euclidean system and not in the Lobachevskian system, so that I may conclude:

No experience will ever be in contradiction to Euclid's postulate; nor, on the other hand, will any experience ever contradict the postulate of Lobachevski.

5. But it is not enough that the Euclidean (or non-Euclidean) geometry can never be directly contradicted by experience. Might it not happen that it can accord with experience only by violating the principle of sufficient reason or that of the relativity of space?

I will explain myself: consider any material system; we shall have to regard, on the one hand, 'the state' of the various bodies of this system (for instance, their temperature, their electric potential, etc.), and, on the other hand, their position in space; and among the data which enable us to define this position we shall, moreover, distinguish the mutual distances of these bodies, which define their relative positions, from the conditions which define the absolute position of the system and its absolute orientation in space.

The laws of the phenomena which will happen in this system will depend on the state of these bodies and their mutual distances; but, because of the relativity and passivity of space, they will not depend on the absolute position and orientation of the system.

In other words, the state of the bodies and their mutual distances at any instant will depend solely on the state of these same bodies and on their mutual distances at the initial instant, but will not at all depend on the absolute initial position of the system or on its absolute initial orientation. This is what for brevity I shall call *the law of relativity*.

Hitherto I have spoken as a Euclidean geometer. As I have said, an experience, whatever it be, admits of an interpretation on the Euclidean hypothesis; but it admits of one equally on the non-Euclidean hypothesis. Well, we have made a series of experiments; we have interpreted them on the Euclidean hypothesis, and we have recognized that these experiments thus interpreted do not violate this 'law of relativity.'

We now interpret them on the non-Euclidean hypothesis: this is always possible; only the non-Euclidean distances of our different bodies in this new interpretation will not generally be the same as the Euclidean distances in the primitive interpretation.

Will our experiments, interpreted in this new manner, still be in accord with our 'law of relativity'? And if there were not this accord, should we not have also the right to say experience had proven the falsity of the non-Euclidean geometry?

It is easy to see that this is an idle fear; in fact, to apply the law of relativity in all rigor, it must be applied to the entire universe. For if only a part of this universe were considered, and if the absolute position of this part happened to vary, the distances to the other bodies of the universe would likewise vary, their influence on the part of the universe considered would consequently augment or diminish, which might modify the laws of the phenomena happening there.

But if our system is the entire universe, experience is powerless to give information about its absolute position and orientation in space. All that our instruments, however perfected they may be, can tell us will be the state of the various parts of the universe and their mutual distances.

So our law of relativity may be thus enunciated:

The readings we shall be able to make on our instruments at any instant will depend only on the readings we could have made on these same instruments at the initial instant.

Now such an enunciation is independent of every interpretation of experimental facts. If the law is true in the Euclidean interpretation, it will also be true in the non-Euclidean interpretation.

Allow me here a short digression. I have spoken above of the data which define the position of the various bodies of the system; I should likewise have spoken of those which define their velocities; I should then have had to distinguish the velocities with which the mutual distances of the different bodies vary;

and, on the other hand, the velocities of translation and rotation of the system, that is to say, the velocities with which its absolute position and orientation vary.

To fully satisfy the mind, the law of relativity should be expressible thus:

The state of bodies and their mutual distances at any instant, as well as the velocities with which these distances vary at this same instant, will depend only on the state of those bodies and their mutual distances at the initial instant, and the velocities with which these distances vary at this initial instant, but they will not depend either upon the absolute initial position of the system, or upon its absolute orientation, or upon the velocities with which this absolute position and orientation varied at the initial instant.

Unhappily the law thus enunciated is not in accord with experiments, at least as they are ordinarily interpreted.

Suppose a man be transported to a planet whose heavens were always covered with a thick curtain of clouds, so that he could never see the other stars; on that planet he would live as if it were isolated in space. Yet this man could become aware that it turned, either by measuring its oblateness (done ordinarily by the aid of astronomic observations, but capable of being done by purely geodetic means), or by repeating the experiment of Foucault's pendulum. The absolute rotation of this planet could therefore be made evident.

That is a fact which shocks the philosopher, but which the physicist is compelled to accept.

We know that from this fact Newton inferred the existence of absolute space; I myself am quite unable to adopt this view. I shall explain why in Part III. For the moment it is not my intention to enter upon this difficulty.

Therefore I must resign myself, in the enunciation of the law of relativity, to including velocities of every kind among the data which define the state of the bodies.

However that may be, this difficulty is the same for Euclid's geometry as for Lobachevski's; I therefore need not trouble myself with it, and have only mentioned it incidentally.

What is important is the conclusion: experiment can not decide between Euclid and Lobachevski.

To sum up, whichever way we look at it, it is impossible to discover in geometric empiricism a rational meaning.

6. Experiments only teach us the relations of bodies to one another; none of them bears or can bear on the relations of bodies with space, or on the mutual relations of different parts of space.

"Yes," you reply, "a single experiment is insufficient, because it gives me only a single equation with several unknowns; but when I shall have made enough experiments I shall have equations enough to calculate all my unknowns."

To know the height of the mainmast does not suffice for calculating the age of the captain. When you have measured every bit of wood in the ship you will have many equations, but you will know his age no better. All your measurements bearing only on your bits of wood can reveal to you nothing except concerning these bits of wood. Just so your experiments, however numerous they may be, bearing only on the relations of bodies to one another, will reveal to us nothing about the mutual relations of the various parts of space.

7. Will you say that if the experiments bear on the bodies, they bear at least upon the geometric properties of the bodies? But, first, what do you understand by geometric properties of the bodies? I assume that it is a question of the relations of the bodies with space; these properties are therefore inaccessible to experiments which bear only on the relations of the bodies to one another. This alone would suffice to show that there can be no question of these properties.

Still let us begin by coming to an understanding about the sense of the phrase: geometric properties of bodies. When I say a body is composed of several parts, I assume that I do not enunciate therein a geometric property, and this would remain true even if I agreed to give the improper name of points to the smallest parts I consider.

When I say that such a part of such a body is in contact with such a part of such another body, I enunciate a proposition which concerns the mutual relations of these two bodies and not their relations with space.

I suppose you will grant me these are not geometric properties; at least I am sure you will grant me these properties are independent of all knowledge of metric geometry.

This presupposed, I imagine that we have a solid body formed of eight slender iron rods, *OA, OB, OC, OD, OE, OF, OG, OH,* united at one of their extremities *O.* Let us besides have a second solid body, for example a bit of wood, to be marked with three little flecks of ink which I shall call *α, β, γ.* I further suppose it ascertained that *αβγ* may be brought into contact with *AGO* (I mean *α* with *A,* and at the same time *β* with *G* and *γ* with *O*), then that we may bring successively into contact *αβγ* with *BGO, CGO, DGO, EGO, FGO,* then with *AHO, BHO, CHO, DHO, EHO, FHO,* then *αγ* successively with *AB, BC, CD, DE, EF, FA.*

These are determinations we may make without having in advance any notion about form or about the metric properties of space. They in no wise bear on the 'geometric properties of bodies.' And these determinations will not be possible if the bodies experimented upon move in accordance with a group having the same structure as the Lobachevskian group (I mean according to the same laws as solid bodies in Lobachevski's geometry). They suffice therefore to prove that these bodies move in accordance with the Euclidean group, or at least that they do not move according to the Lobachevskian group.

That they are compatible with the Euclidean group is easy to see. For they could be made if the body *αβγ* was a rigid solid of our ordinary geometry presenting the form of a right-angled triangle, and if the points *ABCDEFGH* were the summits of a polyhedron formed of two regular hexagonal pyramids of our ordinary geometry, having for common base *ABCDEF* and for apices the one *G* and the other *H.*

Suppose now that in place of the preceding determination it is observed that as above *αβγ* can be successively applied to *AGO, BGO, CGO, DGO, EGO, AHO, BHO, CHO, DHO, EHO, FHO,* then that *αβ* (and no longer *αγ*) can be successively applied to *AB, BC, CD, DE, EF* and *FA.*

These are determinations which could be made if non-Euclidean geometry were true, if the bodies *αβγ* and *OABCDEFGH* were rigid solids, and if the first were a right-angled triangle and the second a double regular hexagonal pyramid of suitable dimensions.

Therefore these new determinations are not possible if the bodies move according to the Euclidean group; but they become so if it be supposed that the bodies move according to the Lobachevskian group. They would suffice, therefore (if one made them), to prove that the bodies in question do not move according to the Euclidean group.

Thus, without making any hypothesis about form, about the nature of space, about the relations of bodies to space, and without attributing to bodies any geometric property, I have made observations which have enabled me to show in one case that the bodies experimented upon move according to a group whose structure is Euclidean, in the other case that they move according to a group whose structure is Lobachevskian.

And one may not say that the first aggregate of determinations would constitute an experiment proving that space is Euclidean, and the second an experiment proving that space is non-Euclidean.

In fact one could imagine (I say imagine) bodies moving so as to render possible the second series of determinations. And the proof is that the first mechanician met could construct such bodies if he cared to take the pains and make the outlay. You will not conclude from that, however, that space is non-Euclidean.

Nay, since the ordinary solid bodies would continue to exist when the mechanician had constructed the strange bodies of which I have just spoken, it would be necessary to conclude that space is at the same time Euclidean and non-Euclidean.

Suppose, for example, that we have a great sphere of radius $R$ and that the temperature decreases from the center to the surface of this sphere according to the law of which I have spoken in describing the non-Euclidean world.

We might have bodies whose expansion would be negligible and which would act like ordinary rigid solids; and, on the other hand, bodies very dilatable and which would act like non-Euclidean solids. We might have two double pyramids $OABCDEFGH$ and $O'A'B'C'D'E'F'G'H'$ and two triangles $\alpha\beta\gamma$ and $\alpha'\beta'\gamma'$  The first double pyramid might be rectilinear and the second curvilinear;

the triangle $\alpha\beta\gamma$ might be made of inexpansible matter and the other of a very dilatable matter.

It would then be possible to make the first observations with the double pyramid $OAH$ and the triangle $\alpha\beta\gamma$, and the second with the double pyramid $O'A'H'$ and the triangle $\alpha'\beta'\gamma'$. And then experiment would seem to prove first that the Euclidean geometry is true and then that it is false.

*Experiments therefore have a bearing, not on space, but on bodies.*

### SUPPLEMENT

8. To complete the matter, I ought to speak of a very delicate question, which would require long development; I shall confine myself to summarizing here what I have expounded in the *Revue de Métaphysique et de Morale* and in *The Monist*. When we say space has three dimensions, what do we mean?

We have seen the importance of those 'internal changes' revealed to us by our muscular sensations. They may serve to characterize the various *attitudes* of our body. Take arbitrarily as origin one of these attitudes $A$. When we pass from this initial attitude to any other attitude $B$, we feel a series of muscular sensations, and this series $S$ will define $B$. Observe, however, that we shall often regard two series $S$ and $S'$ as defining the same attitude $B$ (since the initial and final attitudes $A$ and $B$ remaining the same, the intermediary attitudes and the corresponding sensations may differ). How then shall we recognize the equivalence of these two series? Because they may serve to compensate the same external change, or more generally because, when it is a question of compensating an external change, one of the series can be replaced by the other. Among these series, we have distinguished those which of themselves alone can compensate an external change, and which we have called 'displacements.' As we can not discriminate between two displacements which are too close together, the totality of these displacements presents the characteristics of a physical continuum; experience teaches us that they are those of a physical continuum of six dimensions; but we do not yet know how many dimensions space itself has, we must first solve another question.

What is a point of space? Everybody thinks he knows, but

that is an illusion. What we see when we try to represent to ourselves a point of space is a black speck on white paper, a speck of chalk on a blackboard, always an object. The question should therefore be understood as follows:

What do I mean when I say the object $B$ is at the same point that the object $A$ occupied just now? Or further, what criterion will enable me to apprehend this?

I mean that, *although I have not budged* (which my muscular sense tells me), my first finger which just now touched the object $A$ touches at present the object $B$. I could have used other criteria; for instance another finger or the sense of sight. But the first criterion is sufficient; I know that if it answers yes, all the other criteria will give the same response. I know it *by experience,* I can not know it *a priori.* For the same reason I say that touch can not be exercised at a distance; this is another way of enunciating the same experimental fact. And if, on the contrary, I say that sight acts at a distance, it means that the criterion furnished by sight may respond yes while the others reply no.

And in fact, the object, although moved away, may form its image at the same point of the retina. Sight responds yes, the object has remained at the same point and touch answers no, because my finger which just now touched the object, at present touches it no longer. If experience had shown us that one finger may respond no when the other says yes, we should likewise say that touch acts at a distance.

In short, for each attitude of my body, my first finger determines a point, and this it is, and this alone, which defines a point of space.

To each attitude corresponds thus a point; but it often happens that the same point corresponds to several different attitudes (in this case we say our finger has not budged, but the rest of the body has moved). We distinguish, therefore, among the changes of attitude those where the finger does not budge. How are we led thereto? It is because often we notice that in these changes the object which is in contact with the finger remains in contact with it.

Range, therefore, in the same class all the attitudes obtainable from each other by one of the changes we have thus distinguished.

To all the attitudes of the class will correspond the same point of space. Therefore to each class will correspond a point and to each point a class. But one may say that what experience arrives at is not the point, it is this class of changes or, better, the corresponding class of muscular sensations.

And when we say space has three dimensions, we simply mean that the totality of these classes appears to us with the characteristics of a physical continuum of three dimensions.

One might be tempted to conclude that it is experience which has taught us how many dimensions space has. But in reality here also our experiences have bearing, not on space, but on our body and its relations with the neighboring objects. Moreover they are excessively crude.

In our mind pre-existed the latent idea of a certain number of groups—those whose theory Lie has developed. Which group shall we choose, to make of it a sort of standard with which to compare natural phenomena? And, this group chosen, which of its sub-groups shall we take to characterize a point of space? Experience has guided us by showing us which choice best adapts itself to the properties of our body. But its rôle is limited to that.

### Ancestral Experience

It has often been said that if individual experience could not create geometry the same is not true of ancestral experience. But what does that mean? Is it meant that we could not experimentally demonstrate Euclid's postulate, but that our ancestors have been able to do it? Not in the least. It is meant that by natural selection our mind has *adapted* itself to the conditions of the external world, that it has adopted the geometry *most advantageous* to the species: or in other words *the most convenient*. This is entirely in conformity with our conclusions; geometry is not true, it is advantageous.

# PART III

## FORCE

---

### CHAPTER VI

#### The Classic Mechanics

The English teach mechanics as an experimental science; on the continent it is always expounded as more or less a deductive and *a priori* science. The English are right, that goes without saying; but how could the other method have been persisted in so long? Why have the continental savants who have sought to get out of the ruts of their predecessors been usually unable to free themselves completely?

On the other hand, if the principles of mechanics are only of experimental origin, are they not therefore only approximate and provisional? Might not new experiments some day lead us to modify or even to abandon them?

Such are the questions which naturally obtrude themselves, and the difficulty of solution comes principally from the fact that the treatises on mechanics do not clearly distinguish between what is experiment, what is mathematical reasoning, what is convention, what is hypothesis.

That is not all:

1° There is no absolute space and we can conceive only of relative motions; yet usually the mechanical facts are enunciated as if there were an absolute space to which to refer them.

2° There is no absolute time; to say two durations are equal is an assertion which has by itself no meaning and which can acquire one only by convention.

3° Not only have we no direct intuition of the equality of two durations, but we have not even direct intuition of the simultaneity of two events occurring in different places: this I have explained in an article entitled *La mesure du temps*.[1]

4° Finally, our Euclidean geometry is itself only a sort of convention of language; mechanical facts might be enunciated with reference to a non-Euclidean space which would be a guide less convenient than, but just as legitimate as, our ordinary space; the enunciation would thus become much more complicated, but it would remain possible.

Thus absolute space, absolute time, geometry itself, are not conditions which impose themselves on mechanics; all these things are no more antecedent to mechanics than the French language is logically antecedent to the verities one expresses in French.

We might try to enunciate the fundamental laws of mechanics in a language independent of all these conventions; we should thus without doubt get a better idea of what these laws are in themselves; this is what M. Andrade has attempted to do, at least in part, in his *Leçons de mécanique physique*.

The enunciation of these laws would become of course much more complicated, because all these conventions have been devised expressly to abridge and simplify this enunciation.

As for me, save in what concerns absolute space, I shall ignore all these difficulties; not that I fail to appreciate them, far from that; but we have sufficiently examined them in the first two parts of the book.

I shall therefore admit, *provisionally*, absolute time and Euclidean geometry.

THE PRINCIPLE OF INERTIA.—A body acted on by no force can only move uniformly in a straight line.

Is this a truth imposed *a priori* upon the mind? If it were so, how should the Greeks have failed to recognize it? How could they have believed that motion stops when the cause which gave birth to it ceases? Or again that every body if nothing prevents, will move in a circle, the noblest of motions?

If it is said that the velocity of a body can not change if there is no reason for it to change, could it not be maintained just as well that the position of this body can not change, or that the

[1] *Revue de Métaphysique et de Morale*, t. VI., pp. 1–13 (January, 1898).

curvature of its trajectory can not change, if no external cause
intervenes to modify them?

Is the principle of inertia, which is not an *a priori* truth,
therefore an experimental fact? But has any one ever experimented on bodies withdrawn from the action of every force? and,
if so, how was it known that these bodies were subjected to no
force? The example ordinarily cited is that of a ball rolling a
very long time on a marble table; but why do we say it is subjected to no force? Is this because it is too remote from all other
bodies to experience any appreciable action from them? Yet it
is not farther from the earth than if it were thrown freely into
the air; and every one knows that in this case it would experience
the influence of gravity due to the attraction of the earth.

Teachers of mechanics usually pass rapidly over the example
of the ball; but they add that the principle of inertia is verified
indirectly by its consequences. They express themselves badly;
they evidently mean it is possible to verify various consequences
of a more general principle, of which that of inertia is only a
particular case.

I shall propose for this general principle the following enunciation:

The acceleration of a body depends only upon the position
of this body and of the neighboring bodies and upon their
velocities.

Mathematicians would say the movements of all the material
molecules of the universe depend on differential equations of the
second order.

To make it clear that this is really the natural generalization
of the law of inertia, I shall beg you to permit me a bit of fiction.
The law of inertia, as I have said above, is not imposed upon us
*a priori;* other laws would be quite as compatible with the principle of sufficient reason. If a body is subjected to no force, in
lieu of supposing its velocity not to change, it might be supposed
that it is its position or else its acceleration which is not to change.

Well, imagine for an instant that one of these two hypothetical
laws is a law of nature and replaces our law of inertia. What
would be its natural generalization? A moment's thought will
show us.

In the first case, we must suppose that the velocity of a body depends only upon its position and upon that of the neighboring bodies; in the second case that the change of acceleration of a body depends only upon the position of this body and of the neighboring bodies, upon their velocities and upon their accelerations.

Or to speak the language of mathematics, the differential equations of motion would be of the first order in the first case, and of the third order in the second case.

Let us slightly modify our fiction. Suppose a world analogous to our solar system, but where, by a strange chance, the orbits of all the planets are without eccentricity and without inclination. Suppose further that the masses of these planets are too slight for their mutual perturbations to be sensible. Astronomers inhabiting one of these planets could not fail to conclude that the orbit of a star can only be circular and parallel to a certain plane; the position of a star at a given instant would then suffice to determine its velocity and its whole path. The law of inertia which they would adopt would be the first of the two hypothetical laws I have mentioned.

Imagine now that this system is some day traversed with great velocity by a body of vast mass, coming from distant constellations. All the orbits would be profoundly disturbed. Still our astronomers would not be too greatly astonished; they would very well divine that this new star was alone to blame for all the mischief. "But," they would say, "when it is gone, order will of itself be reestablished; no doubt the distances of the planets from the sun will not revert to what they were before the cataclysm, but when the perturbing star is gone, the orbits will again become circular."

It would only be when the disturbing body was gone and when nevertheless the orbits, in lieu of again becoming circular, became elliptic, that these astronomers would become conscious of their error and the necessity of remaking all their mechanics.

I have dwelt somewhat upon these hypotheses because it seems to me one can clearly comprehend what our generalized law of inertia really is only in contrasting it with a contrary hypothesis.

Well, now, has this generalized law of inertia been verified by experiment, or can it be? When Newton wrote the *Principia* he quite regarded this truth as experimentally acquired and demonstrated. It was so in his eyes, not only through the anthropomorphism of which we shall speak further on, but through the work of Galileo. It was so even from Kepler's laws themselves; in accordance with these laws, in fact, the path of a planet is completely determined by its initial position and initial velocity; this is just what our generalized law of inertia requires.

For this principle to be only in appearance true, for one to have cause to dread having some day to replace it by one of the analogous principles I have just now contrasted with it, would be necessary our having been misled by some amazing chance, like that which, in the fiction above developed, led into error our imaginary astronomers.

Such a hypothesis is too unlikely to delay over. No one will believe that such coincidences can happen; no doubt the probability of two eccentricities being both precisely null, to within errors of observation, is not less than the probability of one being precisely equal to 0.1, for instance, and the other to 0.2, to within errors of observation. The probability of a simple event is not less than that of a complicated event; and yet, if the first happens, we shall not consent to attribute it to chance; we should not believe that nature had acted expressly to deceive us. The hypothesis of an error of this sort being discarded, it may therefore be admitted that in so far as astronomy is concerned, our law has been verified by experiment.

But astronomy is not the whole of physics.

May we not fear lest some day a new experiment should come to falsify the law in some domain of physics? An experimental law is always subject to revision; one should always expect to see it replaced by a more precise law.

Yet no one seriously thinks that the law we are speaking of will ever be abandoned or amended. Why? Precisely because it can never be subjected to a decisive test.

First of all, in order that this trial should be complete, it would be necessary that after a certain time all the bodies in the universe should revert to their initial positions with their initial velocities. It might then be seen whether, starting from this moment, they would resume their original paths.

But this test is impossible, it can be only partially applied, and, however well it is made, there will always be some bodies which will not revert to their initial positions; thus every derogation of the law will easily find its explanation.

This is not all; in astronomy we *see* the bodies whose motions we study and we usually assume that they are not subjected to the action of other invisible bodies. Under these conditions our law must indeed be either verified or not verified.

But it is not the same in physics; if the physical phenomena are due to motions, it is to the motions of molecules which we do not see. If then the acceleration of one of the bodies we see appears to us to depend on *something else* besides the positions or velocities of other visible bodies or of invisible molecules whose existence we have been previously led to admit, nothing prevents our supposing that this *something else* is the position or the velocity of other molecules whose presence we have not before suspected. The law will find itself safeguarded.

Permit me to employ mathematical language a moment to express the same thought under another form. Suppose we observe $n$ molecules and ascertain that their $3n$ coordinates satisfy a system of $3n$ differential equations of the fourth order (and not of the second order as the law of inertia would require). We know that by introducing $3n$ auxiliary variables, a system of $3n$ equations of the fourth order can be reduced to a system of $6n$ equations of the second order. If then we suppose these $3n$ auxiliary variables represent the coordinates of $n$ invisible molecules, the result is again in conformity with the law of inertia.

To sum up, this law, verified experimentally in some particular cases, may unhesitatingly be extended to the most general cases, since we know that in these general cases experiment no longer is able either to confirm or to contradict it.

THE LAW OF ACCELERATION.—The acceleration of a body is equal to the force acting on it divided by its mass. Can this law be verified by experiment? For that it would be necessary to measure the three magnitudes which figure in the enunciation: acceleration, force and mass.

I assume that acceleration can be measured, for I pass over the difficulty arising from the measurement of time. But how measure force, or mass? We do not even know what they are.

What is *mass?* According to Newton, it is the product of the volume by the density. According to Thomson and Tait, it would be better to say that density is the quotient of the mass by the volume. What is *force?* It is, replies Lagrange, that which moves or tends to move a body. It is, Kirchhoff will say, the product of the mass by the *acceleration*. But then, why not say the mass is the quotient of the force by the acceleration?

These difficulties are inextricable.

When we say force is the cause of motion, we talk metaphysics, and this definition, if one were content with it, would be absolutely sterile. For a definition to be of any use, it must teach us to *measure* force; moreover that suffices; it is not at all necessary that it teach us what force is *in itself,* nor whether it is the cause or the effect of motion.

We must therefore first define the equality of two forces. When shall we say two forces are equal? It is, we are told, when, applied to the same mass, they impress upon it the same acceleration, or when, opposed directly one to the other, they produce equilibrium. This definition is only a sham. A force applied to a body can not be uncoupled to hook it up to another body, as one uncouples a locomotive to attach it to another train. It is therefore impossible to know what acceleration such a force, applied to such a body, would impress upon such another body, *if* it were applied to it. It is impossible to know how two forces which are not directly opposed would act, *if* they were directly opposed.

It is this definition we try to materialize, so to speak, when we measure a force with a dynamometer, or in balancing it with a weight. Two forces $F$ and $F''$, which for simplicity I will suppose vertical and directed upward, are applied respectively to two bodies $C$ and $C'$; I suspend the same heavy body $P$ first to the body $C$, then to the body $C'$; if equilibrium is produced in both cases, I shall conclude that the two forces $F$ and $F''$ are equal to one another, since they are each equal to the weight of the body *P*.

But am I sure the body *P* has retained the same weight when I have transported it from the first body to the second? Far from it; *I am sure of the contrary;* I know the intensity of gravity varies from one point to another, and that it is stronger, for instance, at the pole than at the equator. No doubt the difference is very slight and, in practise, I shall take no account of it; but a properly constructed definition should have mathematical rigor; this rigor is lacking. What I say of weight would evidently apply to the force of the resiliency of a dynamometer, which the temperature and a multitude of circumstances may cause to vary.

This is not all; we can not say the weight of the body *P* may be applied to the body *C* and directly balance the force *F*. What is applied to the body *C* is the action *A* of the body *P* on the body *C*; the body *P* is submitted on its part, on the one hand, to its weight; on the other hand, to the reaction *R* of the body *C* on *P*. Finally, the force *F* is equal to the force *A*, since it balances it; the force *A* is equal to *R*, in virtue of the principle of the equality of action and reaction; lastly, the force *R* is equal to the weight of *P*, since it balances it. It is from these three equalities we deduce as consequence the equality of *F* and the weight of *P*.

We are therefore obliged in the definition of the equality of the two forces to bring in the principle of the equality of action and reaction; *on this account, this principle must no longer be regarded as an experimental law, but as a definition.*

For recognizing the equality of two forces here, we are then in possession of two rules: equality of two forces which balance; equality of action and reaction. But, as we have seen above, these two rules are insufficient; we are obliged to have recourse to a third rule and to assume that certain forces, as, for instance, the weight of a body, are constant in magnitude and direction. But this third rule, as I have said, is an experimental law; it is only approximately true; *it is a bad definition.*

We are therefore reduced to Kirchhoff's definition; *force is equal to the mass multipled by the acceleration.* This 'law of Newton' in its turn ceases to be regarded as an experimental law, it is now only a definition. But this definition is still insufficient, for we do not know what mass is. It enables us doubtless to calculate the relation of two forces applied to the same body at different instants; it teaches us nothing about the relation of two forces applied to two different bodies.

To complete it, it is necessary to go back anew to Newton's third law (equality of action and reaction), regarded again, not as an experimental law, but as a definition. Two bodies $A$ and $B$ act one upon the other; the acceleration of $A$ multiplied by the mass of $A$ is equal to the action of $B$ upon $A$; in the same way, the product of the acceleration of $B$ by its mass is equal to the reaction of $A$ upon $B$. As, by definition, action is equal to reaction, the masses of $A$ and $B$ are in the inverse ratio of their accelerations. Here we have the ratio of these two masses defined, and it is for experiment to verify that this ratio is constant.

That would be all very well if the two bodies $A$ and $B$ alone were present and removed from the action of the rest of the world. This is not at all the case; the acceleration of $A$ is not due merely to the action of $B$, but to that of a multitude of other bodies $C, D, \ldots$ To apply the preceding rule, it is therefore necessary to separate the acceleration of $A$ into many components, and discern which of these components is due to the action of $B$.

This separation would still be possible, if we *should assume* that the action of $C$ upon $A$ is simply adjoined to that of $B$ upon $A$, without the presence of the body $C$ modifying the action of $B$ upon $A$; or the presence of $B$ modifying the action of $C$ upon $A$; if we should assume, consequently, that any two bodies attract each other, that their mutual action is along their join and depends only upon their distance apart; if, in a word, we assume *the hypothesis of central forces.*

You know that to determine the masses of the celestial bodies we use a wholly different principle. The law of gravitation teaches us that the attraction of two bodies is proportional to their masses; if $r$ is their distance apart $m$ and $m'$ their masses, $k$ a constant, their attraction will be $kmm'/r^2$.

What we are measuring then is not mass, the ratio of force to acceleration, but the attracting mass; it is not the inertia of the body, but its attracting force.

This is an indirect procedure, whose employment is not theoretically indispensable. It might very well have been that attraction was inversely proportional to the square of the distance without being proportional to the product of the masses, that it was equal to $f/r^2$, but without our having $f = kmm'$.

If it were so, we could nevertheless, by observation of the *relative* motions of the heavenly bodies, measure the masses of these bodies.

But have we the right to admit the hypothesis of central forces? Is this hypothesis rigorously exact? Is it certain it will never be contradicted by experiment? Who would dare affirm that? And if we must abandon this hypothesis, the whole edifice so laboriously erected will crumble.

We have no longer the right to speak of the component of the acceleration of $A$ due to the action of $B$. We have no means of distinguishing it from that due to the action of $C$ or of another body. The rule for the measurement of masses becomes inapplicable.

What remains then of the principle of the equality of action and reaction? If the hypothesis of central forces is rejected, this principle should evidently be enunciated thus: the geometric resultant of all the forces applied to the various bodies of a system isolated from all external action will be null. Or, in other words, *the motion of the center of gravity of this system will be rectilinear and uniform.*

There it seems we have a means of defining mass; the position of the center of gravity evidently depends on the values attributed to the masses; it will be necessary to dispose of these values in such a way that the motion of the center of gravity may be rectilinear and uniform; this will always be possible if Newton's third law is true, and possible in general only in a single way.

But there exists no system isolated from all external action; all the parts of the universe are subject more or less to the action of all the other parts. *The law of the motion of the center of gravity is rigorously true only if applied to the entire universe.*

But then, to get from it the values of the masses, it would be necessary to observe the motion of the center of gravity of the universe. The absurdity of this consequence is manifest; we know only relative motions; the motion of the center of gravity of the universe will remain for us eternally unknown.

Therefore nothing remains and our efforts have been fruitless; we are driven to the following definition, which is only an avowal of powerlessness: *masses are coefficients it is convenient to introduce into calculations.*

We could reconstruct all mechanics by attributing different values to all the masses. This new mechanics would not be in contradiction either with experience or with the general principles of dynamics (principle of inertia, proportionality of forces to masses and to accelerations, equality of action and reaction, rectilinear and uniform motion of the center of gravity, principle of areas).

Only the equations of this new mechanics would be *less simple.* Let us understand clearly: it would only be the first terms which would be less simple, that is those experience has already made us acquainted with; perhaps one could alter the masses by small quantities without the *complete* equations gaining or losing in simplicity.

Hertz has raised the question whether the principles of mechanics are rigorously true. "In the opinion of many physicists," he says, "it is inconceivable that the remotest experience should ever change anything in the immovable principles of mechanics; and yet, what comes from experience may always be rectified by experience." After what we have just said, these fears will appear groundless.

The principles of dynamics at first appeared to us as experimental truths; but we have been obliged to use them as definitions. It is *by definition* that force is equal to the product of mass by acceleration; here, then, is a principle which is henceforth beyond the reach of any further experiment. It is in the same way by definition that action is equal to reaction.

But then, it will be said, these unverifiable principles are absolutely devoid of any significance; experiment can not contradict them; but they can teach us nothing useful; then what is the use of studying dynamics?

This over-hasty condemnation would be unjust. There is not in nature any system *perfectly* isolated, perfectly removed from all external action; but there are systems *almost* isolated.

If such a system be observed, one may study not only the

relative motion of its various parts one in reference to another, but also the motion of its center of gravity in reference to the other parts of the universe. We ascertain then that the motion of this center of gravity is *almost* rectilinear and uniform, in conformity with Newton's third law.

That is an experimental truth, but it can not be invalidated by experience; in fact, what would a more precise experiment teach us? It would teach us that the law was only almost true; but that we knew already.

*We can now understand how experience has been able to serve as basis for the principles of mechanics and yet will never be able to contradict them.*

ANTHROPOMORPHIC MECHANICS.—"Kirchhoff," it will be said, "has only acted in obedience to the general tendency of mathematicians toward nominalism; from this his ability as a physicist has not saved him. He wanted a definition of force, and he took for it the first proposition that presented itself; but we need no definition of force: the idea of force is primitive, irreducible, indefinable; we know all that it is, we have a direct intuition of it. This direct intuition comes from the notion of effort, which is familiar to us from infancy."

But first, even though this direct intuition made known to us the real nature of force in itself, it would be insufficient as a foundation for mechanics; it would besides be wholly useless. What is of importance is not to know what force is, but to know how to measure it.

Whatever does not teach us to measure it is as useless to mechanics as is, for instance, the subjective notion of warmth and cold to the physicist who is studying heat. This subjective notion can not be translated into numbers, therefore it is of no use; a scientist whose skin was an absolutely bad conductor of heat and who, consequently, would never have felt either sensations of cold or sensations of warmth, could read a thermometer just as well as any one else, and that would suffice him for constructing the whole theory of heat.

Now this immediate notion of effort is of no use to us for measuring force; it is clear, for instance, that I should feel more fatigue in lifting a weight of fifty kilos than a man accustomed to carry burdens.

But more than that: this notion of effort does not teach us the real nature of force; it reduces itself finally to a remembrance of muscular sensations, and it will hardly be maintained that the sun feels a muscular sensation when it draws the earth.

All that can there be sought is a symbol, less precise and less convenient than the arrows the geometers use, but just as remote from the reality.

Anthropomorphism has played a considerable historic rôle in the genesis of mechanics; perhaps it will still at times furnish a symbol which will appear convenient to some minds; but it can not serve as foundation for anything of a truly scientific or philosophic character.

'THE SCHOOL OF THE THREAD.'—M. Andrade, in his *Leçons de méchanique physique,* has rejuvenated anthropomorphic mechanics. To the school of mechanics to which Kirchhoff belongs, he opposes that which he bizarrely calls the school of the thread.

This school tries to reduce everything to "the consideration of certain material systems of negligible mass, envisaged in the state of tension and capable of transmitting considerable efforts to distant bodies, systems of which the ideal type is the *thread.*"

A thread which transmits any force is slightly elongated under the action of this force; the direction of the thread tells us the direction of the force, whose magnitude is measured by the elongation of the thread.

One may then conceive an experiment such as this. A body $A$ is attached to a thread; at the other extremity of the thread any force acts which varies until the thread takes an elongation $\alpha$; the acceleration of the body $A$ is noted; $A$ is detached and the body $B$ attached to the same thread; the same force or another force acts anew, and is made to vary until the thread takes again the elongation $\alpha$; the acceleration of the body $B$ is noted. The experiment is then renewed with both $A$ and $B$, but so that the thread takes the elongation $\beta$. The four observed accelerations should be proportional. We have thus an experimental verification of the law of acceleration above enunciated.

Or still better, a body is submitted to the simultaneous action of several identical threads in equal tension, and by experiment it is sought what must be the orientations of all these threads that the body may remain in equilibrium. We have then an experimental verification of the law of the composition of forces.

But, after all, what have we done? We have defined the force to which the thread is subjected by the deformation undergone by this thread, which is reasonable enough; we have further assumed that if a body is attached to this thread, the effort transmitted to it by the thread is equal to the action this body exercises on this thread; after all, we have therefore used the principle of the equality of action and reaction, in considering it, not as an experimental truth, but as the very definition of force.

This definition is just as conventional as Kirchhoff's, but far less general.

All forces are not transmitted by threads (besides, to be able to compare them, they would all have to be transmitted by identical threads). Even if it should be conceded that the earth is attached to the sun by some invisible thread, at least it would be admitted that we have no means of measuring its elongation.

Nine times out of ten, consequently, our definition would be at fault; no sort of sense could be attributed to it, and it would be necessary to fall back on Kirchhoff's.

Why then take this détour? You admit a certain definition of force which has a meaning only in certain particular cases. In these cases you verify by experiment that it leads to the law of acceleration. On the strength of this experiment, you then take the law of acceleration as a definition of force in all the other cases.

Would it not be simpler to consider the law of acceleration as a definition in all cases, and to regard the experiments in question, not as verifications of this law, but as verifications of the principle of reaction, or as demonstrating that the deformations of an elastic body depend only on the forces to which this body is subjected?

And this is without taking into account that the conditions under which your definition could be accepted are never fulfilled except imperfectly, that a thread is never without mass, that it is never removed from every force except the reaction of the bodies attached to its extremities.

Andrade's ideas are nevertheless very interesting; if they do not satisfy our logical craving, they make us understand better the historic genesis of the fundamental ideas of mechanics. The reflections they suggest show us how the human mind has raised itself from a naïve anthropomorphism to the present conceptions of science.

We see at the start a very particular and in sum rather crude experiment; at the finish, a law perfectly general, perfectly precise, the certainty of which we regard as absolute. This certainty we ourselves have bestowed upon it voluntarily, so to speak, by looking upon it as a convention.

Are the law of acceleration, the rule of the composition of forces then only arbitrary conventions? Conventions, yes; arbitrary, no; they would be if we lost sight of the experiments which led the creators of the science to adopt them, and which, imperfect as they may be, suffice to justify them. It is well that from time to time our attention is carried back to the experimental origin of these conventions.

# CHAPTER VII

## RELATIVE MOTION AND ABSOLUTE MOTION

THE PRINCIPLE OF RELATIVE MOTION.—The attempt has sometimes been made to attach the law of acceleration to a more general principle. The motion of any system must obey the same laws, whether it be referred to fixed axes, or to movable axes carried along in a rectilinear and uniform motion. This is the principle of relative motion, which forces itself upon us for two reasons: first, the commonest experience confirms it, and second, the contrary hypothesis is singularly repugnant to the mind.

Assume it then, and consider a body subjected to a force; the relative motion of this body, in reference to an observer moving with a uniform velocity equal to the initial velocity of the body, must be identical to what its absolute motion would be if it started from rest. We conclude hence that its acceleration can not depend upon its absolute velocity; the attempt has even been made to derive from this a demonstration of the law of acceleration.

There long were traces of this demonstration in the regulations for the degree B. ès Sc. It is evident that this attempt is idle. The obstacle which prevented our demonstrating the law of acceleration is that we had no definition of force; this obstacle subsists in its entirety, since the principle invoked has not furnished us the definition we lacked.

The principle of relative motion is none the less highly interesting and deserves study for its own sake. Let us first try to enunciate it in a precise manner.

We have said above that the accelerations of the different bodies forming part of an isolated system depend only on their relative velocities and positions, and not on their absolute velocities and positions, provided the movable axes to which the relative motion is referred move uniformly in a straight line. Or, if we prefer, their accelerations depend only on the differences of their velocities and the differences of their coordinates, and not on the absolute values of these velocities and coordinates.

If this principle is true for relative accelerations, or rather for differences of acceleration, in combining it with the law of reaction we shall thence deduce that it is still true of absolute accelerations.

It then remains to be seen how we may demonstrate that the differences of the accelerations depend only on the differences of the velocities and of the coordinates, or, to speak in mathematical language, that these differences of coordinates satisfy differential equations of the second order.

Can this demonstration be deduced from experiments or from *a priori* considerations?

Recalling what we have said above, the reader can answer for himself.

Thus enunciated, in fact, the principle of relative motion singularly resembles what I called above the generalized principle of inertia; it is not altogether the same thing, since it is a question of the differences of coordinates and not of the coordinates themselves. The new principle teaches us therefore something more than the old, but the same discussion is applicable and would lead to the same conclusions; it is unnecessary to return to it.

Newton's Argument.—Here we encounter a very important and even somewhat disconcerting question. I have said the principle of relative motion was for us not solely a result of experiment and that *a priori* every contrary hypothesis would be repugnant to the mind.

But then, why is the principle true only if the motion of the movable axes is rectilinear and uniform? It seems that it ought to impose itself upon us with the same force, if this motion is varied, or at any rate if it reduces to a uniform rotation. Now, in these two cases, the principle is not true. I will not dwell long on the case where the motion of the axes is rectilinear without being uniform; the paradox does not bear a moment's examination. If I am on board, and if the train, striking any obstacle, stops suddenly, I shall be thrown against the seat in front of me, although I have not been directly subjected to any force. There is nothing mysterious in that; if I have undergone the action of no external force, the train itself has experienced an external impact. There can be nothing paradoxical in the relative motion of two bodies being disturbed when the motion of one or the other is modified by an external cause.

I will pause longer on the case of relative motions referred to axes which rotate uniformly. If the heavens were always covered with clouds, if we had no means of observing the stars, we nevertheless might conclude that the earth turns round; we could learn this from its flattening or again by the Foucault pendulum experiment.

And yet, in this case, would it have any meaning, to say the earth turns round? If there is no absolute space, can one turn without turning in reference to something else? and, on the other hand, how could we admit Newton's conclusion and believe in absolute space?

But it does not suffice to ascertain that all possible solutions are equally repugnant to us; we must analyze, in each case, the reasons for our repugnance, so as to make our choice intelligently. The long discussion which follows will therefore be excused.

Let us resume our fiction: thick clouds hide the stars from men, who can not observe them and are ignorant even of their existence; how shall these men know the earth turns round?

Even more than our ancestors, no doubt, they will regard the ground which bears them as fixed and immovable; they will await much longer the advent of a Copernicus. But in the end the Copernicus would come—how?

The students of mechanics in this world would not at first be confronted with an absolute contradiction. In the theory of relative motion, besides real forces, two fictitious forces are met which are called ordinary and compound centrifugal force. Our imaginary scientists could therefore explain everything by regarding these two forces as real, and they would not see therein any contradiction of the generalized principle of inertia, for these forces would depend, the one on the relative positions of the various parts of the system, as real attractions do, the other on their relative velocities, as real frictions do.

Many difficulties, however, would soon awaken their attention; if they succeeded in realizing an isolated system, the center of gravity of this system would not have an almost rectilinear path. They would invoke, to explain this fact, the centrifugal forces which they would regard as real, and which they would attribute no doubt to the mutual actions of the bodies. Only they would not see these forces become null at great distances, that is to say in proportion as the isolation was better realized; far from it; centrifugal force increases indefinitely with the distance.

This difficulty would seem to them already sufficiently great; and yet it would not stop them long; they would soon imagine some very subtile medium, analogous to our ether, in which all bodies would be immersed and which would exert a repellent action upon them.

But this is not all. Space is symmetric, and yet the laws of motion would not show any symmetry; they would have to distinguish between right and left. It would be seen for instance that cyclones turn always in the same sense, whereas by reason of symmetry these winds should turn indifferently in one sense and in the other. If our scientists by their labor had succeeded in rendering their universe perfectly symmetric, this symmetry would not remain, even though there was no apparent reason why it should be disturbed in one sense rather than in the other.

They would get themselves out of the difficulty doubtless, they would invent something which would be no more extraordinary than the glass spheres of Ptolemy, and so it would go on, complications accumulating, until the long-expected Copernicus sweeps them all away at a single stroke, saying: It is much simpler to assume the earth turns round.

And just as our Copernicus said to us: It is more convenient to suppose the earth turns round, since thus the laws of astronomy are expressible in a much simpler language; this one would say: It is more convenient to suppose the earth turns round, since thus the laws of mechanics are expressible in a much simpler language.

This does not preclude maintaining that absolute space, that is to say the mark to which it would be necessary to refer the earth to know whether it really moves, has no objective existence. Hence, this affirmation: 'the earth turns round' has no meaning, since it can be verified by no experiment; since such an experiment, not only could not be either realized or dreamed by the boldest Jules Verne, but can not be conceived of without contradiction; or rather these two propositions: 'the earth turns round,' and, 'it is more convenient to suppose the earth turns round' have the same meaning; there is nothing more in the one than in the other.

Perhaps one will not be content even with that, and will find it already shocking that among all the hypotheses, or rather all the conventions we can make on this subject, there is one more convenient than the others.

But if it has been admitted without difficulty when it was a question of the laws of astronomy, why should it be shocking in that which concerns mechanics?

We have seen that the coordinates of bodies are determined by differential equations of the second order, and that so are the differences of these coordinates. This is what we have called the generalized principle of inertia and the principle of relative motion. If the distances of these bodies were determined likewise by equations of the second order, it seems that the mind ought to be entirely satisfied. In what measure does the mind get this satisfaction and why is it not content with it?

To account for this, we had better take a simple example. I suppose a system analogous to our solar system, but where one can not perceive fixed stars foreign to this system, so that astronomers can observe only the mutual distances of the planets and the sun, and not the absolute longitudes of the planets. If we deduce directly from Newton's law the differential equations which define the variation of these distances, these equations will not be of the second order. I mean that if, besides Newton's law, one knew the initial values of these distances and of their derivatives with respect to the time, that would not suffice to determine the values of these same distances at a subsequent instant. There would still be lacking one datum, and this datum might be for instance what astronomers call the area-constant.

But here two different points of view may be taken; we may distinguish two sorts of constants. To the eyes of the physicist the world reduces to a series of phenomena, depending, on the one hand, solely upon the initial phenomena; on the other hand, upon the laws which bind the consequents to the antecedents. If then observation teaches us that a certain quantity is a constant, we shall have the choice between two conceptions.

Either we shall assume that there is a law requiring this quantity not to vary, but that by chance, at the beginning of the ages, it had, rather than another, this value it has been forced to keep ever since. This quantity might then be called an *accidental* constant.

Or else we shall assume, on the contrary, that there is a law of nature which imposes upon this quantity such a value and not such another.

We shall then have what we may call an *essential* constant.

For example, in virtue of Newton's laws, the duration of the revolution of the earth must be constant. But if it is 366 sidereal days and something over, and not 300 or 400, this is in consequence of I know not what initial chance. This is an accidental constant. If, on the contrary, the exponent of the distance which figures in the expression of the attractive force is equal to —2 and not to —3, this is not by chance, but because Newton's law requires it. This is an essential constant.

I know not whether this way of giving chance its part is legitimate in itself, and whether this distinction is not somewhat artificial; it is certain at least that, so long as nature shall have secrets, this distinction will be in application extremely arbitrary and always precarious.

As to the area-constant, we are accustomed to regard it as accidental. Is it certain our imaginary astronomers would do the same? If they could have compared two different solar systems, they would have the idea that this constant may have several different values; but my very supposition in the beginning was that their system should appear as isolated, and that they should observe no star foreign to it. Under these conditions, they would see only one single constant which would have a single value absolutely invariable; they would be led without any doubt to regard it as an essential constant.

A word in passing to forestall an objection: the inhabitants of this imaginary world could neither observe nor define the area-constant as we do, since the absolute longitudes escape them; that would not preclude their being quickly led to notice a certain constant which would introduce itself naturally into their equations and which would be nothing but what we call the area-constant.

But then see what would happen. If the area-constant is regarded as essential, as depending upon a law of nature, to calculate the distances of the planets at any instant it will suffice to know the initial values of these distances and those of their first derivatives. From this new point of view, the distances will be determined by differential equations of the second order.

Yet would the mind of these astronomers be completely satisfied? I do not believe so; first, they would soon perceive that in differentiating their equations and thus raising their order, these equations became much simpler. And above all they would be struck by the difficulty which comes from symmetry. It would be necessary to assume different laws, according as the aggregate of the planets presented the figure of a certain polyhedron or of the symmetric polyhedron, and one would escape from this consequence only by regarding the area-constant as accidental.

I have taken a very special example, since I have supposed astronomers who did not at all consider terrestrial mechanics, and whose view was limited to the solar system. Our universe is more extended than theirs, as we have fixed stars, but still it too is limited, and so we might reason on the totality of our universe as the astronomers on their solar system.

Thus we see that finally we should be led to conclude that the equations which define distances are of an order superior to the second. Why should we be shocked at that, why do we find it perfectly natural for the series of phenomena to depend upon the initial values of the first derivatives of these distances, while we hesitate to admit that they may depend on the initial values of the second derivatives? This can only be because of the habits of mind created in us by the constant study of the generalized principle of inertia and its consequences.

The values of the distances at any instant depend upon their initial values, upon those of their first derivatives and also upon something else.   What is this *something else?*

If we will not admit that this may be simply one of the second derivatives, we have only the choice of hypotheses.   Either it may be supposed, as is ordinarily done, that this something else is the absolute orientation of the universe in space, or the rapidity with which this orientation varies; and this supposition may be correct; it is certainly the most convenient solution for geometry; it is not the most satisfactory for the philosopher, because this orientation does not exist.

Or it may be supposed that this something else is the position or the velocity of some invisible body; this has been done by certain persons who have even called it the body alpha, although we are doomed never to know anything of this body but its name.   This is an artifice entirely analogous to that of which I spoke at the end of the paragraph devoted to my reflections on the principle of inertia.

But, after all, the difficulty is artificial.   Provided the future indications of our instruments can depend only on the indications they have given us or would have given us formerly, this is all that is necessary.   Now as to this we may rest easy.

# CHAPTER VIII

## ENERGY AND THERMODYNAMICS

ENERGETICS.—The difficulties inherent in the classic mechanics have led certain minds to prefer a new system they call *energetics*.

Energetics took its rise as an outcome of the discovery of the principle of the conservation of energy. Helmholtz gave it its final form.

It begins by defining two quantities which play the fundamental rôle in this theory. They are *kinetic energy*, or *vis viva*, and *potential energy.*

All the changes which bodies in nature can undergo are regulated by two experimental laws:

1° The sum of kinetic energy and potential energy is constant. This is the principle of the conservation of energy.

2° If a system of bodies is at $A$ at the time $t_0$ and at $B$ at the time $t_1$, it always goes from the first situation to the second in such a way that the *mean* value of the difference between the two sorts of energy, in the interval of time which separates the two epochs $t_0$ and $t_1$, may be as small as possible.

This is Hamilton's principle, which is one of the forms of the principle of least action.

The energetic theory has the following advantages over the classic theory:

1° It is less incomplete; that is to say, Hamilton's principle and that of the conservation of energy teach us more than the fundamental principles of the classic theory, and exclude certain motions not realized in nature and which would be compatible with the classic theory:

2° It saves us the hypothesis of atoms, which it was almost impossible to avoid with the classic theory.

But it raises in its turn new difficulties:

The definitions of the two sorts of energy would raise difficulties almost as great as those of force and mass in the first

115

system. Yet they may be gotten over more easily, at least in the simplest cases.

Suppose an isolated system formed of a certain number of material points; suppose these points subjected to forces depending only on their relative position and their mutual distances, and independent of their velocities. In virtue of the principle of the conservation of energy, a function of forces must exist.

In this simple case the enunciation of the principle of the conservation of energy is of extreme simplicity. A certain quantity, accessible to experiment, must remain constant. This quantity is the sum of two terms; the first depends only on the position of the material points and is independent of their velocities; the second is proportional to the square of these velocities. This resolution can take place only in a single way.

The first of these terms, which I shall call $U$, will be the potential energy; the second, which I shall call $T$, will be the kinetic energy.

It is true that if $T + U$ is a constant, so is any function of $T + U$,

$$\phi(T + U).$$

But this function $\phi(T + U)$ will not be the sum of two terms the one independent of the velocities, the other proportional to the square of these velocities. Among the functions which remain constant there is only one which enjoys this property, that is $T + U$ (or a linear function of $T + U$, which comes to the same thing, since this linear function may always be reduced to $T + U$ by change of unit and of origin). This then is what we shall call energy; the first term we shall call potential energy and the second kinetic energy. The definition of the two sorts of energy can therefore be carried through without any ambiguity.

It is the same with the definition of the masses. Kinetic energy, or *vis viva*, is expressed very simply by the aid of the masses and the relative velocities of all the material points with reference to one of them. These relative velocities are accessible to observation, and, when we know the expression of the kinetic energy as function of these relative velocities, the coefficients of this expression will give us the masses.

Thus, in this simple case, the fundamental ideas may be defined without difficulty. But the difficulties reappear in the more complicated cases and, for instance, if the forces, in lieu of depending only on the distances, depend also on the velocities. For example, Weber supposes the mutual action of two electric molecules to depend not only on their distance, but on their velocity and their acceleration. If material points should attract each other according to an analogous law, $U$ would depend on the velocity, and might contain a term proportional to the square of the velocity.

Among the terms proportional to the squares of the velocities, how distinguish those which come from $T$ or from $U$? Consequently, how distinguish the two parts of energy?

But still more; how define energy itself? We no longer have any reason to take as definition $T + U$ rather than any other function of $T + U$, when the property which characterized $T + U$ has disappeared, that, namely, of being the sum of two terms of a particular form.

But this is not all; it is necessary to take account, not only of mechanical energy properly so called, but of the other forms of energy, heat, chemical energy, electric energy, etc. The principle of the conservation of energy should be written:

$$T + U + Q = \text{const.}$$

where $T$ would represent the sensible kinetic energy, $U$ the potential energy of position, depending only on the position of the bodies, $Q$ the internal molecular energy, under the thermal, chemic or electric form.

All would go well if these three terms were absolutely distinct, if $T$ were proportional to the square of the velocities, $U$ independent of these velocities and of the state of the bodies, $Q$ independent of the velocities and of the positions of the bodies and dependent only on their internal state.

The expression for the energy could be resolved only in one single way into three terms of this form.

But this is not the case; consider electrified bodies; the electrostatic energy due to their mutual action will evidently depend upon their charge, that is to say, on their state; but it will equally depend upon their position. If these bodies are in motion, they will act one upon another electrodynamically and the electro-dynamic energy will depend not only upon their state and their position, but upon their velocities.

We therefore no longer have any means of making the sepa-ration of the terms which should make part of $T$, of $U$ and of $Q$, and of separating the three parts of energy.

If $(T + U + Q)$ is constant so is any function $\phi(T + U + Q)$.

If $T + U + Q$ were of the particular form I have above considered, no ambiguity would result; among the functions $\phi(T + U + Q)$ which remain constant, there would only be one of this particular form, and that I should convene to call energy.

But as I have said, this is not rigorously the case; among the functions which remain constant, there is none which can be put rigorously under this particular form; hence, how choose among them the one which should be called energy? We no longer have anything to guide us in our choice.

There only remains for us one enunciation of the principle of the conservation of energy: *There is something which remains constant.* Under this form it is in its turn out of the reach of experiment and reduces to a sort of tautology. It is clear that if the world is governed by laws, there will be quantities which will remain constant. Like Newton's laws, and, for an analogous reason, the principle of the conservation of energy, founded on experiment, could no longer be invalidated by it.

This discussion shows that in passing from the classic to the energetic system progress has been made; but at the same time it shows this progress is insufficient.

Another objection seems to me still more grave: the prin-ciple of least action is applicable to reversible phenomena; but it is not at all satisfactory in so far as irreversible phenomena are concerned; the attempt by Helmholtz to extend it to this kind of phenomena did not succeed and could not succeed; in this regard everything remains to be done. The very statement of the prin-ciple of least action has something about it repugnant to the mind. To go from one point to another, a material molecule, acted upon by no force, but required to move on a surface, will take the geodesic line, that is to say, the shortest path.

This molecule seems to know the point whither it is to go, to foresee the time it would take to reach it by such and such a route, and then to choose the most suitable path. The statement presents the molecule to us, so to speak, as a living and free being. Clearly it would be better to replace it by an enunciation less objectionable, and where, as the philosophers would say, final causes would not seem to be substituted for efficient causes.

THERMODYNAMICS.[1]—The rôle of the two fundamental principles of thermodynamics in all branches of natural philosophy becomes daily more important. Abandoning the ambitious theories of forty years ago, which were encumbered by molecular hypotheses, we are trying to-day to erect upon thermodynamics alone the entire edifice of mathematical physics. Will the two principles of Mayer and of Clausius assure to it foundations solid enough for it to last some time? No one doubts it; but whence comes this confidence?

An eminent physicist said to me one day *à propos* of the law of errors: "All the world believes it firmly, because the mathematicians imagine that it is a fact of observation, and the observers that it is a theorem of mathematics." It was long so for the principle of the conservation of energy. It is no longer so to-day; no one is ignorant that this is an experimental fact.

But then what gives us the right to attribute to the principle itself more generality and more precision than to the experiments which have served to demonstrate it? This is to ask whether it is legitimate, as is done every day, to generalize empirical data, and I shall not have the presumption to discuss this question, after so many philosophers have vainly striven to solve it. One thing is certain; if this power were denied us, science could not exist or, at least, reduced to a sort of inventory, to the ascertaining of isolated facts, it would have no value for us, since it could give no satisfaction to our craving for order and harmony and since it would be at the same time incapable of foreseeing. As the circumstances which have preceded any fact will probably never be simultaneously reproduced, a first general-

[1] The following lines are a partial reproduction of the preface of my book *Thermodynamique*.

ization is already necessary to foresee whether this fact will be reproduced again after the least of these circumstances shall be changed.

But every proposition may be generalized in an infinity of ways. Among all the generalizations possible, we must choose, and we can only choose the simplest. We are therefore led to act as if a simple law were, other things being equal, more probable than a complicated law.

Half a century ago this was frankly confessed, and it was proclaimed that nature loves simplicity; she has since too often given us the lie. To-day we no longer confess this tendency, and we retain only so much of it as is indispensable if science is not to become impossible.

In formulating a general, simple and precise law on the basis of experiments relatively few and presenting certain divergences, we have therefore only obeyed a necessity from which the human mind can not free itself.

But there is something more, and this is why I dwell upon the point.

No one doubts that Mayer's principle is destined to survive all the particular laws from which it was obtained, just as Newton's law has survived Kepler's laws, from which it sprang, and which are only approximative if account be taken of perturbations.

Why does this principle occupy thus a sort of privileged place among all the physical laws? There are many little reasons for it.

First of all it is believed that we could not reject it or even doubt its absolute rigor without admitting the possibility of perpetual motion; of course we are on our guard at such a prospect, and we think ourselves less rash in affirming Mayer's principle than in denying it.

That is perhaps not wholly accurate; the impossibility of perpetual motion implies the conservation of energy only for reversible phenomena.

The imposing simplicity of Mayer's principle likewise contributes to strengthen our faith. In a law deduced immediately from experiment, like Mariotte's, this simplicity would rather seem to us a reason for distrust; but here this is no longer the case; we see elements, at first sight disparate, arrange themselves in an unexpected order and form a harmonious whole; and we refuse to believe that an unforeseen harmony may be a simple effect of chance. It seems that our conquest is the dearer to us the more effort it has cost us, or that we are the surer of having wrested her true secret from nature the more jealously she has hidden it from us.

But those are only little reasons; to establish Mayer's law as an absolute principle, a more profound discussion is necessary. But if this be attempted, it is seen that this absolute principle is not even easy to state.

In each particular case it is clearly seen what energy is and at least a provisional definition of it can be given; but it is impossible to find a general definition for it.

If we try to enunciate the principle in all its generality and apply it to the universe, we see it vanish, so to speak, and nothing is left but this: *There is something which remains constant.*

But has even this any meaning? In the determinist hypothesis, the state of the universe is determined by an extremely great number $n$ of parameters which I shall call $x_1$, $x_2$, $\ldots x_n$. As soon as the values of these $n$ parameters at any instant are known, their derivatives with respect to the time are likewise known and consequently the values of these same parameters at a preceding or subsequent instant can be calculated. In other words, these $n$ parameters satisfy $n$ differential equations of the first order.

These equations admit of $n-1$ integrals and consequently there are $n-1$ functions of $x_1$, $x_2$, $\ldots x_n$, which remain constant. *If then we say there is something which remains constant,* we only utter a tautology. We should even be puzzled to say which among all our integrals should retain the name of energy.

Besides, Mayer's principle is not understood in this sense when it is applied to a limited system. It is then assumed that $p$ of our parameters vary independently, so that we only have $n-p$ relations, generally linear, between our $n$ parameters and their derivatives.

To simplify the enunciation, suppose that the sum of the work of the external forces is null, as well as that of the quantities of heat given off to the outside. Then the signification of our principle will be:

*There is a combination of these $n-p$ relations whose first member is an exact differential;* and then this differential vanishing in virtue of our $n-p$ relations, its integral is a constant and this integral is called energy.

But how can it be possible that there are several parameters whose variations are independent? That can only happen under the influence of external forces (although we have supposed, for simplicity, that the algebraic sum of the effects of these forces is null). In fact, if the system were completely isolated from all external action, the values of our $n$ parameters at a given instant would suffice to determine the state of the system at any subsequent instant, provided always we retain the determinist hypothesis; we come back therefore to the same difficulty as above.

If the future state of the system is not entirely determined by its present state, this is because it depends besides upon the state of bodies external to the system. But then is it probable that there exist between the parameters $x$, which define the state of the system, equations independent of this state of the external bodies? and if in certain cases we believe we can find such, is this not solely in consequence of our ignorance and because the influence of these bodies is too slight for our experimenting to detect it?

If the system is not regarded as completely isolated, it is probable that the rigorously exact expression of its internal energy will depend on the state of the external bodies. Again, I have above supposed the sum of the external work was null, and if we try to free ourselves from this rather artificial restriction, the enunciation becomes still more difficult.

To formulate Mayer's principle in an absolute sense, it is therefore necessary to extend it to the whole universe, and then we find ourselves face to face with the very difficulty we sought to avoid.

In conclusion, using ordinary language, the law of the conservation of energy can have only one signification, which is that there is a property common to all the possibilities; but on the determinist hypothesis there is only a single possibility, and then the law has no longer any meaning.

On the indeterminist hypothesis, on the contrary, it would have a meaning, even if it were taken in an absolute sense; it would appear as a limitation imposed upon freedom.

But this word reminds me that I am digressing and am on the point of leaving the domain of mathematics and physics. I check myself therefore and will stress of all this discussion only one impression, that Mayer's law is a form flexible enough for us to put into it almost whatever we wish. By that I do not mean it corresponds to no objective reality, nor that it reduces itself to a mere tautology, since, in each particular case, and provided one does not try to push to the absolute, it has a perfectly clear meaning.

This flexibility is a reason for believing in its permanence, and as, on the other hand, it will disappear only to lose itself in a higher harmony, we may work with confidence, supporting ourselves upon it, certain beforehand that our labor will not be lost.

Almost everything I have just said applies to the principle of Clausius. What distinguishes it is that it is expressed by an inequality. Perhaps it will be said it is the same with all physical laws, since their precision is always limited by errors of observation. But they at least claim to be first approximations, and it is hoped to replace them little by little by laws more and more precise. If, on the other hand, the principle of Clausius reduces to an inequality, this is not caused by the imperfection of our means of observation, but by the very nature of the question.

## General Conclusions on Part Third

The principles of mechanics, then, present themselves to us under two different aspects. On the one hand, they are truths founded on experiment and approximately verified so far as concerns almost isolated systems. On the other hand, they are postulates applicable to the totality of the universe and regarded as rigorously true.

If these postulates possess a generality and a certainty which are lacking to the experimental verities whence they are drawn, this is because they reduce in the last analysis to a mere convention which we have the right to make, because we are certain beforehand that no experiment can ever contradict it.

This convention, however, is not absolutely arbitrary; it does not spring from our caprice; we adopt it because certain experiments have shown us that it would be convenient.

Thus is explained how experiment could make the principles of mechanics, and yet why it can not overturn them.

Compare with geometry: The fundamental propositions of geometry, as for instance Euclid's postulate, are nothing more than conventions, and it is just as unreasonable to inquire whether they are true or false as to ask whether the metric system is true or false.

Only, these conventions are convenient, and it is certain experiments which have taught us that.

At first blush, the analogy is complete; the rôle of experiment seems the same. One will therefore be tempted to say: Either mechanics must be regarded as an experimental science, and then the same must hold for geometry; or else, on the contrary, geometry is a deductive science, and then one may say as much of mechanics.

Such a conclusion would be illegitimate. The experiments which have led us to adopt as more convenient the fundamental conventions of geometry bear on objects which have nothing in common with those geometry studies; they bear on the properties of solid bodies, on the rectilinear propagation of light. They are experiments of mechanics, experiments of optics; they can not in any way be regarded as experiments of geometry. And even the principal reason why our geometry seems convenient to us is that the different parts of our body, our eye, our limbs, have the properties of solid bodies. On this account, our fundamental experiments are preeminently physiological experiments, which bear, not on space which is the object the geometer must study, but on his body, that is to say, on the instrument he must use for this study.

On the contrary, the fundamental conventions of mechanics, and the experiments which prove to us that they are convenient, bear on exactly the same objects or on analogous objects. The conventional and general principles are the natural and direct generalization of the experimental and particular principles.

Let it not be said that thus I trace artificial frontiers between the sciences; that if I separate by a barrier geometry properly so called from the study of solid bodies, I could just as well erect one between experimental mechanics and the conventional mechanics of the general principles. In fact, who does not see that in separating these two sciences I mutilate them both, and that what will remain of conventional mechanics when it shall be isolated will be only a very small thing and can in no way be compared to that superb body of doctrine called geometry?

One sees now why the teaching of mechanics should remain experimental.

Only thus can it make us comprehend the genesis of the science, and that is indispensable for the complete understanding of the science itself.

Besides, if we study mechanics, it is to apply it; and we can apply it only if it remains objective. Now, as we have seen, what the principles gain in generality and certainty they lose in objectivity. It is, therefore, above all with the objective side of the principles that we must be familiarized early, and that can be done only by going from the particular to the general, instead of the inverse.

The principles are conventions and disguised definitions. Yet they are drawn from experimental laws; these laws have, so to speak, been exalted into principles to which our mind attributes an absolute value.

Some philosophers have generalized too far; they believed the principles were the whole science and consequently that the whole science was conventional.

This paradoxical doctrine, called nominalism, will not bear examination.

How can a law become a principle? It expressed a relation between two real terms $A$ and $B$. But it was not rigorously true, it was only approximate. We introduce arbitrarily an intermediary term $C$ more or less fictitious, and $C$ is *by definition* that which has with $A$ *exactly* the relation expressed by the law.

Then our law is separated into an absolute and rigorous principle which expresses the relation of $A$ to $C$ and an experimental law, approximate and subject to revision, which expresses the relation of $C$ to $B$. It is clear that, however far this partition is pushed, some laws will always be left remaining.

We go to enter now the domain of laws properly so called.

# PART IV

## NATURE

### CHAPTER IX

#### HYPOTHESES IN PHYSICS

THE RÔLE OF EXPERIMENT AND GENERALIZATION.—Experiment is the sole source of truth. It alone can teach us anything new; it alone can give us certainty. These are two points that can not be questioned.

But then, if experiment is everything, what place will remain for mathematical physics? What has experimental physics to do with such an aid, one which seems useless and perhaps even dangerous?

And yet mathematical physics exists, and has done unquestionable service. We have here a fact that must be explained.

The explanation is that merely to observe is not enough. We must use our observations, and to do that we must generalize. This is what men always have done; only as the memory of past errors has made them more and more careful, they have observed more and more, and generalized less and less.

Every age has ridiculed the one before it, and accused it of having generalized too quickly and too naïvely. Descartes pitied the Ionians; Descartes, in his turn, makes us smile. No doubt our children will some day laugh at us.

But can we not then pass over immediately to the goal? Is not this the means of escaping the ridicule that we foresee? Can we not be content with just the bare experiment?

No, that is impossible; it would be to mistake utterly the true nature of science. The scientist must set in order. Science is built up with facts, as a house is with stones. But a collection of facts is no more a science than a heap of stones is a house.

127

And above all the scientist must foresee. Carlyle has somewhere said something like this: "Nothing but facts are of importance. John Lackland passed by here. Here is something
that is admirable. Here is a reality for which I would give all
the theories in the world." Carlyle was a fellow countryman of
Bacon; but Bacon would not have said that. That is the language
of the historian. The physicist would say rather: "John Lackland passed by here; that makes no difference to me, for he
never will pass this way again."

We all know that there are good experiments and poor ones.
The latter will accumulate in vain; though one may have made a
hundred or a thousand, a single piece of work by a true master,
by a Pasteur, for example, will suffice to tumble them into oblivion.
Bacon would have well understood this; it is he who invented the
phrase *Experimentum crucis*. But Carlyle would not have understood it. A fact is a fact. A pupil has read a certain number on
his thermometer; he has taken no precaution; no matter, he has
read it, and if it is only the fact that counts, here is a reality of
the same rank as the peregrinations of King John Lackland. Why
is the fact that this pupil has made this reading of no interest,
while the fact that a skilled physicist had made another reading
might be on the contrary very important? It is because from the
first reading we could not infer anything. What then is a good
experiment? It is that which informs us of something besides
an isolated fact; it is that which enables us to foresee, that is, that
which enables us to generalize.

For without generalization foreknowledge is impossible. The
circumstances under which one has worked will never reproduce
themselves all at once. The observed action then will never recur;
the only thing that can be affirmed is that under analogous circumstances an analogous action will be produced. In order to
foresee, then, it is necessary to invoke at least analogy, that is to
say, already then to generalize.

No matter how timid one may be, still it is necessary to interpolate. Experiment gives us only a certain number of isolated
points. We must unite these by a continuous line. This is a
veritable generalization. But we do more; the curve that we shall
trace will pass between the observed points and near these points;

it will not pass through these points themselves. Thus one does not restrict himself to generalizing the experiments, but corrects them; and the physicist who should try to abstain from these corrections and really be content with the bare experiment, would be forced to enunciate some very strange laws.

The bare facts, then, would not be enough for us; and that is why we must have science ordered, or rather organized.

It is often said experiments must be made without a preconceived idea. That is impossible. Not only would it make all experiment barren, but that would be attempted which could not be done. Every one carries in his mind his own conception of the world, of which he can not so easily rid himself. We must, for instance, use language; and our language is made up only of preconceived ideas and can not be otherwise. Only these are unconscious preconceived ideas, a thousand times more dangerous than the others.

Shall we say that if we introduce others, of which we are fully conscious, we shall only aggravate the evil? I think not. I believe rather that they will serve as counterbalances to each other—I was going to say as antidotes; they will in general accord ill with one another—they will come into conflict with one another, and thereby force us to regard things under different aspects. This is enough to emancipate us. He is no longer a slave who can choose his master.

Thus, thanks to generalization, each fact observed enables us to foresee a great many others; only we must not forget that the first alone is certain, that all others are merely probable. No matter how solidly founded a prediction may appear to us, we are never *absolutely* sure that experiment will not contradict it, if we undertake to verify it. The probability, however, is often so great that practically we may be content with it. It is far better to foresee even without certainty than not to foresee at all.

One must, then, never disdain to make a verification when opportunity offers. But all experiment is long and difficult; the workers are few; and the number of facts that we need to foresee is immense. Compared with this mass the number of direct verifications that we can make will never be anything but a negligible quantity.

10

Of this few that we can directly attain, we must make the best use; it is very necessary to get from every experiment the greatest possible number of predictions, and with the highest possible degree of probability. The problem is, so to speak, to increase the yield of the scientific machine.

Let us compare science to a library that ought to grow continually. The librarian has at his disposal for his purchases only insufficient funds. He ought to make an effort not to waste them.

It is experimental physics that is entrusted with the purchases. It alone, then, can enrich the library.

As for mathematical physics, its task will be to make out the catalogue. If the catalogue is well made, the library will not be any richer, but the reader will be helped to use its riches.

And even by showing the librarian the gaps in his collections, it will enable him to make a judicious use of his funds; which is all the more important because these funds are entirely inadequate.

Such, then, is the rôle of mathematical physics. It must direct generalization in such a manner as to increase what I just now called the yield of science. By what means it can arrive at this, and how it can do it without danger, is what remains for us to investigate.

THE UNITY OF NATURE.—Let us notice, first of all, that every generalization implies in some measure the belief in the unity and simplicity of nature. As to the unity there can be no difficulty. If the different parts of the universe were not like the members of one body, they would not act on one another, they would know nothing of one another; and we in particular would know only one of these parts. We do not ask, then, if nature is one, but how it is one.

As for the second point, that is not such an easy matter. It is not certain that nature is simple. Can we without danger act as if it were?

There was a time when the simplicity of Mariotte's law was an argument invoked in favor of its accuracy; when Fresnel himself, after having said in a conversation with Laplace that nature was not concerned about analytical difficulties, felt himself obliged to make explanations, in order not to strike too hard at prevailing opinion.

To-day ideas have greatly changed; and yet, those who do not believe that natural laws have to be simple, are still often obliged to act as if they did. They could not entirely avoid this necessity without making impossible all generalization, and consequently all science.

It is clear that any fact can be generalized in an infinity of ways, and it is a question of choice. The choice can be guided only by considerations of simplicity. Let us take the most commonplace case, that of interpolation. We pass a continuous line, as regular as possible, between the points given by observation. Why do we avoid points making angles and too abrupt turns? Why do we not make our curve describe the most capricious zigzags? It is because we know beforehand, or believe we know, that the law to be expressed can not be so complicated as all that.

We may calculate the mass of Jupiter from either the movements of its satellites, or the perturbations of the major planets, or those of the minor planets. If we take the averages of the determinations obtained by these three methods, we find three numbers very close together, but different. We might interpret this result by supposing that the coefficient of gravitation is not the same in the three cases. The observations would certainly be much better represented. Why do we reject this interpretation? Not because it is absurd, but because it is needlessly complicated. We shall only accept it when we are forced to, and that is not yet.

To sum up, ordinarily every law is held to be simple till the contrary is proved.

This custom is imposed upon physicists by the causes that I have just explained. But how shall we justify it in the presence of discoveries that show us every day new details that are richer and more complex? How shall we even reconcile it with the belief in the unity of nature? For if everything depends on everything, relationships where so many diverse factors enter can no longer be simple.

If we study the history of science, we see happen two inverse phenomena, so to speak. Sometimes simplicity hides under complex appearances; sometimes it is the simplicity which is apparent, and which disguises extremely complicated realities.

What is more complicated than the confused movements of

the planets? What simpler than Newton's law? Here nature, making sport, as Fresnel said, of analytical difficulties, employs only simple means, and by combining them produces I know not what inextricable tangle. Here it is the hidden simplicity which must be discovered.

Examples of the opposite abound. In the kinetic theory of gases, one deals with molecules moving with great velocities, whose paths, altered by incessant collisions, have the most capricious forms and traverse space in every direction. The observable result is Mariotte's simple law. Every individual fact was complicated. The law of great numbers has reestablished simplicity in the average. Here the simplicity is merely apparent, and only the coarseness of our senses prevents our perceiving the complexity.

Many phenomena obey a law of proportionality. But why? Because in these phenomena there is something very small. The simple law observed, then, is only a result of the general analytical rule that the infinitely small increment of a function is proportional to the increment of the variable. As in reality our increments are not infinitely small, but very small, the law of proportionality is only approximate, and the simplicity is only apparent. What I have just said applies to the rule of the superposition of small motions, the use of which is so fruitful, and which is the basis of optics.

And Newton's law itself? Its simplicity, so long undetected, is perhaps only apparent. Who knows whether it is not due to some complicated mechanism, to the impact of some subtile matter animated by irregular movements, and whether it has not become simple only through the action of averages and of great numbers? In any case, it is difficult not to suppose that the true law contains complementary terms, which would become sensible at small distances. If in astronomy they are negligible as modifying Newton's law, and if the law thus regains its simplicity, it would be only because of the immensity of celestial distances.

No doubt, if our means of investigation should become more and more penetrating, we should discover the simple under the complex, then the complex under the simple, then again the simple under the complex, and so on, without our being able to foresee what will be the last term.

We must stop somewhere, and that science may be possible, we must stop when we have found simplicity. This is the only ground on which we can rear the edifice of our generalizations. But this simplicity being only apparent, will the ground be firm enough? This is what must be investigated.

For that purpose, let us see what part is played in our generalizations by the belief in simplicity. We have verified a simple law in a good many particular cases; we refuse to admit that this agreement, so often repeated, is simply the result of chance, and conclude that the law must be true in the general case.

Kepler notices that a planet's positions, as observed by Tycho, are all on one ellipse. Never for a moment does he have the thought that by a strange play of chance Tycho never observed the heavens except at a moment when the real orbit of the planet happened to cut this ellipse.

What does it matter then whether the simplicity be real, or whether it covers a complex reality? Whether it is due to the influence of great numbers, which levels down individual differences, or to the greatness or smallness of certain quantities, which allows us to neglect certain terms, in no case is it due to chance. This simplicity, real or apparent, always has a cause. We can always follow, then, the same course of reasoning, and if a simple law has been observed in several particular cases, we can legitimately suppose that it will still be true in analogous cases. To refuse to do this would be to attribute to chance an inadmissible rôle.

There is, however, a difference. If the simplicity were real and essential, it would resist the increasing precision of our means of measure. If then we believe nature to be essentially simple, we must, from a simplicity that is approximate, infer a simplicity that is rigorous. This is what was done formerly; and this is what we no longer have a right to do.

The simplicity of Kepler's laws, for example, is only apparent. That does not prevent their being applicable, very nearly, to all systems analogous to the solar system; but it does prevent their being rigorously exact.

THE RÔLE OF HYPOTHESIS.—All generalization is a hypothesis. Hypothesis, then, has a necessary rôle that no one has ever contested. Only, it ought always, as soon as possible and as often as possible, to be subjected to verification. And, of course, if it does not stand this test, it ought to be abandoned without reserve. This is what we generally do, but sometimes with rather an ill humor.

Well, even this ill humor is not justified. The physicist who has just renounced one of his hypotheses ought, on the contrary, to be full of joy; for he has found an unexpected opportunity for discovery. His hypothesis, I imagine, had not been adopted without consideration; it took account of all the known factors that it seemed could enter into the phenomenon. If the test does not support it, it is because there is something unexpected and extraordinary; and because there is going to be something found that is unknown and new.

Has the discarded hypothesis, then, been barren? Far from that, it may be said it has rendered more service than a true hypothesis. Not only has it been the occasion of the decisive experiment, but, without having made the hypothesis, the experiment would have been made by chance, so that nothing would have been derived from it. One would have seen nothing extraordinary; only one fact the more would have been catalogued without deducing from it the least consequence.

Now on what condition is the use of hypothesis without danger?

The firm determination to submit to experiment is not enough; there are still dangerous hypotheses; first, and above all, those which are tacit and unconscious. Since we make them without knowing it, we are powerless to abandon them. Here again, then, is a service that mathematical physics can render us. By the precision that is characteristic of it, it compels us to formulate all the hypotheses that we should make without it, but unconsciously.

Let us notice besides that it is important not to multiply hypotheses beyond measure, and to make them only one after the other. If we construct a theory based on a number of hypotheses, and if experiment condemns it, which of our premises is it necessary to change? It will be impossible to know. And inversely, if the experiment succeeds, shall we believe that we have demonstrated all the hypotheses at once? Shall we believe that with one single equation we have determined several unknowns?

We must equally take care to distinguish between the different kinds of hypotheses. There are first those which are perfectly natural and from which one can scarcely escape. It is difficult not to suppose that the influence of bodies very remote is quite negligible, that small movements follow a linear law, that the effect is a continuous function of its cause. I will say as much of the conditions imposed by symmetry. All these hypotheses form, as it were, the common basis of all the theories of mathematical physics. They are the last that ought to be abandoned.

There is a second class of hypotheses, that I shall term neutral. In most questions the analyst assumes at the beginning of his calculations either that matter is continuous or, on the contrary, that it is formed of atoms. He might have made the opposite assumption without changing his results. He would only have had more trouble to obtain them; that is all. If, then, experiment confirms his conclusions, will he think that he has demonstrated, for instance, the real existence of atoms?

In optical theories two vectors are introduced, of which one is regarded as a velocity, the other as a vortex. Here again is a neutral hypothesis, since the same conclusions would have been reached by taking precisely the opposite. The success of the experiment, then, can not prove that the first vector is indeed a velocity; it can only prove one thing, that it is a vector. This is the only hypothesis that has really been introduced in the premises. In order to give it that concrete appearance which the weakness of our minds requires, it has been necessary to consider it either as a velocity or as a vortex, in the same way that it has been necessary to represent it by a letter, either $x$ or $y$. The result, however, whatever it may be, will not prove that it was right or wrong to regard it as a velocity any more than it will prove that it was right or wrong to call it $x$ and not $y$.

These neutral hypotheses are never dangerous, if only their character is not misunderstood. They may be useful, either as devices for computation, or to aid our understanding by concrete images, to fix our ideas as the saying is. There is, then, no occasion to exclude them.

The hypotheses of the third class are the real generalizations. They are the ones that experiment must confirm or invalidate. Whether verified or condemned, they will always be fruitful. But for the reasons that I have set forth, they will only be fruitful if they are not too numerous.

ORIGIN OF MATHEMATICAL PHYSICS.—Let us penetrate further, and study more closely the conditions that have permitted the development of mathematical physics. We observe at once that the efforts of scientists have always aimed to resolve the complex phenomenon directly given by experiment into a very large number of elementary phenomena.

This is done in three different ways: first, in time. Instead of embracing in its entirety the progressive development of a phenomenon, the aim is simply to connect each instant with the instant immediately preceding it. It is admitted that the actual state of the world depends only on the immediate past, without being directly influenced, so to speak, by the memory of a distant past. Thanks to this postulate, instead of studying directly the whole succession of phenomena, it is possible to confine ourselves to writing its 'differential equation.' For Kepler's laws we substitute Newton's law.

Next we try to analyze the phenomenon in space. What experiment gives us is a confused mass of facts presented on a stage of considerable extent. We must try to discover the elementary phenomenon, which will be, on the contrary, localized in a very small region of space.

Some examples will perhaps make my thought better understood. If we wished to study in all its complexity the distribution of temperature in a cooling solid, we should never succeed. Everything becomes simple if we reflect that one point of the solid can not give up its heat directly to a distant point; it will give up its heat only to the points in the immediate neighborhood, and it is by degrees that the flow of heat can reach other parts of the solid. The elementary phenomenon is the exchange of heat between two contiguous points. It is strictly localized, and is relatively simple, if we admit, as is natural, that it is not influenced by the temperature of molecules whose distance is sensible.

I bend a rod. It is going to take a very complicated form, the direct study of which would be impossible. But I shall be able, however, to attack it, if I observe that its flexure is a result only of the deformation of the very small elements of the rod, and that the deformation of each of these elements depends only on the forces that are directly applied to it, and not at all on those which may act on the other elements.

In all these examples, which I might easily multiply, we admit that there is no action at a distance, or at least at a great distance. This is a hypothesis. It is not always true, as the law of gravitation shows us. It must, then, be submitted to verification. If it is confirmed, even approximately, it is precious, for it will enable us to make mathematical physics, at least by successive approximations.

If it does not stand the test, we must look for something else analogous; for there are still other means of arriving at the elementary phenomenon. If several bodies act simultaneously, it may happen that their actions are independent and are simply added to one another, either as vectors or as scalars. The elementary phenomenon is then the action of an isolated body. Or again, we have to deal with small movements, or more generally with small variations, which obey the well-known law of superposition. The observed movement will then be decomposed into simple movements, for example, sound into its harmonics, white light into its monochromatic components.

When we have discovered in what direction it is advisable to look for the elementary phenomenon, by what means can we reach it?

First of all, it will often happen that in order to detect it, or rather to detect the part of it useful to us, it will not be necessary to penetrate the mechanism; the law of great numbers will suffice.

Let us take again the instance of the propagation of heat. Every molecule emits rays toward every neighboring molecule. According to what law, we do not need to know. If we should make any supposition in regard to this, it would be a neutral hypothesis and consequently useless and incapable of verification. And, in fact, by the action of averages and thanks to the symmetry of the medium, all the differences are leveled down, and whatever hypothesis may be made, the result is always the same.

The same circumstance is presented in the theory of electricity and in that of capillarity. The neighboring molecules attract and repel one another. We do not need to know according to what law; it is enough for us that this attraction is sensible only at small distances, that the molecules are very numerous, that the medium is symmetrical, and we shall only have to let the law of great numbers act.

Here again the simplicity of the elementary phenomenon was hidden under the complexity of the resultant observable phenomenon; but, in its turn, this simplicity was only apparent, and concealed a very complex mechanism.

The best means of arriving at the elementary phenomenon would evidently be experiment. We ought by experimental contrivance to dissociate the complex sheaf that nature offers to our researches, and to study with care the elements as much isolated as possible. For example, natural white light would be decomposed into monochromatic lights by the aid of the prism, and into polarized light by the aid of the polarizer.

Unfortunately that is neither always possible nor always sufficient, and sometimes the mind must outstrip experiment. I shall cite only one example, which has always struck me forcibly.

If I decompose white light, I shall be able to isolate a small part of the spectrum, but however small it may be, it will retain a certain breadth. Likewise the natural lights, called *monochromatic,* give us a very narrow line, but not, however, infinitely narrow. It might be supposed that by studying experimentally the properties of these natural lights, by working with finer and finer lines of the spectrum, and by passing at last to the limit, so to speak, we should succeed in learning the properties of a light strictly monochromatic.

That would not be accurate. Suppose that two rays emanate from the same source, that we polarize them first in two perpendicular planes, then bring them back to the same plane of polarization, and try to make them interfere. If the light were *strictly* monochromatic, they would interfere. With our lights, which are nearly monochromatic, there will be no interference, and that no matter how narrow the line. In order to be otherwise it would have to be several million times as narrow as the finest known lines.

Here, then, the passage to the limit would have deceived us. The mind must outstrip the experiment, and if it has done so with success, it is because it has allowed itself to be guided by the instinct of simplicity.

The knowledge of the elementary fact enables us to put the problem in an equation. Nothing remains but to deduce from this by combination the complex fact that can be observed and verified. This is what is called *integration*, and is the business of the mathematician.

It may be asked why, in physical sciences, generalization so readily takes the mathematical form. The reason is now easy to see. It is not only because we have numerical laws to express; it is because the observable phenomenon is due to the superposition of a great number of elementary phenomena *all alike*. Thus quite naturally are introduced differential equations.

It is not enough that each elementary phenomenon obeys simple laws; all those to be combined must obey the same law. Then only can the intervention of mathematics be of use; mathematics teaches us in fact to combine like with like. Its aim is to learn the result of a combination without needing to go over the combination piece by piece. If we have to repeat several times the same operation, it enables us to avoid this repetition by telling us in advance the result of it by a sort of induction. I have explained this above, in the chapter on mathematical reasoning.

But, for this, all the operations must be alike. In the opposite case, it would evidently be necessary to resign ourselves to doing them in reality one after another, and mathematics would become useless.

It is then thanks to the approximate homogeneity of the matter studied by physicists, that mathematical physics could be born.

In the natural sciences, we no longer find these conditions: homogeneity, relative independence of remote parts, simplicity of the elementary fact; and this is why naturalists are obliged to resort to other methods of generalization.

# CHAPTER X

## The Theories of Modern Physics

MEANING OF PHYSICAL THEORIES.—The laity are struck to see how ephemeral scientific theories are. After some years of prosperity, they see them successively abandoned; they see ruins accumulate upon ruins; they foresee that the theories fashionable to-day will shortly succumb in their turn and hence they conclude that these are absolutely idle. This is what they call the *bankruptcy of science.*

Their scepticism is superficial; they give no account to themselves of the aim and the rôle of scientific theories; otherwise they would comprehend that the ruins may still be good for something.

No theory seemed more solid than that of Fresnel which attributed light to motions of the ether. Yet now Maxwell's is preferred. Does this mean the work of Fresnel was in vain? No, because the aim of Fresnel was not to find out whether there is really an ether, whether it is or is not formed of atoms, whether these atoms really move in this or that sense; his object was to foresee optical phenomena.

Now, Fresnel's theory always permits of this, to-day as well as before Maxwell. The differential equations are always true; they can always be integrated by the same procedures and the results of this integration always retain their value.

And let no one say that thus we reduce physical theories to the rôle of mere practical recipes; these equations express relations, and if the equations remain true it is because these relations preserve their reality. They teach us, now as then, that there is such and such a relation between some thing and some other thing; only this something formerly we called *motion;* we now call it *electric current.* But these appellations were only images substituted for the real objects which nature will eternally hide from us. The true relations between these real objects are the only reality we can attain to, and the only condition is that the same relations exist between these objects as between the images by which we are forced to replace them. If these relations are known to us, what matter if we deem it convenient to replace one image by another.

That some periodic phenomenon (an electric oscillation, for instance) is really due to the vibration of some atom which, acting like a pendulum, really moves in this or that sense, is neither certain nor interesting. But that between electric oscillation, the motion of the pendulum and all periodic phenomena there exists a close relationship which corresponds to a profound reality; that this relationship, this similitude, or rather this parallelism extends into details; that it is a consequence of more general principles, that of energy and that of least action; this is what we can affirm; this is the truth which will always remain the same under all the costumes in which we may deem it useful to deck it out.

Numerous theories of dispersion have been proposed; the first was imperfect and contained only a small part of truth. Afterwards came that of Helmholtz; then it was modified in various ways, and its author himself imagined another founded on the principles of Maxwell. But, what is remarkable, all the scientists who came after Helmholtz reached the same equations, starting from points of departure in appearance very widely separated. I will venture to say these theories are all true at the same time, not only because they make us foresee the same phenomena, but because they put in evidence a true relation, that of absorption and anomalous dispersion. What is true in the premises of these theories is what is common to all the authors; this is the affirmation of this or that relation between certain things which some call by one name, others by another.

The kinetic theory of gases has given rise to many objections, which we could hardly answer if we pretended to see in it the absolute truth. But all these objections will not preclude its having been useful, and particularly so in revealing to us a relation true and but for it profoundly hidden, that of the gaseous pressure and the osmotic pressure. In this sense, then, it may be said to be true.

When a physicist finds a contradiction between two theories

equally dear to him, he sometimes says: "We will not bother about that, but hold firmly the two ends of the chain, though the intermediate links are hidden from us." This argument of an embarrassed theologian would be ridiculous if it were necessary to attribute to physical theories the sense the laity give them. In case of contradiction, one of them at least must then be regarded as false. It is no longer the same if in them be sought only what should be sought. May be they both express true relations and the contradiction is only in the images wherewith we have clothed the reality.

To those who find we restrict too much the domain accessible to the scientist, I answer: These questions which we interdict to you and which you regret, are not only insoluble, they are illusory and devoid of meaning.

Some philosopher pretends that all physics may be explained by the mutual impacts of atoms. If he merely means there are between physical phenomena the same relations as between the mutual impacts of a great number of balls, well and good, that is verifiable, that is perhaps true. But he means something more; and we think we understand it because we think we know what impact is in itself; why? Simply because we have often seen games of billiards. Shall we think God, contemplating his work, feels the same sensations as we in watching a billiard match? If we do not wish to give this bizarre sense to his assertion, if neither do we wish the restricted sense I have just explained, which is good sense, then it has none.

Hypotheses of this sort have therefore only a metaphorical sense. The scientist should no more interdict them than the poet does metaphors; but he ought to know what they are worth. They may be useful to give a certain satisfaction to the mind, and they will not be injurious provided they are only indifferent hypotheses.

These considerations explain to us why certain theories, supposed to be abandoned and finally condemned by experiment, suddenly arise from their ashes and recommence a new life. It is because they expressed true relations; and because they had not ceased to do so when, for one reason or another, we felt it necessary to enunciate the same relations in another language. So they retained a sort of latent life.

Scarcely fifteen years ago was there anything more ridiculous, more naïvely antiquated, than Coulomb's fluids? And yet here they are reappearing under the name of *electrons*. Wherein do these permanently electrified molecules differ from Coulomb's electric molecules? It is true that in the electrons the electricity is supported by a little, a very little matter; in other words, they have a mass (and yet this is now contested); but Coulomb did not deny mass to his fluids, or, if he did, it was only with reluctance. It would be rash to affirm that the belief in electrons will not again suffer eclipse; it was none the less curious to note this unexpected resurrection.

But the most striking example is Carnot's principle. Carnot set it up starting from false hypotheses; when it was seen that heat is not indestructible, but may be transformed into work, his ideas were completely abandoned; afterwards Clausius returned to them and made them finally triumph. Carnot's theory, under its primitive form, expressed, aside from true relations, other inexact relations, *débris* of antiquated ideas; but the presence of these latter did not change the reality of the others. Clausius had only to discard them as one lops off dead branches.

The result was the second fundamental law of thermodynamics. There were always the same relations; though these relations no longer subsisted, at least in appearance, between the same objects. This was enough for the principle to retain its value. And even the reasonings of Carnot have not perished because of that; they were applied to a material tainted with error; but their form (that is to say, the essential) remained correct.

What I have just said illuminates at the same time the rôle of general principles such as the principle of least action, or that of the conservation of energy.

These principles have a very high value; they were obtained in seeking what there was in common in the enunciation of numerous physical laws; they represent therefore, as it were, the quintessence of innumerable observations.

However, from their very generality a consequence results to which I have called attention in Chapter VIII., namely, that they can no longer be verified. As we can not give a general definition of energy, the principle of the conservation of energy signifies simply that there is *something* which remains constant. Well, whatever be the new notions that future experiments shall give us about the world, we are sure in advance that there will be something there which will remain constant and which may be called *energy*.

Is this to say that the principle has no meaning and vanishes in a tautology? Not at all; it signifies that the different things to which we give the name of *energy* are connected by a true kinship; it affirms a real relation between them. But then if this principle has a meaning, it may be false; it may be that we have not the right to extend indefinitely its applications, and yet it is certain beforehand to be verified in the strict acceptation of the term; how then shall we know when it shall have attained all the extension which can legitimately be given it? Just simply when it shall cease to be useful to us, that is, to make us correctly foresee new phenomena. We shall be sure in such a case that the relation affirmed is no longer real; for otherwise it would be fruitful; experiment, without directly contradicting a new extension of the principle, will yet have condemned it.

PHYSICS AND MECHANISM.—Most theorists have a constant predilection for explanations borrowed from mechanics or dynamics. Some would be satisfied if they could explain all phenomena by motions of molecules attracting each other according to certain laws. Others are more exacting; they would suppress attractions at a distance; their molecules should follow rectilinear paths from which they could be made to deviate only by impacts. Others again, like Hertz, suppress forces also, but suppose their molecules subjected to geometric attachments analogous, for instance, to those of our linkages; they try thus to reduce dynamics to a sort of kinematics.

In a word, all would bend nature into a certain form outside of which their mind could not feel satisfied. Will nature be sufficiently flexible for that?

We shall examine this question in Chapter XII., *à propos* of Maxwell's theory. Whenever the principles of energy and of least action are satisfied, we shall see not only that there is always one possible mechanical explanation, but that there is always an infinity of them. Thanks to a well-known theorem of König's on linkages, it could be shown that we can, in an infinity of ways, explain everything by attachments after the manner of Hertz, or also by central forces. Without doubt it could be demonstrated just as easily that everything can always be explained by simple impacts.

For that, of course, we need not be content with ordinary matter, with that which falls under our senses and whose motions we observe directly. Either we shall suppose that this common matter is formed of atoms whose internal motions elude us, the displacement of the totality alone remaining accessible to our senses. Or else we shall imagine some one of those subtile fluids which under the name of *ether* or under other names, have at all times played so great a rôle in physical theories.

Often one goes further and regards the ether as the sole primitive matter or even as the only true matter. The more moderate consider common matter as condensed ether, which is nothing startling; but others reduce still further its importance and see in it nothing more than the geometric locus of the ether's singularities. For instance, what we call *matter* is for Lord Kelvin only the locus of points where the ether is animated by vortex motions; for Riemann, it was the locus of points where ether is constantly destroyed; for other more recent authors, Wiechert or Larmor, it is the locus of points where the ether undergoes a sort of torsion of a very particular nature. If the attempt is made to occupy one of these points of view, I ask myself by what right shall we extend to the ether, under pretext that this is the true matter, mechanical properties observed in ordinary matter, which is only false matter.

The ancient fluids, caloric, electricity, etc., were abandoned when it was perceived that heat is not indestructible. But they were abandoned for another reason also. In materializing them, their individuality was, so to speak, emphasized, a sort of abyss was opened between them. This had to be filled up on the coming of a more vivid feeling of the unity of nature, and the perception of the intimate relations which bind together all its parts. Not only did the old physicists, in multiplying fluids, create entities unnecessarily, but they broke real ties.

It is not sufficient for a theory to affirm no false relations, it must not hide true relations.

And does our ether really exist? We know the origin of our belief in the ether. If light reaches us from a distant star, during several years it was no longer on the star and not yet on the earth; it must then be somewhere and sustained, so to speak, by some material support.

The same idea may be expressed under a more mathematical and more abstract form. What we ascertain are the changes undergone by material molecules; we see, for instance, that our photographic plate feels the consequences of phenomena of which the incandescent mass of the star was the theater several years before. Now, in ordinary mechanics the state of the system studied depends only on its state at an instant immediately anterior; therefore the system satisfies differential equations. On the contrary, if we should not believe in the ether, the state of the material universe would depend not only on the state immediately preceding, but on states much older; the system would satisfy equations of finite differences. It is to escape this derogation of the general laws of mechanics that we have invented the ether.

That would still only oblige us to fill up, with the ether, the interplanetary void, but not to make it penetrate the bosom of the material media themselves. Fizeau's experiment goes further. By the interference of rays which have traversed air or water in motion, it seems to show us two different media interpenetrating and yet changing place one with regard to the other.

We seem to touch the ether with the finger.

Yet experiments may be conceived which would make us touch it still more nearly. Suppose Newton's principle, of the equality of action and reaction, no longer true if applied to matter *alone,* and that we have established it. The geometric sum of all the forces applied to all the material molecules would no longer be null. It would be necessary then, if we did not wish to change all mechanics, to introduce the ether, in order that this action which matter appeared to experience should be counterbalanced by the reaction of matter on something.

Or again, suppose we discover that optical and electrical phenomena are influenced by the motion of the earth. We should be led to conclude that these phenomena might reveal to us not only the relative motions of material bodies, but what would seem to be their absolute motions. Again, an ether would be necessary, that these so-called absolute motions should not be their displacements with regard to a void space, but their displacements with regard to something concrete.

Shall we ever arrive at that? I have not this hope, I shall soon say why, and yet it is not so absurd, since others have had it.

For instance, if the theory of Lorentz, of which I shall speak in detail further on in Chapter XIII., were true, Newton's principle would not apply to matter *alone,* and the difference would not be very far from being accessible to experiment.

On the other hand, many researches have been made on the influence of the earth's motion. The results have always been negative. But these experiments were undertaken because the outcome was not sure in advance, and, indeed, according to the ruling theories, the compensation would be only approximate, and one might expect to see precise methods give positive results.

I believe that such a hope is illusory; it was none the less interesting to show that a success of this sort would open to us, in some sort, a new world.

And now I must be permitted a digression; I must explain, in fact, why I do not believe, despite Lorentz, that more precise observations can ever put in evidence anything else than the relative displacements of material bodies. Experiments have been made which should have disclosed the terms of the first order; the results have been negative; could that be by chance? No one has assumed that; a general explanation has been sought, and Lorentz has found it; he has shown that the terms of the first order must destroy each other, but not those of the second. Then more precise experiments were made; they also were negative; neither could this be the effect of chance; an explanation was necessary; it was found; they always are found; of hypotheses there is never lack.

But this is not enough; who does not feel that this is still to leave to chance too great a rôle? Would not that also be a chance, this singular coincidence which brought it about that a certain circumstance should come just in the nick of time to destroy the terms of the first order, and that another circumstance, wholly different, but just as opportune, should take upon itself to destroy those of the second order?  No, it is necessary to find an explanation the same for the one as for the other, and then everything leads us to think that this explanation will hold good equally well for the terms of higher order, and that the mutual destruction of these terms will be rigorous and absolute.

PRESENT STATE OF THE SCIENCE.—In the history of the development of physics we distinguish two inverse tendencies.

On the one hand, new bonds are continually being discovered between objects which had seemed destined to remain forever unconnected; scattered facts cease to be strangers to one another; they tend to arrange themselves in an imposing synthesis. Science advances toward unity and simplicity.

On the other hand, observation reveals to us every day new phenomena; they must long await their place and sometimes, to make one for them, a corner of the edifice must be demolished. In the known phenomena themselves, where our crude senses showed us uniformity, we perceive details from day to day more varied; what we believed simple becomes complex, and science appears to advance toward variety and complexity.

Of these two inverse tendencies, which seem to triumph turn about, which will win?  If it be the first, science is possible; but nothing proves this *a priori*, and it may well be feared that after having made vain efforts to bend nature in spite of herself to our ideal of unity, submerged by the ever-rising flood of our new riches, we must renounce classifying them, abandon our ideal, and reduce science to the registration of innumerable recipes.

To this question we can not reply.  All we can do is to observe the science of to-day and compare it with that of yesterday. From this examination we may doubtless draw some encouragement.

Half a century ago, hope ran high.  The discovery of the conservation of energy and of its transformations had revealed to us the unity of force.  Thus it showed that the phenomena of heat could be explained by molecular motions.  What was the nature of these motions was not exactly known, but no one doubted that it soon would be. For light, the task seemed completely accomplished. In what concerns electricity, things were less advanced. Electricity had just annexed magnetism. This was a considerable step toward unity, and a decisive step.

But how should electricity in its turn enter into the general unity, how should it be reduced to the universal mechanism?

Of that no one had any idea. Yet the possibility of this reduction was doubted by none, there was faith. Finally, in what concerns the molecular properties of material bodies, the reduction seemed still easier, but all the detail remained hazy. In a word, the hopes were vast and animated, but vague. To-day, what do we see? First of all, a prime progress, immense progress. The relations of electricity and light are now known; the three realms, of light, of electricity and of magnetism, previously separated, form now but one; and this annexation seems final.

This conquest, however, has cost us some sacrifices. The optical phenomena subordinate themselves as particular cases under the electrical phenomena; so long as they remained isolated, it was easy to explain them by motions that were supposed to be known in all their details, that was a matter of course; but now an explanation, to be acceptable, must be easily capable of extension to the entire electric domain. Now that is a matter not without difficulties.

The most satisfactory theory we have is that of Lorentz, which, as we shall see in the last chapter, explains electric currents by the motions of little electrified particles; it is unquestionbly the one which best explains the known facts, the one which illuminates the greatest number of true relations, the one of which most traces will be found in the final construction. Nevertheless, it still has a serious defect, which I have indicated above; it is contrary to Newton's law of the equality of action and reaction; or rather, this principle, in the eyes of Lorentz, would not be applicable to matter alone; for it to be true, it would be necessary to take account of the action of the ether on matter and of the reaction of matter on the ether.

Now, from what we know at present, it seems probable that things do not happen in this way.

However that may be, thanks to Lorentz, Fizeau's results on

the optics of moving bodies, the laws of normal and anomalous dispersion and of absorption find themselves linked to one another and to the other properties of the ether by bonds which beyond any doubt will never more be broken. See the facility with which the new Zeeman effect has found its place already and has even aided in classifying Faraday's magnetic rotation which had defied Maxwell's efforts; this facility abundantly proves that the theory of Lorentz is not an artificial assemblage destined to fall asunder. It will probably have to be modified, but not destroyed.

But Lorentz had no aim beyond that of embracing in one totality all the optics and electrodynamics of moving bodies; he never pretended to give a mechanical explanation of them. Larmor goes further; retaining the theory of Lorentz in essentials, he grafts upon it, so to speak, MacCullagh's ideas on the direction of the motions of the ether.

According to him, the velocity of the ether would have the same direction and the same magnitude as the magnetic force. However ingenious this attempt may be, the defect of the theory of Lorentz remains and is even aggravated. With Lorentz, we do not know what are the motions of the ether; thanks to this ignorance, we may suppose them such that, compensating those of matter, they reestablish the equality of action and reaction. With Larmor, we know the motions of the ether, and we can ascertain that the compensation does not take place.

If Larmor has failed, as it seems to me he has, does that mean that a mechanical explanation is impossible? Far from it: I have said above that when a phenomenon obeys the two principles of energy and of least action, it admits of an infinity of mechanical explanations; so it is, therefore, with the optical and electrical phenomena.

But this is not enough: for a mechanical explanation to be good, it must be simple; for choosing it among all which are possible, there should be other reasons besides the necessity of making a choice. Well, we have not as yet a theory satisfying this condition and consequently good for something. Must we lament this? That would be to forget what is the goal sought; this is not mechanism; the true, the sole aim is unity.

We must therefore set bounds to our ambition; let us not try to formulate a mechanical explanation; let us be content with showing that we could always find one if we wished to. In this regard we have been successful; the principle of the conservation of energy has received only confirmations; a second principle has come to join it, that of least action, put under the form which is suitable for physics. It also has always been verified, at least in so far as concerns reversible phenomena which thus obey the equations of Lagrange, that is to say, the most general laws of mechanics.

Irreversible phenomena are much more rebellious. Yet these also are being coordinated, and tend to come into unity; the light which has illuminated them has come to us from Carnot's principle. Long did thermodynamics confine itself to the study of the dilatation of bodies and their changes of state. For some time past it has been growing bolder and has considerably extended its domain. We owe to it the theory of the galvanic battery, and that of the thermoelectric phenomena; there is not in all physics a corner that it has not explored, and it has attacked chemistry itself.

Everywhere the same laws reign; everywhere, under the diversity of appearances, is found again Carnot's principle; everywhere also is found that concept so prodigiously abstract of entropy, which is as universal as that of energy and seems like it to cover a reality. Radiant heat seemed destined to escape it; but recently we have seen that submit to the same laws.

In this way fresh analogies are revealed to us, which may often be followed into detail; ohmic resistance resembles the viscosity of liquids; hysteresis would resemble rather the friction of solids. In all cases, friction would appear to be the type which the most various irreversible phenomena copy, and this kinship is real and profound.

Of these phenomena a mechanical explanation, properly so called, has also been sought. They hardly lent themselves to it. To find it, it was necessary to suppose that the irreversibility is only apparent, that the elementary phenomena are reversible and obey the known laws of dynamics. But the elements are extremely numerous and blend more and more, so that to our crude sight all appears to tend toward uniformity, that is, everything seems to go forward in the same sense without hope of return. The apparent irreversibility is thus only an effect of the law of great numbers. But, only a being with infinitely subtle senses, like Maxwell's imaginary demon, could disentangle this inextricable skein and turn back the course of the universe.

This conception, which attaches itself to the kinetic theory of gases, has cost great efforts and has not, on the whole, been fruitful; but it may become so. This is not the place to examine whether it does not lead to contradictions and whether it is in conformity with the true nature of things.

We signalize, however, M. Gouy's original ideas on the Brownian movement. According to this scientist, this singular motion should escape Carnot's principle. The particles which it puts in swing would be smaller than the links of that so compacted skein; they would therefore be fitted to disentangle them and hence to make the world go backward. We should almost see Maxwell's demon at work.

To summarize, the previously known phenomena are better and better classified, but new phenomena come to claim their place; most of these, like the Zeeman effect, have at once found it.

But we have the cathode rays, the X-rays, those of uranium and of radium. Herein is a whole world which no one suspected. How many unexpected guests must be stowed away!

No one can yet foresee the place they will occupy. But I do not believe they will destroy the general unity; I think they will rather complete it. On the one hand, in fact, the new radiations seem connected with the phenomena of luminescence; not only do they excite fluorescence, but they sometimes take birth in the same conditions as it.

Nor are they without kinship with the causes which produce the electric spark under the action of the ultra-violet light.

Finally, and above all, it is believed that in all these phenomena are found true ions, animated, it is true, by velocities incomparably greater than in the electrolytes.

That is all very vague, but it will all become more precise.

Phosphorescence, the action of light on the spark, these were regions rather isolated, and consequently somewhat neglected by investigators. One may now hope that a new path will be constructed which will facilitate their communications with the rest of science.

Not only do we discover new phenomena, but in those we thought we knew, unforeseen aspects reveal themselves. In the free ether, the laws retain their majestic simplicity; but matter, properly so called, seems more and more complex; all that is said of it is never more than approximate, and at each instant our formulas require new terms.

Nevertheless the frames are not broken; the relations that we have recognized between objects we thought simple still subsist between these same objects when we know their complexity, and it is that alone which is of importance. Our equations become, it is true, more and more complicated, in order to embrace more closely the complexity of nature; but nothing is changed in the relations which permit the deducing of these equations one from another. In a word, the form of these equations has persisted.

Take, for example, the laws of reflection: Fresnel had established them by a simple and seductive theory which experiment seemed to confirm. Since then more precise researches have proved that this verification was only approximate; they have shown everywhere traces of elliptic polarization. But, thanks to the help that the first approximation gave us, we found forthwith the cause of these anomalies, which is the presence of a transition layer; and Fresnel's theory has subsisted in its essentials.

But there is a reflection we can not help making: All these relations would have remained unperceived if one had at first suspected the complexity of the objects they connect. It has long been said: If Tycho had had instruments ten times more precise neither Kepler, nor Newton, nor astronomy would ever have been. It is a misfortune for a science to be born too late, when the means of observation have become too perfect. This is to-day the case with physical chemistry; its founders are embarrassed in their general grasp by third and fourth decimals; happily they are men of a robust faith.

The better one knows the properties of matter the more one sees continuity reign. Since the labors of Andrews and of van der Wals, we get an idea of how the passage is made from the liquid to the gaseous state and that this passage is not abrupt. Similarly, there is no gap between the liquid and solid states, and in the proceedings of a recent congress is to be seen, alongside of a work on the rigidity of liquids, a memoir on the flow of solids.

By this tendency no doubt simplicity loses; some phenomenon was formerly represented by several straight lines, now these straights must be joined by curves more or less complicated. In compensation unity gains notably. Those cut-off categories quieted the mind, but they did not satisfy it.

Finally the methods of physics have invaded a new domain, that of chemistry; physical chemistry is born. It is still very young, but we already see that it will enable us to connect such phenomena as electrolysis, osmosis and the motions of ions.

From this rapid exposition, what shall we conclude?

Everything considered, we have approached unity; we have not been as quick as was hoped fifty years ago, we have not always taken the predicted way; but, finally, we have gained ever so much ground.

# CHAPTER XI

## THE CALCULUS OF PROBABILITIES

DOUBTLESS it will be astonishing to find here thoughts about the calculus of probabilities. What has it to do with the method of the physical sciences? And yet the questions I shall raise without solving present themselves naturally to the philosopher who is thinking about physics. So far is this the case that in the two preceding chapters I have often been led to use the words 'probability' and 'chance.'

'Predicted facts,' as I have said above, 'can only be probable.' ''However solidly founded a prediction may seem to us to be, we are never absolutely sure that experiment will not prove it false. But the probability is often so great that practically we may be satisfied with it.'' And a little further on I have added: ''See what a rôle the belief in simplicity plays in our generalizations. We have verified a simple law in a great number of particular cases; we refuse to admit that this coincidence, so often repeated, can be a mere effect of chance. . . .''

Thus in a multitude of circumstances the physicist is in the same position as the gambler who reckons up his chances. As often as he reasons by induction, he requires more or less consciously the calculus of probabilities, and this is why I am obliged to introduce a parenthesis, and interrupt our study of method in the physical sciences in order to examine a little more closely the value of this calculus, and what confidence it merits.

The very name calculus of probabilities is a paradox. Probability opposed to certainty is what we do not know, and how can we calculate what we do not know? Yet many eminent savants have occupied themselves with this calculus, and it can not be denied that science has drawn therefrom no small advantage.

How can we explain this apparent contradiction?

Has probability been defined? Can it even be defined? And if it can not, how dare we reason about it? The definition, it will

155

be said, is very simple: the probability of an event is the ratio of the number of cases favorable to this event to the total number of possible cases.

A simple example will show how incomplete this definition is. I throw two dice. What is the probability that one of the two at least turns up a six? Each die can turn up in six different ways; the number of possible cases is $6 \times 6 = 36$; the number of favorable cases is 11; the probability is 11/36.

That is the correct solution. But could I not just as well say: The points which turn up on the two dice can form $6 \times 7/2 = 21$ different combinations? Among these combinations 6 are favorable; the probability is 6/21.

Now why is the first method of enumerating the possible cases more legitimate than the second? In any case it is not our definition that tells us.

We are therefore obliged to complete this definition by saying: '. . . to the total number of possible cases provided these cases are equally probable.' So, therefore, we are reduced to defining the probable by the probable.

How can we know that two possible cases are equally probable? Will it be by a convention? If we place at the beginning of each problem an explicit convention, well and good. We shall then have nothing to do but apply the rules of arithmetic and of algebra, and we shall complete our calculation without our result leaving room for doubt. But if we wish to make the slightest application of this result, we must prove our convention was legitimate, and we shall find ourselves in the presence of the very difficulty we thought to escape.

Will it be said that good sense suffices to show us what convention should be adopted? Alas! M. Bertrand has amused himself by discussing the following simple problem: "What is the probability that a chord of a circle may be greater than the side of the inscribed equilateral triangle?" The illustrious geometer successively adopted two conventions which good sense seemed equally to dictate and with one he found 1/2, with the other 1/3.

The conclusion which seems to follow from all this is that the calculus of probabilities is a useless science, and that the obscure instinct which we may call good sense, and to which we are wont to appeal to legitimatize our conventions, must be distrusted.

But neither can we subscribe to this conclusion; we can not do without this obscure instinct. Without it science would be impossible, without it we could neither discover a law nor apply it. Have we the right, for instance, to enunciate Newton's law? Without doubt, numerous observations are in accord with it; but is not this a simple effect of chance? Besides how do we know whether this law, true for so many centuries, will still be true next year? To this objection, you will find nothing to reply, except: 'That is very improbable.'

But grant the law. Thanks to it, I believe myself able to calculate the position of Jupiter a year from now. Have I the right to believe this? Who can tell if a gigantic mass of enormous velocity will not between now and that time pass near the solar system, and produce unforeseen perturbations? Here again the only answer is: 'It is very improbable.'

From this point of view, all the sciences would be only unconscious applications of the calculus of probabilities. To condemn this calculus would be to condemn the whole of science.

I'shall dwell lightly on the scientific problems in which the intervention of the calculus of probabilities is more evident. In the forefront of these is the problem of interpolation, in which, knowing a certain number of values of a function, we seek to divine the intermediate values.

I shall likewise mention: the celebrated theory of errors of observation, to which I shall return later; the kinetic theory of gases, a well-known hypothesis, wherein each gaseous molecule is supposed to describe an extremely complicated trajectory; but in which, through the effect of great numbers, the mean phenomena, alone observable, obey the simple laws of Mariotte and Gay-Lussac.

All these theories are based on the laws of great numbers, and the calculus of probabilities would evidently involve them in its ruin. It is true that they have only a particular interest, and that, save as far as interpolation is concerned, these are sacrifices to which we might readily be resigned.

But, as I have said above, it would not be only these partial sacrifices that would be in question; it would be the legitimacy of the whole of science that would be challenged.

I quite see that it might be said: "We are ignorant, and yet we must act. For action, we have not time to devote ourselves to an inquiry sufficient to dispel our ignorance. Besides, such an inquiry would demand an infinite time. We must therefore decide without knowing; we are obliged to do so, hit or miss, and we must follow rules without quite believing them. What I know is not that such and such a thing is true, but that the best course for me is to act as if it were true." The calculus of probabilities, and consequently science itself, would thenceforth have merely a practical value.

Unfortunately the difficulty does not thus disappear. A gambler wants to try a *coup;* he asks my advice. If I give it to him, I shall use the calculus of probabilities, but I shall not guarantee success. This is what I shall call *subjective probability.* In this case, we might be content with the explanation of which I have just given a sketch. But suppose that an observer is present at the game, that he notes all its *coups,* and that the game goes on a long time. When he makes a summary of his book, he will find that events have taken place in conformity with the laws of the calculus of probabilities. This is what I shall call *objective probability,* and it is this phenomenon which has to be explained.

There are numerous insurance companies which apply the rules of the calculus of probabilities, and they distribute to their shareholders dividends whose objective reality can not be contested. To invoke our ignorance and the necessity to act does not suffice to explain them.

Thus absolute skepticism is not admissible. We may distrust, but we can not condemn *en bloc.* Discussion is necessary.

I. CLASSIFICATION OF THE PROBLEMS OF PROBABILITY.—In order to classify the problems which present themselves *à propos* of probabilities, we may look at them from many different points of view, and, first, from the *point of view of generality.* I have said above that probability is the ratio of the number of favorable cases to the number of possible cases. What for want of a better term I call the generality will increase with the number of possible cases. This number may be finite, as, for instance, if we take a throw of the dice in which the number of possible cases is 36. That is the first degree of generality.

But if we ask, for example, what is the probability that a point within a circle is within the inscribed square, there are as many possible cases as there are points in the circle, that is to say, an infinity. This is the second degree of generality. Generality can be pushed further still. We may ask the probability that a function will satisfy a given condition. There are then as many possible cases as one can imagine different functions. This is the third degree of generality, to which we rise,˙for instance, when we seek to find the most probable law in conformity with a finite number of observations.

We may place ourselves at a point of view wholly different. If we were not ignorant, there would be no probability, there would be room for nothing but certainty. But our ignorance can not be absolute, for then there would no longer be any probability at all, since a little light is necessary to attain even this uncertain science. Thus the problems of probability may be classed according to the greater or less depth of this ignorance.

In mathematics even we may set ourselves problems of probability. What is the probability that the fifth decimal of a logarithm taken at random from a table is a '9'? There is no hesitation in answering that this probability is 1/10; here we possess all the data of the problem. We can calculate our logarithm without recourse to the table, but we do not wish to give ourselves the trouble. This is the first degree of ignorance.

In the physical sciences our ignorance becomes greater. The state of a system at a given instant depends on two things: Its initial state, and the law according to which that state varies. If we know both this law and this initial state, we shall have then only a mathematical problem to solve, and we fall back upon the first degree of ignorance.

But it often happens that we know the law, and do not know the initial state. It may be asked, for instance, what is the present distribution of the minor planets? We know that from all time they have obeyed the laws of Kepler, but we do not know what was their initial distribution.

In the kinetic theory of gases, we assume that the gaseous molecules follow rectilinear trajectories, and obey the laws of impact of elastic bodies. But, as we know nothing of their initial velocities, we know nothing of their present velocities.

The calculus of probabilities only enables us to predict the mean phenomena which will result from the combination of these velocities. This is the second degree of ignorance.

Finally it is possible that not only the initial conditions but the laws themselves are unknown. We then reach the third degree of ignorance and in general we can no longer affirm anything at all as to the probability of a phenomenon.

It often happens that instead of trying to guess an event, by means of a more or less imperfect knowledge of the law, the events may be known and we want to find the law; or that instead of deducing effects from causes, we wish to deduce the causes from the effects. These are the problems called *probability of causes*, the most interesting from the point of view of their scientific applications.

I play écarté with a gentleman I know to be perfectly honest. He is about to deal. What is the probability of his turning up the king? It is 1/8. This is a problem of the probability of effects.

I play with a gentleman whom I do not know. He has dealt ten times, and he has turned up the king six times. What is the probability that he is a sharper? This is a problem in the probability of causes.

It may be said that this is the essential problem of the experimental method. I have observed $n$ values of $x$ and the corresponding values of $y$. I have found that the ratio of the latter to the former is practically constant. There is the event, what is the cause?

Is it probable that there is a general law according to which $y$ would be proportional to $x$, and that the small divergencies are due to errors of observation? This is a type of question that one is ever asking, and which we unconsciously solve whenever we are engaged in scientific work.

I am now going to pass in review these different categories of problems, discussing in succession what I have called above subjective and objective probability.

II. PROBABILITY IN MATHEMATICS.—The impossibility of squaring the circle has been proved since 1882; but even before that date all geometers considered that impossibility as so 'probable,' that the Academy of Sciences rejected without examination the alas! too numerous memoirs on this subject, that some unhappy madmen sent in every year.

Was the Academy wrong? Evidently not, and it knew well that in acting thus it did not run the least risk of stifling a discovery of moment. The Academy could not have proved that it was right; but it knew quite well that its instinct was not mistaken. If you had asked the Academicians, they would have answered: ''We have compared the probability that an unknown savant should have found out what has been vainly sought for so long, with the probability that there is one madman the more on the earth; the second appears to us the greater.'' These are very good reasons, but there is nothing mathematical about them; they are purely psychological.

And if you had pressed them further they would have added: ''Why do you suppose a particular value of a transcendental function to be an algebraic number; and if $\pi$ were a root of an algebraic equation, why do you suppose this root to be a period of the function sin $2x$, and not the same about the other roots of this same equation?'' To sum up, they would have invoked the principle of sufficient reason in its vaguest form.

But what could they deduce from it? At most a rule of conduct for the employment of their time, more usefully spent at their ordinary work than in reading a lucubration that inspired in them a legitimate distrust. But what I call above objective probability has nothing in common with this first problem.

It is otherwise with the second problem.

Consider the first 10,000 logarithms that we find in a table. Among these 10,000 logarithms I take one at random. What is the probability that its third decimal is an even number? You will not hesitate to answer 1/2; and in fact if you pick out in a table the third decimals of these 10,000 numbers, you will find nearly as many even digits as odd.

12

Or if you prefer, let us write 10,000 numbers corresponding to our 10,000 logarithms, each of these numbers being $+1$ if the third decimal of the corresponding logarithm is even, and $-1$ if odd. Then take the mean of these 10,000 numbers.

I do not hesitate to say. that the mean of these 10,000 numbers is probably 0, and if I were actually to calculate it I should verify that it is extremely small.

But even this verification is needless. I might have rigorously proved that this mean is less than 0.003. To prove this result, I should have had to make a rather long calculation for which there is no room here, and for which I confine myself to citing an article I published in the *Revue générale des Sciences*, April 15, 1899. The only point to which I wish to call attention is the following: in this calculation, I should have needed only to rest my case on two facts, to wit, that the first and second derivatives of the logarithm remain, in the interval considered, between certain limits.

Hence this important consequence that the property is true not only of the logarithm, but of any continuous function whatever, since the derivatives of every continuous function are limited.

If I was certain beforehand of the result, it is first, because I had often observed analogous facts for other continuous functions; and next, because I made in my mind, in a more or less unconscious and imperfect manner, the reasoning which led me to the preceding inequalities, just as a skilled calculator before finishing his multiplication takes into account what it should come to approximately.

And besides, since what I call my intuition was only an incomplete summary of a piece of true reasoning, it is clear why observation has confirmed my predictions, and why the objective probability has been in agreement with the subjective probability.

As a third example I shall choose the following problem: A number $u$ is taken at random, and $n$ is a given very large integer. What is the probable value of $\sin nu$? This problem has no meaning by itself. To give it one a convention is needed. We *shall agree* that the probability for the number $u$ to lie between $a$ and $a + da$ is equal to $\phi(a)\, da$; that it is therefore proportional to the infinitely small interval $da$, and equal to this multiplied by a function $\phi(a)$ depending only on $a$. As for this function, I choose it arbitrarily, but I must assume it to be continuous. The value of sin $nu$ remaining the same when $u$ increases by $2\pi$, I may without loss of generality assume that $u$ lies between 0 and $2\pi$, and I shall thus be led to suppose that $\phi(a)$ is a periodic function whose period is $2\pi$.

The probable value sought is readily expressed by a simple integral, and it is easy to show that this integral is less than

$$2\pi M_k/n^k,$$

$M_k$ being the maximum value of the $k^{\text{th}}$ derivative of $\phi(u)$. We see then that if the $k^{\text{th}}$ derivative is finite, our probable value will tend toward 0 when $n$ increases indefinitely, and that more rapidly than $1/n^{k-1}$.

The probable value of sin $nu$ when $n$ is very large is therefore naught. To define this value I required a convention; but the result remains the same *whatever that convention may be.* I have imposed upon myself only slight restrictions in assuming that the function $\phi(a)$ is continuous and periodic, and these hypotheses are so natural that we may ask ourselves how they can be escaped.

Examination of the three preceding examples, so different in all respects, has already given us a glimpse, on the one hand, of the rôle of what philosophers call the principle of sufficient reason, and, on the other hand, of the importance of the fact that certain properties are common to all continuous functions. The study of probability in the physical sciences will lead us to the same result.

III. PROBABILITY IN THE PHYSICAL SCIENCES.—We come now to the problems connected with what I have called the second degree of ignorance, those, namely, in which we know the law, but do not know the initial state of the system. I could multiply examples, but will take only one. What is the probable present distribution of the minor planets on the zodiac?

We know they obey the laws of Kepler. We may even, without at all changing the nature of the problem, suppose that their orbits are all circular, and situated in the same plane, and that we know this plane. On the other hand, we are in absolute ignorance as to what was their initial distribution. However, we do not hesitate to affirm that their distribution is now nearly uniform. Why?

Let $b$ be the longitude of a minor planet in the initial epoch, that is to say, the epoch zero. Let $a$ be its mean motion. Its longitude at the present epoch, that is to say, at the epoch $t$, will be $at + b$. To say that the present distribution is uniform is to say that the mean value of the sines and cosines of multiples of $at + b$ is zero. Why do we assert this?

Let us represent each minor planet by a point in a plane, to wit, by a point whose coordinates are precisely $a$ and $b$. All these representative points will be contained in a certain region of the plane, but as they are very numerous, this region will appear dotted with points. We know nothing else about the distribution of these points.

What do we do when we wish to apply the calculus of probabilities to such a question? What is the probability that one or more representative points may be found in a certain portion of the plane? In our ignorance, we are reduced to making an arbitrary hypothesis. To explain the nature of this hypothesis, allow me to use, in lieu of a mathematical formula, a crude but concrete image. Let us suppose that over the surface of our plane has been spread an imaginary substance, whose density is variable, but varies continuously. We shall then agree to say that the probable number of representative points to be found on a portion of the plane is proportional to the quantity of fictitious matter found there. If we have then two regions of the plane of the same extent, the probabilities that a representative point of one of our minor planets is found in one or the other of these regions will be to one another as the mean densities of the fictitious matter in the one and the other region.

Here then are two distributions, one real, in which the representative points are very numerous, very close together, but discrete like the molecules of matter in the atomic hypothesis; the other remote from reality, in which our representative points are replaced by continuous fictitious matter. We know that the latter can not be real, but our ignorance forces us to adopt it.

If again we had some idea of the real distribution of the representative points, we could arrange it so that in a region of some extent the density of this imaginary continuous matter would be nearly proportional to the number of the representative points, or, if you wish, to the number of atoms which are contained in that region. Even that is impossible, and our ignorance is so great that we are forced to choose arbitrarily the function which defines the density of our imaginary matter. Only we shall be forced to a hypothesis from which we can hardly get away, we shall suppose that this function is continuous. That is sufficient, as we shall see, to enable us to reach a conclusion.

What is at the instant $t$ the probable distribution of the minor planets? Or rather what is the probable value of the sine of the longitude at the instant $t$, that is to say of $\sin (at + b)$? We made at the outset an arbitrary convention, but if we adopt it, this probable value is entirely defined. Divide the plane into elements of surface. Consider the value of $\sin (at + b)$ at the center of each of these elements; multiply this value by the surface of the element, and by the corresponding density of the imaginary matter. Take then the sum for all the elements of the plane. This sum, by definition, will be the probable mean value we seek, which will thus be expressed by a double integral. It may be thought at first that this mean value depends on the choice of the function which defines the density of the imaginary matter, and that, as this function $\phi$ is arbitrary, we can, according to the arbitrary choice which we make, obtain any mean value. This is not so.

A simple calculation shows that our double integral decreases very rapidly when $t$ increases. Thus I could not quite tell what hypothesis to make as to the probability of this or that initial distribution; but whatever the hypothesis made, the result will be the same, and this gets me out of my difficulty.

Whatever be the function $\phi$, the mean value tends toward zero as $t$ increases, and as the minor planets have certainly accomplished a very great number of revolutions, I may assert that this mean value is very small.

I may choose $\phi$ as I wish, save always one restriction: this function must be continuous; and, in fact, from the point of view of subjective probability, the choice of a discontinuous function would have been unreasonable. For instance, what reason could

I have for supposing that the initial longitude might be exactly 0°, but that it could not lie between 0° and 1°?

But the difficulty reappears if we take the point of view of objective probability, if we pass from our imaginary distribution in which the fictitious matter was supposed continuous, to the real distribution in which our representative points form, as it were, discrete atoms.

The mean value of sin $(at + b)$ will be represented quite simply by

$$\frac{1}{n} \Sigma \sin (at + b),$$

$n$ being the number of minor planets. In lieu of a double integral referring to a continuous function, we shall have a sum of discrete terms. And yet no one will seriously doubt that this mean value is practically very small.

Our representative points being very close together, our discrete sum will in general differ very little from an integral.

An integral is the limit toward which a sum of terms tends when the number of these terms is indefinitely increased. If the terms are very numerous, the sum will differ very little from its limit, that is to say from the integral, and what I said of this latter will still be true of the sum itself.

Nevertheless, there are exceptions. If, for instance, for all the minor planets,

$$b = \frac{\pi}{2} - at,$$

the longitude for all the planets at the time $t$ would be $\pi/2$, and the mean value would evidently be equal to unity. For this to be the case, it would be necessary that at the epoch 0, the minor planets must have all been lying on a spiral of peculiar form, with its spires very close together. Every one will admit that such an initial distribution is extremely improbable (and, even supposing it realized, the distribution would not be uniform at the present time, for example, on January 1, 1913, but it would become so a few years later).

Why then do we think this initial distribution improbable? This must be explained, because if we had no reason for rejecting as improbable this absurd hypothesis everything would break down, and we could no longer make any affirmation about the probability of this or that present distribution.

Once more we shall invoke the principle of sufficient reason to which we must always recur. We might admit that at the beginning the planets were distributed almost in a straight line. We might admit that they were irregularly distributed. But it seems to us that there is no sufficient reason for the unknown cause that gave them birth to have acted along a curve so regular and yet so complicated, which would appear to have been expressely chosen so that the present distribution would not be uniform.

IV. Rouge et Noir.—The questions raised by games of chance, such as roulette, are, fundamentally, entirely analogous to those we have just treated. For example, a wheel is partitioned into a great number of equal subdivisions, alternately red and black. A needle is whirled with force, and after having made a great number of revolutions, it stops before one of these subdivisions. The probability that this division is red is evidently 1/2. The needle describes an angle $\theta$, including several complete revolutions. I do not know what is the probability that the needle may be whirled with a force such that this angle should lie between $\theta$ and $\theta + d\theta$; but I can make a convention. I can suppose that this probability is $\phi(\theta) d\theta$. As for the function $\phi(\theta)$, I can choose it in an entirely arbitrary manner. There is nothing that can guide me in my choice, but I am naturally led to suppose this function continuous.

Let $\epsilon$ be the length (measured on the circumference of radius 1) of each red and black subdivision. We have to calculate the integral of $\phi(\theta) d\theta$, extending it, on the one hand, to all the red divisions, and, on the other hand, to all the black divisions, and to compare the results.

Consider an interval $2\epsilon$, comprising a red division and a black division which follows it. Let M and $m$ be the greatest and least values of the function $\phi(\theta)$ in this interval. The integral extended to the red divisions will be smaller than $\Sigma M\epsilon$; the integral extended to the black divisions will be greater than $\Sigma m\epsilon$; the difference will therefore be less than $\Sigma (M - m)\epsilon$. But, if the function $\theta$ is supposed continuous; if, besides, the interval $\epsilon$ is very small with respect to the total angle described by the needle, the difference M — $m$ will be very small. The difference of the two integrals will therefore be very small, and the probability will be very nearly 1/2.

We see that without knowing anything of the function $\theta$, I must act as if the probability were 1/2. We understand, on the other hand, why, if, placing myself at the objective point of view, I observe a certain number of coups, observation will give me about as many black coups as red.

All players know this objective law; but it leads them into a remarkable error, which has been often exposed, but into which they always fall again. When the red has won, for instance, six times running, they bet on the black, thinking they are playing a safe game; because, say they, it is very rare that red wins seven times running.

In reality their probability of winning remains 1/2. Observation shows, it is true, that series of seven consecutive reds are very rare, but series of six reds followed by a black are just as rare.

They have noticed the rarity of the series of seven reds; if they have not remarked the rarity of six reds and a black, it is only because such series strike the attention less.

V. The Probability of Causes.—We now come to the problems of the probability of causes, the most important from the point of view of scientific applications. Two stars, for instance, are very close together on the celestial sphere. Is this apparent contiguity a mere effect of chance? Are these stars, although on almost the same visual ray, situated at very different distances from the earth, and consequently very far from one another? Or, perhaps, does the apparent correspond to a real contiguity? This is a problem on the probability of causes.

I recall first that at the outset of all problems of the probability of effects that have hitherto occupied us, we have always had to make a convention, more or less justified. And if in most cases the result was, in a certain measure, independent of this convention, this was only because of certain hypotheses which permitted us to reject *a priori* discontinuous functions, for example, or certain absurd conventions.

We shall find something analogous when we deal with the

probability of causes. An effect may be produced by the cause *A* or by the cause *B*. The effect has just been observed. We ask the probability that it is due to the cause *A*. This is an *a posteriori* probability of cause. But I could not calculate it, if a convention more or less justified did not tell me *in advance* what is the *a priori* probability for the cause *A* to come into play; I mean the probability of this event for some one who had not observed the effect.

The better to explain myself I go back to the example of the game of écarté mentioned above. My adversary deals for the first time and he turns up a king. What is the probability that he is a sharper? The formulas ordinarily taught give 8/9, a result evidently rather surprising. If we look at it closer, we see that the calculation is made as if, *before sitting down at the table,* I had considered that there was one chance in two that my adversary was not honest. An absurd hypothesis, because in that case I should have certainly not played with him, and this explains the absurdity of the conclusion.

The convention about the *a priori* probability was unjustified, and that is why the calculation of the *a posteriori* probability led me to an inadmissible result. We see the importance of this preliminary convention. I shall even add that if none were made, the problem of the *a posteriori* probability would have no meaning. It must always be made either explicitly or tacitly.

Pass to an example of a more scientific character. I wish to determine an experimental law. This law, when I know it, can be represented by a curve. I make a certain number of isolated observations; each of these will be represented by a point. When I have obtained these different points, I draw a curve between them, striving to pass as near to them as possible and yet preserve for my curve a regular form, without angular points, or inflections too accentuated, or brusque variation of the radius of curvature. This curve will represent for me the probable law, and I assume not only that it will tell me the values of the function intermediate between those which have been observed, but also that it will give me the observed values themselves more exactly than direct observation. This is why I make it pass near the points, and not through the points themselves.

Here is a problem in the probability of causes. The effects are the measurements I have recorded; they depend on a combination of two causes: the true law of the phenomenon and the errors of observation. Knowing the effects, we have to seek the probability that the phenomenon obeys this law or that, and that the observations have been affected by this or that error. The most probable law then corresponds to the curve traced, and the most probable error of an observation is represented by the distance of the corresponding point from this curve.

But the problem would have no meaning if, before any observation, I had not fashioned an *a priori* idea of the probability of this or that law, and of the chances of error to which I am exposed.

If my instruments are good (and that I knew before making the observations), I shall not permit my curve to depart much from the points which represent the rough measurements. If they are bad, I may go a little further away from them in order to obtain a less sinuous curve; I shall sacrifice more to regularity.

Why then is it that I seek to trace a curve without sinuosities? It is because I consider *a priori* a law represented by a continuous function (or by a function whose derivatives of high order are small), as more probable than a law not satisfying these conditions. Without this belief, the problem of which we speak would have no meaning; interpolation would be impossible; no law could be deduced from a finite number of observations; science would not exist.

Fifty years ago physicists considered, other things being equal, a simple law as more probable than a complicated law. They even invoked this principle in favor of Mariotte's law as against the experiments of Regnault. To-day they have repudiated this belief; and yet, how many times are they compelled to act as though they still held it! However that may be, what remains of this tendency is the belief in continuity, and we have just seen that if this belief were to disappear in its turn, experimental science would become impossible.

VI. THE THEORY OF ERRORS.—We are thus led to speak of the theory of errors, which is directly connected with the problem of the probability of causes. Here again we find *effects*, to wit, a certain number of discordant observations, and we seek to divine the *causes*, which are, on the one hand, the real value of the quantity to be measured; on the other hand, the error made in each isolated observation. It is necessary to calculate what is *a posteriori* the probable magnitude of each error, and consequently the probable value of the quantity to be measured.

But as I have just explained, we should not know how to undertake this calculation if we did not admit *a priori*, that is to say, before all observation, a law of probability of errors. Is there a law of errors?

The law of errors admitted by all calculators is Gauss's law, which is represented by a certain transcendental curve known under the name of 'the bell.'

But first it is proper to recall the classic distinction between systematic and accidental errors. If we measure a length with too long a meter, we shall always find too small a number, and it will be of no use to measure several times; this is a systematic error. If we measure with an accurate meter, we may, however, make a mistake; but we go wrong, now too much, now too little, and when we take the mean of a great number of measurements, the error will tend to grow small. These are accidental errors.

It is evident from the first that systematic errors can not satisfy Gauss's law; but do the accidental errors satisfy it? A great number of demonstrations have been attempted; almost all are crude paralogisms. Nevertheless, we may demonstrate Gauss's law by starting from the following hypotheses: the error committed is the result of a great number of partial and independent errors; each of the partial errors is very little and besides, obeys any law of probability, provided that the probability of a positive error is the same as that of an equal negative error. It is evident that these conditions will be often but not always fulfilled, and we may reserve the name of accidental for errors which satisfy them.

We see that the method of least squares is not legitimate in every case; in general the physicists are more distrustful of it than the astronomers. This is, no doubt, because the latter, besides the systematic errors to which they and the physicists are subject alike, have to contend with an extremely important source of error which is wholly accidental; I mean atmospheric undulations.  So it is very curious to hear a physicist discuss with an astronomer about a method of observation.  The physicist, persuaded that one good measurement is worth more than many bad ones, is before all concerned with eliminating by dint of precautions the least systematic errors, and the astronomer says to him: 'But thus you can observe only a small number of stars; the accidental errors will not disappear.'

What should we conclude?  Must we continue to use the method of least squares?  We must distinguish.  We have eliminated all the systematic errors we could suspect; we know well there are still others, but we can not detect them; yet it is necessary to make up our mind and adopt a definitive value which will be regarded as the probable value; and for that it is evident the best thing to do is to apply Gauss's method.  We have only applied a practical rule referring to subjective probability.  There is nothing more to be said.

But we wish to go farther and affirm that not only is the probable value so much, but that the probable error in the result is so much.  *This is absolutely illegitimate;* it would be true only if we were sure that all the systematic errors were eliminated, and of that we know absolutely nothing.  We have two series of observations; by applying the rule of least squares, we find that the probable error in the first series is twice as small as in the second.  The second series may, however, be better than the first, because the first perhaps is affected by a large systematic error.  All we can say is that the first series is *probably* better than the second, since its accidental error is smaller, and we have no reason to affirm that the systematic error is greater for one of the series than for the other, our ignorance on this point being absolute.

VII. CONCLUSIONS.—In the lines which precede, I have set many problems without solving any of them.  Yet I do not regret having written them, because they will perhaps invite the reader to reflect on these delicate questions.

However that may be, there are certain points which seem well established.  To undertake any calculation of probability, and even for that calculation to have any meaning, it is necessary to admit, as point of departure, a hypothesis or convention which has always something arbitrary about it. In the choice of this convention, we can be guided only by the principle of sufficient reason. Unfortunately this principle is very vague and very elastic, and in the cursory examination we have just made, we have seen it take many different forms. The form under which we have met it most often is the belief in continuity, a belief which it would be difficult to justify by apodeictic reasoning, but without which all science would be impossible. Finally the problems to which the calculus of probabilities may be applied with profit are those in which the result is independent of the hypothesis made at the outset, provided only that this hypothesis satisfies the condition of continuity.

# CHAPTER XII

## Optics and Electricity

Fresnel's Theory.—The best example[1] that can be chosen of physics in the making is the theory of light and its relations to the theory of electricity. Thanks to Fresnel, optics is the best developed part of physics; the so-called wave-theory forms a whole truly satisfying to the mind. We must not, however, ask of it what it can not give us.

The object of mathematical theories is not to reveal to us the true nature of things; this would be an unreasonable pretension. Their sole aim is to coordinate the physical laws which experiment reveals to us, but which, without the help of mathematics, we should not be able even to state.

It matters little whether the ether really exists; that is the affair of metaphysicians. The essential thing for us is that everything happens as if it existed, and that this hypothesis is convenient for the explanation of phenomena. After all, have we any other reason to believe in the existence of material objects? That, too, is only a convenient hypothesis; only this will never cease to be so, whereas, no doubt, some day the ether will be thrown aside as useless. But even at that day, the laws of optics and the equations which translate them analytically will remain true, at least as a first approximation. It will always be useful, then, to study a doctrine that unites all these equations.

The undulatory theory rests on a molecular hypothesis. For those who think they have thus discovered the cause under the law, this is an advantage. For the others it is a reason for distrust. But this distrust seems to me as little justified as the illusion of the former.

These hypotheses play only a secondary part. They might be sacrificed. They usually are not, because then the explanation would lose in clearness; but that is the only reason.

[1] This chapter is a partial reproduction of the prefaces of two of my works: *Théorie mathématique de la lumière* (Paris, Naud, 1889), and *Électricité et optique* (Paris, Naud, 1901).

In fact, if we looked closer we should see that only two things are borrowed from the molecular hypotheses: the principle of the conservation of energy, and the linear form of the equations, which is the general law of small movements, as of all small variations.

This explains why most of Fresnel's conclusions remain unchanged when we adopt the electromagnetic theory of light.

MAXWELL'S THEORY.—Maxwell, we know, connected by a close bond two parts of physics until then entirely foreign to one another, optics and electricity. By blending thus in a vaster whole, in a higher harmony, the optics of Fresnel has not ceased to be alive. Its various parts subsist, and their mutual relations are still the same. Only the language we used to express them has changed; and, on the other hand, Maxwell has revealed to us other relations, before unsuspected, between the different parts of optics and the domain of electricity.

When a French reader first opens Maxwell's book, a feeling of uneasiness and often even of mistrust mingles at first with his admiration. Only after a prolonged acquaintance and at the cost of many efforts does this feeling disappear. There are even some eminent minds that never lose it.

Why are the English scientist's ideas with such difficulty acclimatized among us? It is, no doubt, because the education received by the majority of enlightened Frenchmen predisposes them to appreciate precision and logic above every other quality.

The old theories of mathematical physics gave us in this respect complete satisfaction. All our masters, from Laplace to Cauchy, have proceeded in the same way. Starting from clearly stated hypotheses, they deduced all their consequences with mathematical rigor, and then compared them with experiment. It seemed their aim to give every branch of physics the same precision as celestial mechanics.

A mind accustomed to admire such models is hard to suit with a theory. Not only will it not tolerate the least appearance of contradiction, but it will demand that the various parts be logically connected with one another, and that the number of distinct hypotheses be reduced to minimum.

This is not all; it will have still other demands, which seem to me less reasonable. Behind the matter which our senses can reach, and which experiment tells us of, it will desire to see another, and in its eyes the only real, matter, which will have only purely geometric properties, and whose atoms will be nothing but mathematical points, subject to the laws of dynamics alone. And yet these atoms, invisible and without color, it will seek by an unconscious contradiction to represent to itself and consequently to identify as closely as possible with common matter.

Then only will it be fully satisfied and imagine that it has penetrated the secret of the universe. If this satisfaction is deceitful, it is none the less difficult to renounce.

Thus, on opening Maxwell, a Frenchman expects to find a theoretical whole as logical and precise as the physical optics based on the hypothesis of the ether; he thus prepares for himself a disappointment which I should like to spare the reader by informing him immediately of what he must look for in Maxwell, and what he can not find there.

Maxwell does not give a mechanical explanation of electricity and magnetism; he confines himself to demonstrating that such an explanation is possible.

He shows also that optical phenomena are only a special case of electromagnetic phenomena. From every theory of electricity, one can therefore deduce immediately a theory of light.

The converse unfortunately is not true; from a complete explanation of light, it is not always easy to derive a complete explanation of electric phenomena. This is not easy, in particular, if we wish to start from Fresnel's theory. Doubtless it would not be impossible; but nevertheless we must ask whether we are not going to be forced to renounce admirable results that we thought definitely acquired. That seems a step backward; and many good minds are not willing to submit to it.

When the reader shall have consented to limit his hopes, he will still encounter other difficulties. The English scientist does not try to construct a single edifice, final and well ordered; he seems rather to erect a great number of provisional and independent constructions, between which communication is difficult and sometimes impossible.

Take as example the chapter in which he explains electrostatic attractions by pressures and tensions in the dielectric medium. This chapter might be omitted without making thereby the rest of the book less clear or complete; and, on the other hand, it contains a theory complete in itself which one could understand without having read a single line that precedes or follows. But it is not only independent of the rest of the work; it is difficult to reconcile with the fundamental ideas of the book. Maxwell does not even attempt this reconciliation; he merely says: "I have not been able to make the next step, namely, to account by mechanical considerations for these stresses in the dielectric."

This example will suffice to make my thought understood; I could cite many others. Thus who would suspect, in reading the pages devoted to magnetic rotary polarization, that there is an identity between optical and magnetic phenomena?

One must not then flatter himself that he can avoid all contradiction; to that it is necessary to be resigned. In fact, two contradictory theories, provided one does not mingle them, and if one does not seek in them the basis of things, may both be useful instruments of research; and perhaps the reading of Maxwell would be less suggestive if he had not opened up to us so many new and divergent paths.

The fundamental idea, however, is thus a little obscured. So far is this the case that in the majority of popularized versions it is the only point completely left aside.

I feel, then, that the better to make its importance stand out, I ought to explain in what this fundamental idea consists. But for that a short digression is necessary.

THE MECHANICAL EXPLANATION OF PHYSICAL PHENOMENA.— There is in every physical phenomenon a certain number of parameters which experiment reaches directly and allows us to measure. I shall call these the parameters $q$.

Observation then teaches us the laws of the variations of these parameters; and these laws can generally be put in the form of differential equations, which connect the parameters $q$ with the time.

What is it necessary to do to give a mechanical interpretation of such a phenomenon?

13

One will try to explain it either by the motions of ordinary matter, or by those of one or more hypothetical fluids.

These fluids will be considered as formed of a very great number of isolated molecules $m$.

When shall we say, then, that we have a complete mechanical explanation of the phenomenon? It will be, on the one hand, when we know the differential equations satisfied by the coordinates of these hypothetical molecules $m$, equations which, moreover, must conform to the principles of dynamics; and, on the other hand, when we know the relations that define the coordinates of the molecules $m$ as functions of the parameters $q$ accessible to experiment.

These equations, as I have said, must conform to the principles of dynamics, and, in particular, to the principle of the conservation of energy and the principle of least action.

The first of these two principles teaches us that the total energy is constant and that this energy is divided into two parts:

1° The kinetic energy, or *vis viva*, which depends on the masses of the hypothetical molecules $m$, and their velocities, and which I shall call $T$.

2° The potential energy, which depends only on the coordinates of these molecules and which I shall call $U$. It is the *sum* of the two energies $T$ and $U$ which is constant.

What now does the principle of least action tell us? It tells us that to pass from the initial position occupied at the instant $t_0$ to the final position occupied at the instant $t_1$, the system must take such a path that, in the interval of time that elapses between the two instants $t_0$ and $t_1$, the average value of 'the action' (that is to say, of the *difference* between the two energies $T$ and $U$) shall be as small as possible.

If the two functions $T$ and $U$ are known, this principle suffices to determine the equations of motion.

Among all the possible ways of passing from one position to another, there is evidently one for which the average value of the action is less than for any other. There is, moreover, only one; and it results from this that the principle of least action suffices to determine the path followed and consequently the equations of motion.

Thus we obtain what are called the equations of Lagrange.

In these equations, the independent variables are the coordinates of the hypothetical molecules $m$; but I now suppose that one takes as variables the parameters $q$ directly accessible to experiment.

The two parts of the energy must then be expressed as functions of the parameters $q$ and of their derivatives. They will evidently appear under this form to the experimenter. The latter will naturally try to define the potential and the kinetic energy by the aid of quantities that he can directly observe.[2]

That granted, the system will always go from one position to another by a path such that the average action shall be a minimum.

It matters little that $T$ and $U$ are now expressed by the aid of the parameters $q$ and their derivatives; it matters little that it is also by means of these parameters that we define the initial and final positions; the principle of least action remains always true.

Now here again, of all the paths that lead from one position to another, there is one for which the average action is a minimum, and there is only one. The principle of least action suffices, then, to determine the differential equations which define the variations of the parameters $q$.

The equations thus obtained are another form of the equations of Lagrange.

To form these equations we need to know neither the relations that connect the parameters $q$ with the coordinates of the hypothetical molecules, nor the masses of these molecules, nor the expression of $U$ as a function of the coordinates of these molecules.

All we need to know is the expression of $U$ as a function of the parameters, and that of $T$ as a function of the parameters $q$ and their derivatives, that is, the expressions of the kinetic and of the potential energy as functions of the experimental data.

Then we shall have one of two things: either for a suitable

---

[2] We add that $U$ will depend only on the parameters $q$, that $T$ will depend on the parameters $q$ and their derivatives with respect to the time and will be a homogeneous polynomial of the second degree with respect to these derivatives.

choice of the functions $T$ and $U$, the equations of Lagrange, constructed as we have just said, will be identical with the differential equations deduced from experiments; or else there will exist no functions $T$ and $U$, for which this agreement takes place. In the latter case it is clear that no mechanical explanation is possible.

The *necessary* condition for a mechanical explanation to be possible is therefore that we can choose the functions $T$ and $U$ in such a way as to satisfy the principle of least action, which involves that of the conservation of energy.

This condition, moreover, is *sufficient*. Suppose, in fact, that we have found a function $U$ of the parameters $q$, which represents one of the parts of the energy; that another part of the energy, which we shall represent by $T$, is a function of the parameters $q$ and their derivatives, and that it is a homogeneous polynomial of the second degree with respect to these derivatives; and finally that the equations of Lagrange, formed by means of these two functions, $T$ and $U$, conform to the data of the experiment.

What is necessary in order to deduce from this a mechanical explanation? It is necessary that $U$ can be regarded as the potential energy of a system and $T$ as the *vis viva* of the same system.

There is no difficulty as to $U$, but can $T$ be regarded as the *vis viva* of a material system?

It is easy to show that this is always possible, and even in an infinity of ways. I will confine myself to referring for more details to the preface of my work, 'Électricité et optique.'

Thus if the principle of least action can not be satisfied, no mechanical explanation is possible; if it can be satisfied, there is not only one, but an infinity, whence it follows that as soon as there is one there is an infinity of others.

One more observation.

Among the quantities that experiment gives us directly, we shall regard some as functions of the coordinates of our hypothetical molecules; these are our parameters $q$. We shall look upon the others as dependent not only on the coordinates, but on the velocities, or, what comes to the same thing, on the derivatives of the parameters $q$, or as combinations of these parameters and their derivatives.

And then a question presents itself: among all these quantities measured experimentally, which shall we choose to represent the parameters $q$? Which shall we prefer to regard as the derivatives of these parameters? This choice remains arbitrary to a very large extent; but, for a mechanical explanation to be possible, it suffices if we can make the choice in such a way as to accord with the principle of least action.

And then Maxwell asked himself whether he could make this choice and that of the two energies $T$ and $U$, in such a way that the electrical phenomena would satisfy this principle. Experiment shows us that the energy of an electromagnetic field is decomposed into two parts, the electrostatic energy and the electrodynamic energy. Maxwell observed that if we regard the first as representing the potential energy $U$, the second as representing the kinetic energy $T$; if, moreover, the electrostatic charges of the conductors are considered as parameters $q$ and the intensities of the currents as the derivatives of other parameters $q$; under these conditions, I say, Maxwell observed that the electric phenomena satisfy the principle of least action. Thenceforth he was certain of the possibility of a mechanical explanation.

If he had explained this idea at the beginning of his book instead of relegating it to an obscure part of the second volume, it would not have escaped the majority of readers.

If, then, a phenomenon admits of a complete mechanical explanation, it will admit of an infinity of others, that will render an account equally well of all the particulars revealed by experiment.

And this is confirmed by the history of every branch of physics; in optics, for instance, Fresnel believed vibration to be perpendicular to the plane of polarization; Neumann regarded it as parallel to this plane. An 'experimentum crucis' has long been sought which would enable us to decide between these two theories, but it has not been found.

In the same way, without leaving the domain of electricity, we may ascertain that the theory of two fluids and that of the single fluid both account in a fashion equally satisfactory for all the observed laws of electrostatics.

All these facts are easily explicable, thanks to the properties of the equations of Lagrange which I have just recalled.

It is easy now to comprehend what is Maxwell's fundamental idea.

To demonstrate the possibility of a mechanical explanation of electricity, we need not preoccupy ourselves with finding this explanation itself; it suffices us to know the expression of the two functions $T$ and $U$, which are the two parts of energy, to form with these two functions the equations of Lagrange and then to compare these equations with the experimental laws.

Among all these possible explanations, how make a choice for which the aid of experiment fails us? A day will come perhaps when physicists will not interest themselves in these questions, inaccessible to positive methods, and will abandon them to the metaphysicians. This day has not yet arrived; man does not resign himself so easily to be forever ignorant of the foundation of things.

Our choice can therefore be further guided only by considerations where the part of personal appreciation is very great; there are, however, solutions that all the world will reject because of their whimsicality, and others that all the world will prefer because of their simplicity.

In what concerns electricity and magnetism, Maxwell abstains from making any choice. It is not that he systematically disdains all that is unattainable by positive methods; the time he has devoted to the kinetic theory of gases sufficiently proves that. I will add that if, in his great work, he develops no complete explanation, he had previously attempted to give one in an article in the *Philosophical Magazine*. The strangeness and the complexity of the hypotheses he had been obliged to make had led him afterwards to give this up.

The same spirit is found throughout the whole work. What is essential, that is to say what must remain common to all theories, is made prominent; all that would only be suitable to a particular theory is nearly always passed over in silence. Thus the reader finds himself in the presence of a form almost devoid of matter, which he is at first tempted to take for a fugitive shadow not to be grasped. But the efforts to which he is thus condemned force him to think and he ends by comprehending what was often rather artificial in the theoretic constructs he had previously only wondered at.

# CHAPTER XIII

THE history of electrodynamics is particularly instructive from our point of view.

Ampère entitled his immortal work, 'Théorie des phénomènes électrodynamiques, *uniquement* fondée sur l'expérience.' He therefore imagined that he had made *no* hypothesis, but he had made them, as we shall soon see; only he made them without being conscious of it.

His successors, on the other hand, perceived them, since their attention was attracted by the weak points in Ampère's solution. They made new hypotheses, of which this time they were fully conscious; but how many times it was necessary to change them before arriving at the classic system of to-day which is perhaps not yet final; this we shall see.

I. AMPÈRE'S THEORY.—When Ampère studied experimentally the mutual actions of currents, he operated and he only could operate with closed currents.

It was not that he denied the possibility of open currents. If two conductors are charged with positive and negative electricity and brought into communication by a wire, a current is established going from one to the other, which continues until the two potentials are equal. According to the ideas of Ampère's time this was an open current; the current was known to go from the first conductor to the second, it was not seen to return from the second to the first.

So Ampère considered as open currents of this nature, for example, the currents of discharge of condensers; but he could not make them the objects of his experiments because their duration is too short.

Another sort of open current may also be imagined. I suppose two conductors, A and B, connected by a wire AMB. Small conducting masses in motion first come in contact with the

184

conductor *B,* take from it an electric charge, leave contact with *B* and move along the path *BNA,* and, transporting with them their charge, come into contact with *A* and give to it their charge, which returns then to *B* along the wire *AMB.*

Now there we have in a sense a closed circuit, since the electricity describes the closed circuit *BNAMB*; but the two parts of this current are very different. In the wire *AMB,* the electricity is displaced through a fixed conductor, like a voltaic current, overcoming an ohmic resistance and developing heat; we say that it is displaced by conduction. In the part *BNA,* the electricity is carried by a moving conductor; it is said to be displaced by convection.

If then the current of convection is considered as altogether analogous to the current of conduction, the circuit *BNAMB* is closed; if, on the contrary, the convection current is not 'a true current,' and, for example, does not act on the magnet, there remains only the conduction current *AMB,* which is open.

For example, if we connect by a wire the two poles of a Holtz machine, the charged rotating disc transfers the electricity by convection from one pole to the other, and it returns to the first pole by conduction through the wire.

But currents of this sort are very difficult to produce with appreciable intensity. With the means at Ampère's disposal, we may say that this was impossible.

To sum up, Ampère could conceive of the existence of two kinds of open currents, but he could operate on neither because they were not strong enough or because their duration was too short.

Experiment therefore could only show him the action of a closed current on a closed current, or, more accurately, the action of a closed current on a portion of a current, because a current can be made to describe a closed circuit composed of a moving part and a fixed part. It is possible then to study the displacements of the moving part under the action of another closed current.

On the other hand, Ampère had no means of studying the action of an open current, either on a closed current or another open current.

1. *The Case of Closed Currents.*—In the case of the mutual action of two closed currents, experiment revealed to Ampère remarkably simple laws.

I recall rapidly here those which will be useful to us in the sequel:

1° *If the intensity of the currents is kept constant,* and if the two circuits, after having undergone any deformations and displacements whatsoever, return finally to their initial positions, the total work of the electrodynamic actions will be null.

In other words, there is an *electrodynamic potential* of the two circuits, proportional to the product of the intensities, and depending on the form and relative position of the circuits; the work of the electrodynamic actions is equal to the variation of this potential:

2° The action of a closed solenoid is null.

3° The action of a circuit $C$ on another voltaic circuit $C'$ depends only on the 'magnetic field' developed by this circuit. At each point in space we can in fact define in magnitude and direction a certain force called *magnetic force,* which enjoys the following properties:

(*a*) The force exercised by $C$ on a magnetic pole is applied to that pole and is equal to the magnetic force multiplied by the magnetic mass of that pole;

(*b*) A very short magnetic needle tends to take the direction of the magnetic force, and the couple to which it tends to reduce is proportional to the magnetic force, the magnetic moment of the needle and the sine of the dip of the needle;

(*c*) If the circuit $C$ is displaced, the work of the electrodynamic action exercised by $C$ on $C'$ will be equal to the increment of the 'flow of magnetic force' which passes through the circuit.

2. *Action of a Closed Current on a Portion of Current.*— Ampère not having been able to produce an open current, properly so called, had only one way of studying the action of a closed current on a portion of current.

This was by operating on a circuit $C$ composed of two parts, the one fixed, the other movable. The movable part was, for instance, a movable wire $\alpha\beta$ whose extremities $\alpha$ and $\beta$ could slide along a fixed wire. In one of the positions of the movable wire, the end $\alpha$ rested on the $A$ of the fixed wire and the extremity $\beta$ on the point $B$ of the fixed wire. The current circulated from $\alpha$ to $\beta$, that is to say, from $A$ to $B$ along the movable wire, and then it returned from $B$ to $A$ along the fixed wire. *This current was therefore closed.*

In a second position, the movable wire having slipped, the extremity $\alpha$ rested on another point $A'$ of the fixed wire, and the extremity $\beta$ on another point $B'$ of the fixed wire. The current circulated then from $\alpha$ to $\beta$, that is to say from $A'$ to $B'$ along the movable wire, and it afterwards returned from $B'$ to $B$, then from $B$ to $A$, then finally from $A$ to $A'$, always following the fixed wire. The current was therefore also closed.

If a like current is subjected to the action of a closed current $C$, the movable part will be displaced just as if it were acted upon by a force. Ampère *assumes* that the apparent force to which this movable part $AB$ seems thus subjected, representing the action of the $C$ on the portion $\alpha\beta$ of the current, is the same as if $\alpha\beta$ were traversed by an open current, stopping at $\alpha$ and $\beta$, in place of being traversed by a closed current which after arriving at $\beta$ returns to $\alpha$ through the fixed part of the circuit.

This hypothesis seems natural enough, and Ampère made it unconsciously; nevertheless *it is not necessary,* since we shall see further on that Helmholtz rejected it. However that may be, it permitted Ampère, though he had never been able to produce an open current, to enunciate the laws of the action of a closed current on an open current, or even on an element of current.

The laws are simple:

1° The force which acts on an element of current is applied to this element; it is normal to the element and to the magnetic force, and proportional to the component of this magnetic force which is normal to the element.

2° The action of a closed solenoid on an element of current is null.

But the electrodynamic potential has disappeared, that is to say that, when a closed current and an open current, whose intensities have been maintained constant, return to their initial positions, the total work is not null.

3. *Continuous Rotations.*—Among electrodynamic experiments, the most remarkable are those in which continuous rotations are produced and which are sometimes called *unipolar induction* experiments. A magnet may turn about its axis; a current passes first through a fixed wire, enters the magnet by the pole $N$, for example, passes through half the magnet, emerges by a sliding contact and reenters the fixed wire.

The magnet then begins to rotate continuously without being able ever to attain equilibrium; this is Faraday's experiment.

How is it possible? If it were a question of two circuits of invariable form, the one $C$ fixed, the other $C'$ movable about an axis, this latter could never take on continuous rotation; in fact there is an electrodynamic potential; there must therefore be necessarily a position of equilibrium when this potential is a maximum.

Continuous rotations are therefore possible only when the circuit $C'$ is composed of two parts: one fixed, the other movable about an axis, as is the case in Faraday's experiment. Here again it is convenient to draw a distinction. The passage from the fixed to the movable part, or inversely, may take place either by simple contact (the same point of the movable part remaining constantly in contact with the same point of the fixed part), or by a sliding contact (the same point of the movable part coming successively in contact with diverse points of the fixed part).

It is only in the second case that there can be continuous rotation. This is what then happens: The system tends to take a position of equilibrium; but, when at the point of reaching that position, the sliding contact puts the movable part in communication with a new point of the fixed part; it changes the connections, it changes therefore the conditions of equilibrium, so that the position of equilibrium fleeing, so to say, before the system which seeks to attain it, rotation may take place indefinitely.

Ampère assumes that the action of the circuit on the movable part of $C'$ is the same as if the fixed part of $C'$ did not exist, and therefore as if the current passing through the movable part were open.

He concludes therefore that the action of a closed on an open current, or inversely that of an open current on a closed current, may give rise to a continuous rotation.

But this conclusion depends on the hypothesis I have enunciated and which, as I said above, is not admitted by Helmholtz.

4. *Mutual Action of Two Open Currents.*—In what concerns the mutual actions of two open currents, and in particular that of two elements of current, all experiment breaks down. Ampère has recourse to hypothesis. He supposes:

1° That the mutual action of two elements reduces to a force acting along their join;

2° That the action of two closed currents is the resultant of the mutual actions of their diverse elements, which are besides the same as if these elements were isolated.

What is remarkable is that here again Ampère makes these hypotheses unconsciously.

However that may be, these two hypotheses, together with the experiments on closed currents, suffice to determine completely the law of the mutual action of two elements. But then most of the simple laws we have met in the case of closed currents are no longer true.

In the first place, there is no electrodynamic potential; nor was there any, as we have seen, in the case of a closed current acting on an open current.

Next there is, properly speaking, no magnetic force.

And, in fact, we have given above three different definitions of this force:

1° By the action on a magnetic pole;

2° By the director couple which orientates the magnetic needle;

3° By the action on an element of current.

But in the case which now occupies us, not only these three definitions are no longer in harmony, but each has lost its meaning, and in fact:

1° A magnetic pole is no longer acted upon simply by a single force applied to this pole. We have seen in fact that the force due to the action of an element of current on a pole is not applied to the pole, but to the element; it may moreover be replaced by a force applied to the pole and by a couple;

2° The couple which acts on the magnetic needle is no longer a simple director couple, for its moment with respect to the axis of the needle is not null. It breaks up into a director couple, properly so called, and a supplementary couple which tends to produce the continuous rotation of which we have above spoken;

3° Finally the force acting on an element of current is not normal to this element.

In other words, *the unity of the magnetic force has disappeared*.

Let us see in what this unity consists. Two systems which exercise the same action on a magnetic pole will exert also the same action on an indefinitely small magnetic needle, or on an element of current placed at the same point of space as this pole.

Well, this is true if these two systems contain only closed currents; this would no longer be true if these two systems contained open currents.

It suffices to remark, for instance, that, if a magnetic pole is placed at $A$ and an element at $B$, the direction of the element being along the prolongation of the sect $AB$, this element which will exercise no action on this pole will, on the other hand, exercise an action either on a magnetic needle placed at the point $A$, or on an element of current placed at the point $A$.

5. *Induction.*—We know that the discovery of electrodynamic induction soon followed the immortal work of Ampère.

As long as it is only a question of closed currents there is no difficulty, and Helmholtz has even remarked that the principle of the conservation of energy is sufficient for deducing the laws of induction from the electrodynamic laws of Ampère. But always on one condition, as Bertrand has well shown; that we make besides a certain number of hypotheses.

The same principle again permits this deduction in the case of open currents, although of course we can not submit the result to the test of experiment, since we can not produce such currents.

If we try to apply this mode of analysis to Ampère's theory of open currents, we reach results calculated to surprise us.

In the first place, induction can not be deduced from the variation of the magnetic field by the formula well known to savants and practicians, and, in fact, as we have said, properly speaking there is no longer a magnetic field.

But, further; if a circuit $C$ is subjected to the induction of a variable voltaic system $S$, if this system $S$ be displaced and deformed in any way whatever, so that the intensity of the currents of this system varies according to any law whatever, but that after these variations the system finally returns to its initial situation, it seems natural to suppose that the *mean* electromotive force induced in the circuit $C$ is null.

This is true if the circuit $C$ is closed and if the system $S$ contains only closed currents. This would no longer be true, if one accepts the theory of Ampère, if there were open currents. So that not only induction will no longer be the variation of the flow of magnetic force, in any of the usual senses of the word, but it can not be represented by the variation of anything whatever.

II. THEORY OF HELMHOLTZ.—I have dwelt upon the consequences of Ampère's theory, and of his method of explaining open currents.

It is difficult to overlook the paradoxical and artificial character of the propositions to which we are thus led. One can not help thinking 'that can not be so.'

We understand therefore why Helmholtz was led to seek something else.

Helmholtz rejects Ampère's fundamental hypothesis, to wit, that the mutual action of two elements of current reduces to a force along their join. He assumes that an element of current is not subjected to a single force, but to a force and a couple. It is just this which gave rise to the celebrated polemic between Bertrand and Helmholtz.

Helmholtz replaces Ampère's hypothesis by the following: two elements always admit of an electrodynamic potential depending solely on their position and orientation; and the work of the forces that they exercise, one on the other, is equal to the variation of this potential. Thus Helmholtz can no more do without hypothesis than Ampère; but at least he does not make one without explicitly announcing it.

In the case of closed currents, which are alone accessible to experiment, the two theories agree.

In all other cases they differ.

In the first place, contrary to what Ampère supposed, the force which seems to act on the movable portion of a closed current is not the same as would act upon this movable portion if it were isolated and constituted an open current.

Let us return to the circuit $C'$, of which we spoke above, and which was formed of a movable wire $\alpha\beta$ sliding on a fixed wire. In the only experiment that can be made, the movable portion $\alpha\beta$ is not isolated, but is part of a closed circuit. When it passes from $AB$ to $A'B'$, the total electrodynamic potential varies for two reasons:

1° It undergoes a first increase because the potential of $A'B'$ with respect to the circuit $C$ is not the same as that of $AB$;

2° It takes a second increment because it must be increased by the potentials of the elements $AA'$, $BB'$ with respect to $C$.

It is this *double* increment which represents the work of the force to which the portion $AB$ seems subjected.

If, on the contrary, $\alpha\beta$ were isolated, the potential would undergo·only the first increase, and this first increment alone would measure the work of the force which acts on $AB$.

In the second place, there could be no continuous rotation without sliding contact, and, in fact, that, as we have seen *à propos* of closed currents, is an immediate consequence of the existence of an electrodynamic potential.

In Faraday's experiment, if the magnet is fixed and if the part of the current exterior to the magnet runs along a movable wire, that movable part may undergo a continuous rotation. But this does not mean to say that if the contacts of the wire with the magnet were suppressed, and an *open* current were to run along the wire, the wire would still take a movement of continuous rotation.

I have just said in fact that an *isolated* element is not acted upon in the same way as a movable element making part of a closed circuit.

Another difference: The action of a closed solenoid on a closed current is null according to experiment and according to the two theories. Its action on an open current would be null according to Ampère; it would not be null according to Helmholtz. From this follows an important consequence. We have given above three definitions of magnetic force. The third has no meaning here since an element of current is no longer acted upon by a single force. No more has the first any meaning. What, in fact, is a magnetic pole? It is the extremity of an indefinite linear magnet. This magnet may be replaced by an indefinite solenoid. For the definition of magnetic force to have any meaning, it would be necessary that the action exercised by an open current on an indefinite solenoid should depend only on the position of the extremity of this solenoid, that is to say, that the action on a closed solenoid should be null. Now we have just seen that such is not the case.

On the other hand, nothing prevents our adopting the second definition, which is founded on the measurement of the director couple which tends to orientate the magnetic needle.

But if it is adopted, neither the effects of induction nor the electrodynamic effects will depend solely on the distribution of the lines of force in this magnetic field.

III. DIFFICULTIES RAISED BY THESE THEORIES.—The theory of Helmholtz is in advance of that of Ampère; it is necessary, however, that all the difficulties should be smoothed away. In the one as in the other, the phrase 'magnetic field' has no meaning, or, if we give it one, by a more or less artificial convention, the ordinary laws so familiar to all electricians no longer apply; thus the electromotive force induced in a wire is no longer measured by the number of lines of force met by this wire.

And our repugnance does not come alone from the difficulty of renouncing inveterate habits of language and of thought. There is something more. If we do not believe in action at a distance, electrodynamic phenomena must be explained by a modification of the medium. It is precisely this modification that we call 'magnetic field.' And then the electrodynamic effects must depend only on this field.

All these difficulties arise from the hypothesis of open currents.

IV. MAXWELL'S THEORY.—Such were the difficulties raised by the dominant theories when Maxwell appeared, who with a stroke of the pen made them all vanish. To his mind, in fact, all currents are closed currents. Maxwell assumes that if in a dielectric the electric field happens to vary, this dielectric becomes the seat of a particular phenomenon, acting on the gal-

14

vanometer like a current, and which he calls *current of displacement.*

If then two conductors bearing contrary charges are put in communication by a wire, in this wire during the discharge there is an open current of conduction; but there are produced at the same time in the surrounding dielectric, currents of displacement which close this current of conduction.

We know that Maxwell's theory leads to the explanation of optical phenomena, which would be due to extremely rapid electrical oscillations.

At that epoch such a conception was only a bold hypothesis, which could be supported by no experiment.

At the end of twenty years, Maxwell's ideas received the confirmation of experiment. Hertz succeeded in producing systems of electric oscillations which reproduce all the properties of light, and only differ from it by the length of their wave; that is to say as violet differs from red. In some measure he made the synthesis of light.

It might be said that Hertz has not demonstrated directly Maxwell's fundamental idea, the action of the current of displacement on the galvanometer. This is true in a sense. What he has shown in sum is that electromagnetic induction is not propagated instantaneously as was supposed; but with the speed of light.

But to suppose there is no current of displacement, and induction is propagated with the speed of light; or to suppose that the currents of displacement produce effects of induction, and that the induction is propagated instantaneously, *comes to the same thing.*

This can not be seen at the first glance, but it is proved by an analysis of which I must not think of giving even a summary here.

V. ROWLAND'S EXPERIMENT.—But as I have said above, there are two kinds of open conduction currents. There are first the currents of discharge of a condenser or of any conductor whatever.

There are also the cases in which electric discharges describe a closed contour, being displaced by conduction in one part of the circuit and by convection in the other part.

For open currents of the first sort, the question might be considered as solved; they were closed by the currents of displacement.

For open currents of the second sort, the solution appeared still more simple. It seemed that if the current were closed, it could only be by the current of convection itself. For that it sufficed to assume that a 'convection current,' that is to say a charged conductor in motion, could act on the galvanometer.

But experimental confirmation was lacking. It appeared difficult in fact to obtain a sufficient intensity even by augmenting as much as possible the charge and the velocity of the conductors. It was Rowland, an extremely skillful experimenter, who first triumphed over these difficulties. A disc received a strong electrostatic charge and a very great speed of rotation. An astatic magnetic system placed beside the disc underwent deviations.

The experiment was made twice by Rowland, once in Berlin, once in Baltimore. It was afterwards repeated by Himstedt. These physicists even announced that they had succeeded in making quantitative measurements.

In fact, for twenty years Rowland's law was admitted without objection by all physicists. Besides everything seemed to confirm it. The spark certainly does produce a magnetic effect. Now does it not seem probable that the discharge by spark is due to particles taken from one of the electrodes and transferred to the other electrode with their charge? Is not the very spectrum of the spark, in which we recognize the lines of the metal of the electrode, a proof of it? The spark would then be a veritable current of convection.

On the other hand, it is also admitted that in an electrolyte the electricity is carried by the ions in motion. The current in an electrolyte would therefore be also a current of convection; now, it acts on the magnetic needle.

The same for cathode rays. Crookes attributed these rays to a very subtile matter charged with electricity and moving with a very great velocity. He regarded them, in other words, as currents of convection. Now these cathode rays are deviated by the magnet. In virtue of the principle of action and reaction, they should in turn deviate the magnetic needle. It is true that Hertz believed he had demonstrated that the cathode rays do not carry electricity, and that they do not act on the magnetic needle. But Hertz was mistaken. First of all, Perrin succeeded in collecting the electricity carried by these rays, electricity of which Hertz denied the existence; the German scientist appears to have been deceived by effects due to the action of X-rays, which were not yet discovered. Afterwards, and quite recently, the action of the cathode rays on the magnetic needle has been put in evidence.

Thus all these phenomena regarded as currents of convection, sparks, electrolytic currents, cathode rays, act in the same manner on the galvanometer and in conformity with Rowland's law.

VI. THEORY OF LORENTZ.—We soon went further. According to the theory of Lorentz, currents of conduction themselves would be true currents of convection. Electricity would remain inseparably connected with certain material particles called *electrons*. The circulation of these electrons through bodies would produce voltaic currents. And what would distinguish conductors from insulators would be that the one could be traversed by these electrons while the others would arrest their movements.

The theory of Lorentz is very attractive. It gives a very simple explanation of certain phenomena which the earlier theories, even Maxwell's in its primitive form, could not explain in a satisfactory way; for example, the aberration of light, the partial carrying away of luminous waves, magnetic polarization and the Zeeman effect.

Some objections still remained. The phenomena of an electric system seemed to depend on the absolute velocity of translation of the center of gravity of this system, which is contrary to the idea we have of the relativity of space. Supported by M. Crémieu, M. Lippmann has presented this objection in a striking form. Imagine two charged conductors with the same velocity of translation; they are relatively at rest. However, each of them being equivalent to a current of convection, they ought to attract one another, and by measuring this attraction we could measure their absolute velocity.

"No!" replied the partisans of Lorentz. "What we could measure in that way is not their absolute velocity, but their relative velocity *with respect to the ether*, so that the principle of relativity is safe."

Whatever there may be in these latter objections, the edifice of electrodynamics, at least in its broad lines, seemed definitively constructed. Everything was presented under the most satisfactory aspect. The theories of Ampère and of Helmholtz, made for open currents which no longer existed, seemed to have no longer anything but a purely historic interest, and the inextricable complications to which these theories led were almost forgotten.

This quiescence has been recently disturbed by the experiments of M. Crémieu, which for a moment seemed to contradict the result previously obtained by Rowland.

But fresh researches have not confirmed them, and the theory of Lorentz has victoriously stood the test.

The history of these variations will be none the less instructive; it will teach us to what pitfalls the scientist is exposed, and how he may hope to escape them.

# THE VALUE OF SCIENCE.

# TRANSLATOR'S INTRODUCTION

1. *Does the Scientist create Science?*—Professor Rados of Budapest in his report to the Hungarian Academy of Science on the award to Poincaré of the Bolyai prize of ten thousand crowns, speaking of him as unquestionably the most powerful investigator in the domain of mathematics and mathematical physics, characterized him as the intuitive genius drawing the inspiration for his wide-reaching researches from the exhaustless fountain of geometric and physical intuition, yet working this inspiration out in detail with marvelous logical keenness. With his brilliant creative genius was combined the capacity for sharp and successful generalization, pushing far out the boundaries of thought in the most widely different domains, so that his works must be ranked with the greatest mathematical achievements of all time. "Finally," says Rados, "permit me to make especial mention of his intensely interesting book, 'The Value of Science,' in which he in a way has laid down the scientist's creed." Now what is this creed?

Sense may act as stimulus, as suggestive, yet not to awaken a dormant depiction, or to educe the conception of an archetypal form, but rather to strike the hour for creation, to summon to work a sculptor capable of smoothing a Venus of Milo out of the formless clay. Knowledge is not a gift of bare experience, nor even made solely out of experience. The creative activity of mind is in mathematics particularly clear. The axioms of geometry are conventions, disguised definitions or unprovable hypotheses precreated by auto-active animal and human minds. Bertrand Russell says of projective geometry: "It takes nothing from experience, and has, like arithmetic, a creature of the pure intellect for its object. It deals with an object whose properties are logically deduced from its definition, not empirically discovered from data." Then does the scientist create science? This is a question Poincaré here dissects with a master hand.

The physiologic-psychologic investigation of the space problem must give the meaning of the words *geometric fact, geometric reality*. Poincaré here subjects to the most successful analysis ever made the tridimensionality of our space.

2. *The Mind Dispelling Optical Illusions.*—Actual perception of spatial properties is accompanied by movements corresponding to its character. In the case of optical illusions, with the so-called false perceptions eye-movements are closely related. But though the perceived object and its environment remain constant, the sufficiently powerful mind can, as we say, dispel these illusions, the perception itself being creatively changed. Photographs taken at intervals during the presence of these optical illusions, during the change, perhaps gradual and unconscious, in the perception, and after these illusions have, as the phrase is, finally disappeared, show quite clearly that changes in eye-movements corresponding to those internally created in perception itself successively occur. What is called accuracy of movement is created by what is called correctness of perception. The higher creation in the perception is the determining cause of an improvement, a precision in the motion. Thus we see correct perception in the individual helping to make that cerebral organization and accurate motor adjustment on which its possibility and permanence seem in so far to depend. So-called correct perception is connected with a long-continued process of perceptual education motived and initiated from within. How this may take place is here illustrated at length by our author.

3. *Euclid not Necessary.*—Geometry is a construction of the intellect, in application not certain but convenient. As Schiller says, when we see these facts as clearly as the development of metageometry has compelled us to see them, we must surely confess that the Kantian account of space is hopelessly and demonstrably antiquated. As Royce says in 'Kant's Doctrine of the Basis of Mathematics,' "That very use of intuition which Kant regarded as geometrically ideal, the modern geometer regards as scientifically defective, because surreptitious. No mathematical exactness without explicit proof from assumed principles—such is the motto of the modern geometer. But suppose the reasoning of Euclid purified of this comparatively surreptitious

appeal to intuition. Suppose that the principles of geometry are
made quite explicit at the outset of the treatise, as Pieri and
Hilbert or Professor Halsted or Dr. Veblen makes his principles
explicit in his recent treatment of geometry. Then, indeed, geometry becomes for the modern mathematician a purely rational
science. But very few students of the logic of mathematics at the
present time can see any warrant in the analysis of geometrical
truth for regarding just the Euclidean system of principles as
possessing any discoverable necessity.'' Yet the environmental
and perhaps hereditary premiums on Euclid still make even the
scientist think Euclid most convenient.

4. *Without Hypotheses, no Science.*—Nobody ever observed an
equidistantial, but also nobody ever observed a straight line.
Emerson's Uriel

> ''Gave his sentiment divine
> Against the being of a line.
> Line in Nature is not found.''

Clearly not, being an eject from man's mind. What is called 'a
knowledge of facts' is usually merely a subjective realization that
the old hypotheses are still sufficiently elastic to serve in some
domain; that is, with a sufficiency of conscious or unconscious
omissions and doctorings and fudgings more or less wilful. In
the present book we see the very foundation rocks of science, the
conservation of energy and the indestructibility of matter, beat-
ing against the bars of their cages, seemingly anxious to take
wing away into the empyrean, to chase the once divine parallel
postulate broken loose from Euclid and Kant.

5. *What Outcome?*—What now is the definite, the permanent
outcome? What new islets raise their fronded palms in air within
thought's musical domain? Over what age-gray barriers rise the
fragrant floods of this new spring-tide, redolent of the wolf-
haunted forest of Transylvania, of far Erdély's plunging river,
Maros the bitter, or broad mother Volga at Kazan? What victory
heralded the great rocket for which young Lobachevski, the
widow's son, was cast into prison? What severing of age-old
mental fetters symbolized young Bolyai's cutting-off with his

Damascus blade the spikes driven into his door-post, and strewing over the sod the thirteen Austrian cavalry officers? This book by the greatest mathematician of our time gives weightiest and most charming answer.

GEORGE BRUCE HALSTED.

# INTRODUCTION

THE search for truth should be the goal of our activities; it is the sole end worthy of them. Doubtless we should first bend our efforts to assuage human suffering, but why? Not to suffer is a negative ideal more surely attained by the annihilation of the world. If we wish more and more to free man from material cares, it is that he may be able to employ the liberty obtained in the study and contemplation of truth.

But sometimes truth frightens us. And in fact we know that it is sometimes deceptive, that it is a phantom never showing itself for a moment except to ceaselessly flee, that it must be pursued further and ever further without ever being attained. Yet to work one must stop, as some Greek, Aristotle or another, has said. We also know how cruel the truth often is, and we wonder whether illusion is not more consoling, yea, even more bracing, for illusion it is which gives confidence. When it shall have vanished, will hope remain and shall we have the courage to achieve? Thus would not the horse harnessed to his treadmill refuse to go, were his eyes not bandaged? And then to seek truth it is necessary to be independent, wholly independent. If, on the contrary, we wish to act, to be strong, we should be united. This is why many of us fear truth; we consider it a cause of weakness. Yet truth should not be feared, for it alone is beautiful.

When I speak here of truth, assuredly I refer first to scientific truth; but I also mean moral truth, of which what we call justice is only one aspect. It may seem that I am misusing words, that I combine thus under the same name two things having nothing in common; that scientific truth, which is demonstrated, can in no way be likened to moral truth, which is felt. And yet I can not separate them, and whosoever loves the one can not help loving the other. To find the one, as well as to find the other, it is necessary to free the soul completely from prejudice and from passion; it is necessary to attain absolute sincerity. These two sorts of truth when discovered give the same joy; each when perceived beams with the same splendor, so that we must see it or close our eyes. Lastly, both attract us and flee from us; they are never fixed: when we think to have reached them, we find that we have still to advance, and he who pursues them is condemned never to know repose. It must be added that those who fear the one will also fear the other; for they are the ones who in everything are concerned above all with consequences. In a word, I liken the two truths, because the same reasons make us love them and because the same reasons make us fear them.

If we ought not to fear moral truth, still less should we dread scientific truth. In the first place it can not conflict with ethics. Ethics and science have their own domains, which touch but do not interpenetrate. The one shows us to what goal we should aspire, the other, given the goal, teaches us how to attain it. So they can never conflict since they can never meet. There can no more be immoral science than there can be scientific morals.

But if science is feared, it is above all because it can not give us happiness. Of course it can not. We may even ask whether the beast does not suffer less than man. But can we regret that earthly paradise where man brute-like was really immortal in knowing not that he must die? When we have tasted the apple, no suffering can make us forget its savor. We always come back to it. Could it be otherwise? As well ask if one who has seen and is blind will not long for the light. Man, then, can not be happy through science, but to-day he can much less be happy without it.

But if truth be the sole aim worth pursuing, may we hope to attain it? It may well be doubted. Readers of my little book 'Science and Hypothesis' already know what I think about the question. The truth we are permitted to glimpse is not altogether what most men call by that name. Does this mean that our most legitimate, most imperative aspiration is at the same time the most vain? Or can we, despite all, approach truth on some side? This it is which must be investigated.

In the first place, what instrument have we at our disposal for this conquest? Is not human intelligence, more specifically the intelligence of the scientist, susceptible of infinite variation? Volumes could be written without exhausting this subject; I, in a few brief pages, have only touched it lightly. That the geometer's mind is not like the physicist's or the naturalist's, all the world would agree; but mathematicians themselves do not resemble each other; some recognize only implacable logic, others appeal to intuition and see in it the only source of discovery. And this would be a reason for distrust. To minds so unlike can the mathematical theorems themselves appear in the same light? Truth which is not the same for all, is it truth? But looking at things more closely, we see how these very different workers collaborate in a common task which could not be achieved without their cooperation. And that already reassures us.

Next must be examined the frames in which nature seems enclosed and which are called time and space. In 'Science and Hypothesis' I have already shown how relative their value is; it is not nature which imposes them upon us, it is we who impose them upon nature because we find them convenient. But I have spoken of scarcely more than space, and particularly quantitative space, so to say, that is of the mathematical relations whose aggregate constitutes geometry. I should have shown that it is the same with time as with space and still the same with 'qualitative space'; in particular, I should have investigated why we attribute three dimensions to space. I may be pardoned then for taking up again these important questions.

Is mathematical analysis, then, whose principal object is the study of these empty frames, only a vain play of the mind? It can give to the physicist only a convenient language; is this not a mediocre service, which, strictly speaking, could be done without; and even is it not to be feared that this artificial language may be a veil interposed between reality and the eye of the physicist? Far from it; without this language most of the intimate analogies of things would have remained forever unknown to us; and we should forever have been ignorant of the internal harmony of the world, which is, we shall see, the only true objective reality.

The best expression of this harmony is law. Law is one of the most recent conquests of the human mind; there still are people who live in the presence of a perpetual miracle and are not astonished at it. On the contrary, we it is who should be astonished at nature's regularity. Men demand of their gods to prove their existence by miracles; but the eternal marvel is that there are not miracles without cease. The world is divine because it is a harmony. If it were ruled by caprice, what could prove to us it was not ruled by chance?

This conquest of law we owe to astronomy, and just this makes the grandeur of the science rather than the material grandeur of the objects it considers. It was altogether natural, then, that celestial mechanics should be the first model of mathematical physics; but since then this science has developed; it is still developing, even rapidly developing. And it is already necessary to modify in certain points the scheme from which I drew two chapters of 'Science and Hypothesis.' In an address at the St. Louis exposition, I sought to survey the road traveled; the result of this investigation the reader shall see farther on.

The progress of science has seemed to imperil the best established principles, those even which were regarded as fundamental. Yet nothing shows they will not be saved; and if this comes about only imperfectly, they will still subsist even though they are modified. The advance of science is not comparable to the changes of a city, where old edifices are pitilessly torn down to give place to new, but to the continuous evolution of zoologic types which develop ceaselessly and end by becoming unrecognizable to the common sight, but where an expert eye finds always traces of the prior work of the centuries past. One must not think then that the old-fashioned theories have been sterile and vain.

Were we to stop there, we should find in these pages some reasons for confidence in the value of science, but many more for distrusting it; an impression of doubt would remain; it is needful now to set things to rights.

Some people have exaggerated the rôle of convention in science; they have even gone so far as to say that law, that scientific fact itself, was created by the scientist. This is going much too far in the direction of nominalism. No, scientific laws are not artificial creations; we have no reason to regard them as accidental, though it be impossible to prove they are not.

Does the harmony the human intelligence thinks it discovers in nature exist outside of this intelligence? No, beyond doubt a reality completely independent of the mind which conceives it, sees or feels it, is an impossibility. A world as exterior as that, even if it existed, would for us be forever inaccessible. But what we call objective reality is, in the last analysis, what is common to many thinking beings, and could be common to all; this common part, we shall see, can only be the harmony expressed by mathematical laws. It is this harmony then which is the sole objective reality, the only truth we can attain; and when I add that the universal harmony of the world is the source of all beauty, it will be understood what price we should attach to the slow and difficult progress which little by little enables us to know it better.

# PART I

## THE MATHEMATICAL SCIENCES

---

### CHAPTER I

#### Intuition and Logic in Mathematics

### I

It is impossible to study the works of the great mathematicians, or even those of the lesser, without noticing and distinguishing two opposite tendencies, or rather two entirely different kinds of minds. The one sort are above all preoccupied with logic; to read their works, one is tempted to believe they have advanced only step by step, after the manner of a Vauban who pushes on his trenches against the place besieged, leaving nothing to chance. The other sort are guided by intuition and at the first stroke make quick but sometimes precarious conquests, like bold cavalrymen of the advance guard.

The method is not imposed by the matter treated. Though one often says of the first that they are *analysts* and calls the others *geometers*, that does not prevent the one sort from remaining analysts even when they work at geometry, while the others are still geometers even when they occupy themselves with pure analysis. It is the very nature of their mind which makes them logicians or intuitionalists, and they can not lay it aside when they approach a new subject.

Nor is it education which has developed in them one of the two tendencies and stifled the other. The mathematician is born, not made, and it seems he is born a geometer or an analyst. I should like to cite examples and there are surely plenty; but to accentuate the contrast I shall begin with an extreme example, taking the liberty of seeking it in two living mathematicians.

M. Méray wants to prove that a binomial equation always has a root, or, in ordinary words, that an angle may always be subdivided. If there is any truth that we think we know by direct intuition, it is this. Who could doubt that an angle may always be divided into any number of equal parts? M. Méray does not look at it that way; in his eyes this proposition is not at all evident and to prove it he needs several pages.

On the other hand, look at Professor Klein: he is studying one of the most abstract questions of the theory of functions: to determine whether on a given Riemann surface there always exists a function admitting of given singularities. What does the celebrated German geometer do? He replaces his Riemann surface by a metallic surface whose electric conductivity varies according to certain laws. He connects two of its points with the two poles of a battery. The current, says he, must pass, and the distribution of this current on the surface will define a function whose singularities will be precisely those called for by the enunciation.

Doubtless Professor Klein well knows he has given here only a sketch; nevertheless he has not hesitated to publish it; and he would probably believe he finds in it, if not a rigorous demonstration, at least a kind of moral certainty. A logician would have rejected with horror such a conception, or rather he would not have had to reject it, because in his mind it would never have originated.

Again, permit me to compare two men, the honor of French science, who have recently been taken from us, but who both entered long ago into immortality. I speak of M. Bertrand and M. Hermite. They were scholars of the same school at the same time; they had the same education, were under the same influences; and yet what a difference! Not only does it blaze forth in their writings; it is in their teaching, in their way of speaking, in their very look. In the memory of all their pupils these two faces are stamped in deathless lines; for all who have had the pleasure of following their teaching, this remembrance is still fresh; it is easy for us to evoke it.

While speaking, M. Bertrand is always in motion; now he seems in combat with some outside enemy, now he outlines with a gesture of the hand the figures he studies. Plainly he sees and he is eager to paint, this is why he calls gesture to his aid. With M. Hermite, it is just the opposite; his eyes seem to shun contact with the world; it is not without, it is within he seeks the vision of truth.

Among the German geometers of this century, two names above all are illustrious, those of the two scientists who founded the general theory of functions, Weierstrass and Riemann. Weierstrass leads everything back to the consideration of series and their analytic transformations; to express it better, he reduces analysis to a sort of prolongation of arithmetic; you may turn through all his books without finding a figure. Riemann, on the contrary, at once calls geometry to his aid; each of his conceptions is an image that no one can forget, once he has caught its meaning.

More recently, Lie was an intuitionalist; this might have been doubted in reading his books, no one could doubt it after talking with him; you saw at once that he thought in pictures. Madame Kovalevski was a logician.

Among our students we notice the same differences; some prefer to treat their problems 'by analysis,' others 'by geometry.' The first are incapable of 'seeing in space,' the others are quickly tired of long calculations and become perplexed.

The two sorts of minds are equally necessary for the progress of science; both the logicians and the intuitionalists have achieved great things that others could not have done. Who would venture to say whether he preferred that Weierstrass had never written or that there had never been a Riemann? Analysis and synthesis have then both their legitimate rôles. But it is interesting to study more closely in the history of science the part which belongs to each.

## II

Strange! If we read over the works of the ancients we are tempted to class them all among the intuitionalists. And yet nature is always the same; it is hardly probable that it has begun in this century to create minds devoted to logic. If we could put ourselves into the flow of ideas which reigned in their time, we should recognize that many of the old geometers were in tendency analysts.   Euclid, for example, erected a scientific structure wherein his contemporaries could find no fault.   In this vast construction, of which each piece however is due to intuition, we may still to-day, without much effort, recognize the work of a logician.

It is not minds that have changed, it is ideas; the intuitional minds have remained the same; but their readers have required of them greater concessions.

What is the cause of this evolution?   It is not hard to find. Intuition can not give us rigor, nor even certainty; this has been recognized more and more.   Let us cite some examples.   We know there exist continuous functions lacking derivatives.   Nothing is more shocking to intuition than this proposition which is imposed upon us by logic.   Our fathers would not have failed to say: "It is evident that every continuous function has a derivative, since every curve has a tangent."

How can intuition deceive us on this point?   It is because when we seek to imagine a curve we can not represent it to ourselves without width; just so, when we represent to ourselves a straight line, we see it under the form of a rectilinear band of a certain breadth.   We well know these lines have no width; we try to imagine them narrower and narrower and thus to approach the limit; so we do in a certain measure, but we shall never attain this limit.   And then it is clear we can always picture these two narrow bands, one straight, one curved, in a position such that they encroach slightly one upon the other without crossing.   We shall thus be led, unless warned by a rigorous analysis, to conclude that a curve always has a tangent.

I shall take as second example Dirichlet's principle on which rest so many theorems of mathematical physics; to-day we establish it by reasoning very rigorous but very long; heretofore, on the contrary, we were content with a very summary proof.   A certain integral depending on an arbitrary function can never vanish.   Hence it is concluded that it must have a minimum.   The flaw in this reasoning strikes us immediately, since we use the abstract term *function* and are familiar with all the singularities functions can present when the word is understood in the most general sense.

But it would not be the same had we used concrete images, had we, for example, considered this function as an electric potential; it would have been thought legitimate to affirm that electrostatic equilibrium can be attained. Yet perhaps a physical comparison would have awakened some vague distrust. But if care had been taken to translate the reasoning into the language of geometry, intermediate between that of analysis and that of physics, doubtless this distrust would not have been produced, and perhaps one might thus, even to-day, still deceive many readers not forewarned.

Intuition, therefore, does not give us certainty. This is why the evolution had to happen; let us now see how it happened.

It was not slow in being noticed that rigor could not be introduced in the reasoning unless first made to enter into the definitions. For the most part the objects treated of by mathematicians were long ill defined; they were supposed to be known because represented by means of the senses or the imagination; but one had only a crude image of them and not a precise idea on which reasoning could take hold. It was there first that the logicians had to direct their efforts.

So, in the case of incommensurable numbers. The vague idea of continuity, which we owe to intuition, resolved itself into a complicated system of inequalities referring to whole numbers.

By that means the difficulties arising from passing to the limit, or from the consideration of infinitesimals, are finally removed. To-day in analysis only whole numbers are left or systems, finite or infinite, of whole numbers bound together by a net of equality or inequality relations. Mathematics, as they say, is arithmetized.

## III

A first question presents itself. Is this evolution ended? Have we finally attained absolute rigor? At each stage of the evolution our fathers also thought they had reached it. If they deceived themselves, do we not likewise cheat ourselves?

We believe that in our reasonings we no longer appeal to intuition; the philosophers will tell us this is an illusion. Pure logic could never lead us to anything but tautologies; it could create nothing new; not from it alone can any science issue. In one sense these philosopers are right; to make arithmetic, as to make geometry, or to make any science, something else than pure logic is necessary. To designate this something else we have no word other than *intuition*. But how many different ideas are hidden under this same word?

Compare these four axioms: (1) Two quantities equal to a third are equal to one another; (2) if a theorem is true of the number 1 and if we prove that it is true of $n+1$ if true for $n$, then will it be true of all whole numbers; (3) if on a straight the point $C$ is between $A$ and $B$ and the point $D$ between $A$ and $C$, then the point $D$ will be between $A$ and $B$; (4) through a given point there is not more than one parallel to a given straight.

All four are attributed to intuition, and yet the first is the enunciation of one of the rules of formal logic; the second is a real synthetic *a priori* judgment, it is the foundation of rigorous mathematical induction; the third is an appeal to the imagination; the fourth is a disguised definition.

Intuition is not necessarily founded on the evidence of the senses; the senses would soon become powerless; for example, we can not represent to ourselves a chiliagon, and yet we reason by intuition on polygons in general, which include the chiliagon as a particular case.

You know what Poncelet understood by the *principle of continuity*. What is true of a real quantity, said Poncelet, should be true of an imaginary quantity; what is true of the hyperbola whose asymptotes are real, should then be true of the ellipse whose asymptotes are imaginary. Poncelet was one of the most intuitive minds of this century; he was passionately, almost ostentatiously, so; he regarded the principle of continuity as one of his boldest conceptions, and yet this principle did not rest on the evidence of the senses. To assimilate the hyperbola to the ellipse was rather to contradict this evidence. It was only a sort of precocious and instinctive generalization which, moreover, I have no desire to defend.

We have then many kinds of intuition; first, the appeal to the senses and the imagination; next generalization by induction, copied, so to speak, from the procedures of the experimental sciences; finally, we have the intuition of pure number, whence arose the second of the axioms just enunciated, which is able to create the real mathematical reasoning. I have shown above by examples that the first two can not give us certainty; but who will seriously doubt the third, who will doubt arithmetic?

Now in the analysis of to-day, when one cares to take the trouble to be rigorous, there can be nothing but syllogisms or appeals to this intuition of pure number, the only intuition which can not deceive us. It may be said that to-day absolute rigor is attained.

## IV

The philosophers make still another objection: "What you gain in rigor," they say, "you lose in objectivity. You can rise toward your logical ideal only by cutting the bonds which attach you to reality. Your science is infallible, but it can only remain so by imprisoning itself in an ivory tower and renouncing all relation with the external world. From this seclusion it must go out when it would attempt the slightest application."

For example, I seek to show that some property pertains to some object whose concept seems to me at first indefinable, because it is intuitive. At first I fail or must content myself with approximate proofs; finally I decide to give to my object a precise definition, and this enables me to establish this property in an irreproachable manner.

"And then," say the philosophers, "it still remains to show that the object which corresponds to this definition is indeed the same made known to you by intuition; or else that some real and concrete object whose conformity with your intuitive idea you believe you immediately recognize corresponds to your new definition. Only then could you affirm that it has the property in question. You have only displaced the difficulty."

That is not exactly so; the difficulty has not been displaced, it has been divided. The proposition to be established was in reality composed of two different truths, at first not distinguished. The first was a mathematical truth, and it is now rigorously established. The second was an experimental verity. Experience alone can teach us that some real and concrete object corresponds or does not correspond to some abstract definition. This second verity is not mathematically demonstrated, but neither can it be, no more than can the empirical laws of the physical and natural sciences. It would be unreasonable to ask more.

Well, is it not a great advance to have distinguished what long was wrongly confused? Does this mean that nothing is left of this objection of the philosophers? That I do not intend to say; in becoming rigorous, mathematical science takes a character so artificial as to strike every one; it forgets its historical origins; we see how the questions can be answered, we no longer see how and why they are put.

This shows us that logic is not enough; that the science of demonstration is not all science and that intuition must retain its rôle as complement, I was about to say as counterpoise or as antidote of logic.

I have already had occasion to insist on the place intuition should hold in the teaching of the mathematical sciences. Without it young minds could not make a beginning in the understanding of mathematics; they could not learn to love it and would see in it only a vain logomachy; above all, without intuition they would never become capable of applying mathematics. But now I wish before all to speak of the rôle of intuition in science itself. If it is useful to the student it is still more so to the creative scientist.

## V

We seek reality, but what is reality? The physiologists tell us that organisms are formed of cells; the chemists add that cells themselves are formed of atoms. Does this mean that these atoms or these cells constitute reality, or rather the sole reality? The way in which these cells are arranged and from which results the unity of the individual, is not it also a reality much more interesting than that of the isolated elements, and should a naturalist who had never studied the elephant except by means of the microscope think himself sufficiently acquainted with that animal?

Well, there is something analogous to this in mathematics. The logician cuts up, so to speak, each demonstration into a very great number of elementary operations; when we have examined these operations one after the other and ascertained that each is correct, are we to think we have grasped the real meaning of the demonstration? Shall we have understood it even when, by an effort of memory, we have become able to repeat this proof by reproducing all these elementary operations in just the order in which the inventor had arranged them? Evidently not; we shall not yet possess the entire reality; that I know not what, which makes the unity of the demonstration, will completely elude us.

Pure analysis puts at our disposal a multitude of procedures whose infallibility it guarantees; it opens to us a thousand different ways on which we can embark in all confidence; we are assured of meeting there no obstacles; but of all these ways, which will lead us most promptly to our goal? Who shall tell us which to choose? We need a faculty which makes us see the the end from afar, and intuition is this faculty. It is necessary to the explorer for choosing his route; it is not less so to the one following his trail who wants to know why he chose it.

If you are present at a game of chess, it will not suffice, for the understanding of the game, to know the rules for moving the pieces. That will only enable you to recognize that each move has been made conformably to these rules, and this knowledge will truly have very little value. Yet this is what the reader of a book on mathematics would do if he were a logician only. To understand the game is wholly another matter; it is to know why the player moves this piece rather than that other which he could have moved without breaking the rules of the game. It is to perceive the inward reason which makes of this series of successive moves a sort of organized whole. This faculty is still more necessary for the player himself, that is, for the inventor.

Let us drop this comparison and return to mathematics. For example, see what has happened to the idea of continuous function. At the outset this was only a sensible image, for example, that of a continuous mark traced by the chalk on a blackboard. Then it became little by little more refined; ere long it was used to construct a complicated system of inequalities, which reproduced, so to speak, all the lines of the original image; this construction finished, the centering of the arch, so to say, was removed, that crude representation which had temporarily served as support and which was afterward useless was rejected; there remained only the construction itself, irreproachable in the eyes of the logician. And yet if the primitive image had totally disappeared from our recollection, how could we divine by what caprice all these inequalities were erected in this fashion one upon another?

Perhaps you think I use too many comparisons; yet pardon still another. You have doubtless seen those delicate assemblages of silicious needles which form the skeleton of certain sponges. When the organic matter has disappeared, there remains only a frail and elegant lace-work. True, nothing is there except silica, but what is interesting is the form this silica has taken, and we could not understand it if we did not know the living sponge which has given it precisely this form. Thus it is that the old intuitive notions of our fathers, even when we have abandoned them, still imprint their form upon the logical constructions we have put in their place.

This view of the aggregate is necessary for the inventor; it is equally necessary for whoever wishes really to comprehend the inventor. Can logic give it to us? No; the name mathematicians give it would suffice to prove this. In mathematics logic is called *analysis* and analysis means *division, dissection.* It can have, therefore, no tool other than the scalpel and the microscope.

Thus logic and intuition have each their necessary rôle. Each is indispensable. Logic, which alone can give certainty, is the instrument of demonstration; intuition is the instrument of invention.

## VI

But at the moment of formulating this conclusion I am seized with scruples. At the outset I distinguished two kinds of mathematical minds, the one sort logicians and analysts, the others intuitionalists and geometers. Well, the analysts also have been inventors. The names I have just cited make my insistence on this unnecessary.

Here is a contradiction, at least apparently, which needs explanation. And first, do you think these logicians have always proceeded from the general to the particular, as the rules of formal

logic would seem to require of them? Not thus could they have
extended the boundaries of science; scientific conquest is to be
made only by generalization.

In one of the chapters of 'Science and Hypothesis,' I have had
occasion to study the nature of mathematical reasoning, and I
have shown how this reasoning, without ceasing to be absolutely
rigorous, could lift us from the particular to the general by a
procedure I have called *mathematical induction*. It is by this
procedure that the analysts have made science progress, and if we
examine the detail itself of their demonstrations, we shall find it
there at each instant beside the classic syllogism of Aristotle.
We, therefore, see already that the analysts are not simply
makers of syllogisms after the fashion of the scholastics.

Besides, do you think they have always marched step by step
with no vision of the goal they wished to attain? They must have
divined the way leading thither, and for that they needed a guide.
This guide is, first, analogy. For example, one of the methods of
demonstration dear to analysts is that founded on the employment of dominant functions. We know it has already served to
solve a multitude of problems; in what consists then the rôle of
the inventor who wishes to apply it to a new problem? At the
outset he must recognize the analogy of this question with those
which have already been solved by this method; then he must
perceive in what way this new question differs from the others,
and thence deduce the modifications necessary to apply to the
method.

But how does one perceive these analogies and these differences?
In the example just cited they are almost always evident, but I
could have found others where they would have been much more
deeply hidden; often a very uncommon penetration is necessary
for their discovery. The analysts, not to let these hidden analogies escape them, that is, in order to be inventors, must, without
the aid of the senses and imagination, have a direct sense of what
constitutes the unity of a piece of reasoning, of what makes, so
to speak, its soul and inmost life.

When one talked with M. Hermite, he never evoked a sensuous
image, and yet you soon perceived that the most abstract entities
were for him like living beings. He did not see them, but he per-

ceived that they are not an artificial assemblage, and that they have some principle of internal unity.

But, one will say, that still is intuition. Shall we conclude that the distinction made at the outset was only apparent, that there is only one sort of mind and that all the mathematicians are intuitionalists, at least those who are capable of inventing?

No, our distinction corresponds to something real. I have said above that there are many kinds of intuition. I have said how much the intuition of pure number, whence comes rigorous mathematical induction, differs from sensible intuition to which the imagination, properly so called, is the principal contributor.

Is the abyss which separates them less profound than it at first appeared? Could we recognize with a little attention that this pure intuition itself could not do without the aid of the senses? This is the affair of the psychologist and the metaphysician and I shall not discuss the question. But the thing's being doubtful is enough to justify me in recognizing and affirming an essential difference between the two kinds of intuition; they have not the same object and seem to call into play two different faculties of our soul; one would think of two search-lights directed upon two worlds strangers to one another.

It is the intuition of pure number, that of pure logical forms, which illumines and directs those we have called *analysts*. This it is which enables them not alone to demonstrate, but also to invent. By it they perceive at a glance the general plan of a logical edifice, and that too without the senses appearing to intervene. In rejecting the aid of the imagination, which, as we have seen, is not always infallible, they can advance without fear of deceiving themselves. Happy, therefore, are those who can do without this aid! We must admire them; but how rare they are!

Among the analysts there will then be inventors, but they will be few. The majority of us, if we wished to see afar by pure intuition alone, would soon feel ourselves seized with vertigo. Our weakness has need of a staff more solid, and, despite the exceptions of which we have just spoken, it is none the less true that sensible intuition is in mathematics the most usual instrument of invention.

Apropos of these reflections, a question comes up that I have not the time either to solve or even to enunciate with the developments it would admit of. Is there room for a new distinction, for distinguishing among the analysts those who above all use pure intuition and those who are first of all preoccupied with formal logic?

M. Hermite, for example, whom I have just cited, can not be classed among the geometers who make use of the sensible intuition; but neither is he a logician, properly so called. He does not conceal his aversion to purely deductive procedures which start from the general and end in the particular.

# CHAPTER II

## The Measure of Time

### I

So long as we do not go outside the domain of consciousness, the notion of time is relatively clear. Not only do we distinguish without difficulty present sensation from the remembrance of past sensations or the anticipation of future sensations, but we know perfectly well what we mean when we say that of two conscious phenomena which we remember, one was anterior to the other; or that, of two foreseen conscious phenomena, one will be anterior to the other.

When we say that two conscious facts are simultaneous, we mean that they profoundly interpenetrate, so that analysis can not separate them without mutilating them.

The order in which we arrange conscious phenomena does not admit of any arbitrariness. It is imposed upon us and of it we can change nothing.

I have only a single observation to add. For an aggregate of sensations to have become a remembrance capable of classification in time, it must have ceased to be actual, we must have lost the sense of its infinite complexity, otherwise it would have remained present. It must, so to speak, have crystallized around a center of associations of ideas which will be a sort of label. It is only when they thus have lost all life that we can classify our memories in time as a botanist arranges dried flowers in his herbarium.

But these labels can only be finite in number. On that score, psychologic time should be discontinuous. Whence comes the feeling that between any two instants there are others? We arrange our recollections in time, but we know that there remain empty compartments. How could that be, if time were not a form pre-existent in our minds? How could we know there were empty compartments, if these compartments were revealed to us only by their content?

223

## II

But that is not all; into this form we wish to put not only the phenomena of our own consciousness, but those of which other consciousnesses are the theater. But more, we wish to put there physical facts, these I know not what with which we people space and which no consciousness sees directly. This is necessary because without it science could not exist. In a word, psychologic time is given to us and must needs create scientific and physical time. There the difficulty begins, or rather the difficulties, for there are two.

Think of two consciousnesses, which are like two worlds impenetrable one to the other. By what right do we strive to put them into the same mold, to measure them by the same standard? Is it not as if one strove to measure length with a gram or weight with a meter? And besides, why do we speak of measuring? We know perhaps that some fact is anterior to some other, but not *by how much* it is anterior.

Therefore two difficulties: (1) Can we transform psychologic time, which is qualitative, into a quantitative time? (2) Can we reduce to one and the same measure facts which transpire in different worlds?

## III

The first difficulty has long been noticed; it has been the subject of long discussions and one may say the question is settled. *We have not a direct intuition of the equality of two intervals of time.* The persons who believe they possess this intuition are dupes of an illusion. When I say, from noon to one the same time passes as from two to three, what meaning has this affirmation?

The least reflection shows that by itself it has none at all. It will only have that which I choose to give it, by a definition which will certainly possess a certain degree of arbitrariness. Psychologists could have done without this definition; physicists and astronomers could not; let us see how they have managed.

To measure time they use the pendulum and they suppose by definition that all the beats of this pendulum are of equal duration. But this is only a first approximation; the temperature, the resistance of the air, the barometric pressure, make the pace of the pendulum vary. If we could escape these sources of error, we should obtain a much closer approximation, but it would still be only an approximation. New causes, hitherto neglected, electric, magnetic or others, would introduce minute perturbations.

In fact, the best chronometers must be corrected from time to time, and the corrections are made by the aid of astronomic observations; arrangements are made so that the sidereal clock marks the same hour when the same star passes the meridian. In other words, it is the sidereal day, that is, the duration of the rotation of the earth, which is the constant unit of time. It is supposed, by a new definition substituted for that based on the beats of the pendulum, that two complete rotations of the earth about its axis have the same duration.

However, the astronomers are still not content with this definition. Many of them think that the tides act as a check on our globe, and that the rotation of the earth is becoming slower and slower. Thus would be explained the apparent acceleration of the motion of the moon, which would seem to be going more rapidly than theory permits because our watch, which is the earth, is going slow.

## IV

All this is unimportant, one will say; doubtless our instruments of measurement are imperfect, but it suffices that we can conceive a perfect instrument. This ideal can not be reached, but it is enough to have conceived it and so to have put rigor into the definition of the unit of time.

The trouble is that there is no rigor in the definition. When we use the pendulum to measure time, what postulate do we implicitly admit? *It is that the duration of two identical phenomena is the same;* or, if you prefer, that the same causes take the same time to produce the same effects.

And at first blush, this is a good definition of the equality of two durations. But take care. Is it impossible that experiment may some day contradict our postulate?

Let me explain myself. I suppose that at a certain place in the world the phenomenon $\alpha$ happens, causing as consequence at the end of a certain time the effect $\alpha'$. At another place in the world

16

very far away from the first, happens the phenomenon $\beta$, which causes as consequence the effect $\beta'$. The phenomena $\alpha$ and $\beta$ are simultaneous, as are also the effects $\alpha'$ and $\beta'$.

Later, the phenomenon $\alpha$ is reproduced under approximately the same conditions as before, and *simultaneously* the phenomenon $\beta$ is also reproduced at a very distant place in the world and almost under the same circumstances. The effects $\alpha'$ and $\beta'$ also take place. Let us suppose that the effect $\alpha'$ happens perceptibly before the effect $\beta'$.

If experience made us witness such a sight, our postulate would be contradicted. For experience would tell us that the first duration $\alpha\alpha'$ is equal to the first duration $\beta\beta'$ and that the second duration $\alpha\alpha'$ is less than the second duration $\beta\beta'$. On the other hand, our postulate would require that the two durations $\alpha\alpha'$ should be equal to each other, as likewise the two durations $\beta\beta'$. The equality and the inequality deduced from experience would be incompatible with the two equalities deduced from the postulate.

Now can we affirm that the hypotheses I have just made are absurd? They are in no wise contrary to the principle of contradiction. Doubtless they could not happen without the principle of sufficient reason seeming violated. But to justify a definition so fundamental I should prefer some other guarantee.

## V

But that is not all. In physical reality one cause does not produce a given effect, but a multitude of distinct causes contribute to produce it, without our having any means of discriminating the part of each of them.

Physicists seek to make this distinction; but they make it only approximately, and, however they progress, they never will make it except approximately. It is approximately true that the motion of the pendulum is due solely to the earth's attraction; but in all rigor every attraction, even of Sirius, acts on the pendulum.

Under these conditions, it is clear that the causes which have produced a certain effect will never be reproduced except approximately. Then we should modify our postulate and our definition. Instead of saying: 'The same causes take the same time to produce the same effects,' we should say: 'Causes almost identical take almost the same time to produce almost the same effects.'

Our definition therefore is no longer anything but approximate. Besides, as M. Calinon very justly remarks in a recent memoir:[1]

One of the circumstances of any phenomenon is the velocity of the earth's rotation; if this velocity of rotation varies, it constitutes in the reproduction of this phenomenon a circumstance which no longer remains the same. But to suppose this velocity of rotation constant is to suppose that we know how to measure time.

Our definition is therefore not yet satisfactory; it is certainly not that which the astronomers of whom I spoke above implicitly adopt, when they affirm that the terrestrial rotation is slowing down.

What meaning according to them has this affirmation? We can only understand it by analyzing the proofs they give of their proposition. They say first that the friction of the tides producing heat must destroy *vis viva*. They invoke therefore the principle of *vis viva*, or of the conservation of energy.

They say next that the secular acceleration of the moon, calculated according to Newton's law, would be less than that deduced from observations unless the correction relative to the slowing down of the terrestrial rotation were made. They invoke therefore Newton's law. In other words, they define duration in the following way: time should be so defined that Newton's law and that of *vis viva* may be verified. Newton's law is an experimental truth; as such it is only approximate, which shows that we still have only a definition by approximation.

If now it be supposed that another way of measuring time is adopted, the experiments on which Newton's law is founded would none the less have the same meaning. Only the enunciation of the law would be different, because it would be translated into another language; it would evidently be much less simple. So that the definition implicitly adopted by the astronomers may be summed up thus: Time should be so defined that

[1] *Etude sur les diverses grandeurs,* Paris, Gauthier-Villars, 1897.

the equations of mechanics may be as simple as possible. In other words, there is not one way of measuring time more true than another; that which is generally adopted is only more *convenient*. Of two watches, we have no right to say that the one goes true, the other wrong; we can only say that it is advantageous to conform to the indications of the first.

The difficulty which has just occupied us has been, as I have said, often pointed out; among the most recent works in which it is considered, I may mention, besides M. Calinon's little book, the treatise on mechanics of Andrade.

## VI

The second difficulty has up to the present attracted much less attention; yet it is altogether analogous to the preceding; and even, logically, I should have spoken of it first.

Two psychological phenomena happen in two different consciousnesses; when I say they are simultaneous, what do I mean? When I say that a physical phenomenon, which happens outside of every consciousness, is before or after a psychological phenomenon, what do I mean?

In 1572, Tycho Brahe noticed in the heavens a new star. An immense conflagration had happened in some far distant heavenly body; but it had happened long before; at least two hundred years were necessary for the light from that star to reach our earth. This conflagration therefore happened before the discovery of America. Well, when I say that; when, considering this gigantic phenomenon, which perhaps had no witness, since the satellites of that star were perhaps uninhabited, I say this phenomenon is anterior to the formation of the visual image of the isle of Española in the consciousness of Christopher Columbus, what do I mean?

A little reflection is sufficient to understand that all these affirmations have by themselves no meaning. They can have one only as the outcome of a convention.

## VII

We should first ask ourselves how one could have had the idea of putting into the same frame so many worlds impenetrable to one another. We should like to represent to ourselves the external universe, and only by so doing could we feel that we understood it. We know we never can attain this representation: our weakness is too great. But at least we desire the ability to conceive an infinite intelligence for which this representation could be possible, a sort of great consciousness which should see all, and which should classify all *in its time,* as we classify, *in. our time,* the little we see.

This hypothesis is indeed crude and incomplete, because this supreme intelligence would be only a demigod; infinite in one sense, it would be limited in another, since it would have only an imperfect recollection of the past; and it could have no other, since otherwise all recollections would be equally present to it and for it there would be no time. And yet when we speak of time, for all which happens outside of us, do we not unconsciously adopt this hypothesis; do we not put ourselves in the place of this imperfect god; and do not even the atheists put themselves in the place where god would be if he existed?

What I have just said shows us, perhaps, why we have tried to put all physical phenomena into the same frame. But that can not pass for a definition of simultaneity, since this hypothetical intelligence, even if it existed, would be for us impenetrable. It is therefore necessary to seek something else.

## VIII

The ordinary definitions which are proper for psychologic time would suffice us no more. Two simultaneous psychologic facts are so closely bound together that analysis can not separate without mutilating them. Is it the same with two physical facts? Is not my present nearer my past of yesterday than the present of Sirius?

It has also been said that two facts should be regarded as simultaneous when the order of their succession may be inverted at will. It is evident that this definition would not suit two physical facts which happen far from one another, and that, in what concerns them, we no longer even understand what this reversibility would be; besides, succession itself must first be defined.

## IX

Let us then seek to give an account of what is understood by simultaneity or antecedence, and for this let us analyze some examples.

I write a letter; it is afterward read by the friend to whom I have addressed it. There are two facts which have had for their theater two different consciousnesses. In writing this letter I have had the visual image of it, and my friend has had in his turn this same visual image in reading the letter. Though these two facts happen in impenetrable worlds, I do not hesitate to regard the first as anterior to the second, because I believe it is its cause.

I hear thunder, and I conclude there has been an electric discharge; I do not hesitate to consider the physical phenomenon as anterior to the auditory image perceived in my consciousness, because I believe it is its cause.

Behold then the rule we follow, and the only one we can follow: when a phenomenon appears to us as the cause of another, we regard it as anterior. It is therefore by cause that we define time; but most often, when two facts appear to us bound by a constant relation, how do we recognize which is the cause and which the effect? We assume that the anterior fact, the antecedent, is the cause of the other, of the consequent. It is then by time that we define cause. How save ourselves from this *petitio principii?*

We say now *post hoc, ergo propter hoc;* now *propter hoc, ergo post hoc;* shall we escape from this vicious circle?

## X

Let us see, not how we succeed in escaping, for we do not completely succeed, but how we try to escape.

I execute a voluntary act $A$ and I feel afterward a sensation $D$, which I regard as a consequence of the act $A$; on the other hand, for whatever reason, I infer that this consequence is not immediate, but that outside my consciousness two facts $B$ and $C$, which I have not witnessed, have happened, and in such a way that $B$ is the effect of $A$, that $C$ is the effect of $B$, and $D$ of $C$.

But why? If I think I have reason to regard the four facts $A, B, C, D$, as bound to one another by a causal connection, why range them in the causal order *A B C D*, and at the same time in the chronologic order *A B C D*, rather than in any other order?

I clearly see that in the act *A* I have the feeling of having been active, while in undergoing the sensation *D* I have that of having been passive. This is why I regard *A* as the initial cause and *D* as the ultimate effect; this is why I put *A* at the beginning of the chain and *D* at the end; but why put *B* before *C* rather than *C* before *B*?

If this question is put, the reply ordinarily is: we know that it is *B* which is the cause of *C* because we always see *B* happen before *C*. These two phenomena, when witnessed, happen in a certain order; when analogous phenomena happen without witness, there is no reason to invert this order.

Doubtless, but take care; we never know directly the physical phenomena *B* and *C*. What we know are sensations *B'* and *C'* produced respectively by *B* and *C*. Our consciousness tells us immediately that *B'* precedes *C'* and we suppose that *B* and *C* succeed one another in the same order.

This rule appears in fact very natural, and yet we are often led to depart from it. We hear the sound of the thunder only some seconds after the electric discharge of the cloud. Of two flashes of lightning, the one distant, the other near, can not the first be anterior to the second, even though the sound of the second comes to us before that of the first?

## XI

Another difficulty; have we really the right to speak of the cause of a phenomenon? If all the parts of the universe are interchained in a certain measure, any one phenomenon will not be the effect of a single cause, but the resultant of causes infinitely numerous; it is, one often says, the consequence of the state of the universe a moment before. How enunciate rules applicable to circumstances so complex? And yet it is only thus that these rules can be general and rigorous.

Not to lose ourselves in this infinite complexity, let us make a simpler hypothesis. Consider three stars, for example, the sun, Jupiter and Saturn; but, for greater simplicity, regard them as reduced to material points and isolated from the rest of the world. The positions and the velocities of three bodies at a given instant suffice to determine their positions and velocities at the following instant, and consequently at any instant. Their positions at the instant $t$ determine their positions at the instant $t + h$ as well as their positions at the instant $t - h$.

Even more; the position of Jupiter at the instant $t$, together with that of Saturn at the instant $t + a$, determines the position of Jupiter at any instant and that of Saturn at any instant.

The aggregate of positions occupied by Jupiter at the instant $t + e$ and Saturn at the instant $t + a + e$ is bound to the aggregate of positions occupied by Jupiter at the instant $t$ and Saturn at the instant $t + a$, by laws as precise as that of Newton, though more complicated. Then why not regard one of these aggregates as the cause of the other, which would lead to considering as simultaneous the instant $t$ of Jupiter and the instant $t + a$ of Saturn?

In answer there can only be reasons, very strong, it is true, of convenience and simplicity.

## XII

But let us pass to examples less artificial; to understand the definition implicitly supposed by the savants, let us watch them at work and look for the rules by which they investigate simultaneity.

I will take two simple examples, the measurement of the velocity of light and the determination of longitude.

When an astronomer tells me that some stellar phenomenon, which his telescope reveals to him at this moment, happened, nevertheless, fifty years ago, I seek his meaning, and to that end I shall ask him first how he knows it, that is, how he has measured the velocity of light.

He has begun by *supposing* that light has a constant velocity, and in particular that its velocity is the same in all directions. That is a postulate without which no measurement of this velocity could be attempted. This postulate could never be verified directly by experiment; it might be contradicted by it if the results of different measurements were not concordant. We should think ourselves fortunate that this contradiction has not happened and that the slight discordances which may happen can be readily explained.

The postulate, at all events, resembling the principle of sufficient reason, has been accepted by everybody; what I wish to emphasize is that it furnishes us with a new rule for the investigation of simultaneity, entirely different from that which we have enunciated above.

This postulate assumed, let us see how the velocity of light has been measured. You know that Roemer used eclipses of the satellites of Jupiter, and sought how much the event fell behind its prediction. But how is this prediction made? It is by the aid of astronomic laws; for instance Newton's law.

Could not the observed facts be just as well explained if we attributed to the velocity of light a little different value from that adopted, and supposed Newton's law only approximate? Only this would lead to replacing Newton's law by another more complicated. So for the velocity of light a value is adopted, such that the astronomic laws compatible with this value may be as simple as possible. When navigators or geographers determine a longitude, they have to solve just the problem we are discussing; they must, without being at Paris, calculate Paris time. How do they accomplish it? They carry a chronometer set for Paris. The qualitative problem of simultaneity is made to depend upon the quantitative problem of the measurement of time. I need not take up the difficulties relative to this latter problem, since above I have emphasized them at length.

Or else they observe an astronomic phenomenon, such as an eclipse of the moon, and they suppose that this phenomenon is perceived simultaneously from all points of the earth. That is not altogether true, since the propagation of light is not instantaneous; if absolute exactitude were desired, there would be a correction to make according to a complicated rule.

Or else finally they use the telegraph. It is clear first that the reception of the signal at Berlin, for instance, is after the sending of this same signal from Paris. This is the rule of cause and effect analyzed above. But how much after? In general, the duration of the transmission is neglected and the two events are regarded as simultaneous. But, to be rigorous, a little correction would still have to be made by a complicated calculation; in practise it is not made, because it would be well within the errors of observation; its theoretic necessity is none the less from our point of view, which is that of a rigorous definition. From this discussion, I wish to emphasize two things: (1) The rules applied are exceedingly various. (2) It is difficult to separate the qualitative problem of simultaneity from the quantitative problem of the measurement of time; no matter whether a chronometer is used, or whether account must be taken of a velocity of transmission, as that of light, because such a velocity could not be measured without *measuring* a time.

## XIII

To conclude: We have not a direct intuition of simultaneity, nor of the equality of two durations. If we think we have this intuition, this is an illusion. We replace it by the aid of certain rules which we apply almost always without taking count of them.

But what is the nature of these rules? No general rule, no rigorous rule; a multitude of little rules applicable to each particular case.

These rules are not imposed upon us and we might amuse ourselves in inventing others; but they could not be cast aside without greatly complicating the enunciation of the laws of physics, mechanics and astronomy.

We therefore choose these rules, not because they are true, but because they are the most convenient, and we may recapitulate them as follows: " The simultaneity of two events, or the order of their succession, the equality of two durations, are to be so defined that the enunciation of the natural laws may be as simple as possible. In other words, all these rules, all these definitions are only the fruit of an unconscious opportunism.''

# CHAPTER III

## THE NOTION OF SPACE

### 1. *Introduction*

IN the articles I have heretofore devoted to space I have above all emphasized the problems raised by non-Euclidean geometry, while leaving almost completely aside other questions more difficult of approach, such as those which pertain to the number of dimensions. All the geometries I considered had thus a common basis, that tridimensional continuum which was the same for all and which differentiated itself only by the figures one drew in it or when one aspired to measure it.

In this continuum, primitively amorphous, we may imagine a network of lines and surfaces, we may then convene to regard the meshes of this net as equal to one another, and it is only after this convention that this continuum, become measurable, becomes Euclidean or non-Euclidean space. From this amorphous continuum can therefore arise indifferently one or the other of the two spaces, just as on a blank sheet of paper may be traced indifferently a straight or a circle.

In space we know rectilinear triangles the sum of whose angles is equal to two right angles; but equally we know curvilinear triangles the sum of whose angles is less than two right angles. The existence of the one sort is not more doubtful than that of the other. To give the name of straights to the sides of the first is to adopt Euclidean geometry; to give the name of straights to the sides of the latter is to adopt the non-Euclidean geometry. So that to ask what geometry it is proper to adopt is to ask, to what line is it proper to give the name straight?

It is evident that experiment can not settle such a question; one would not ask, for instance, experiment to decide whether I should call $AB$ or $CD$ a straight. On the other hand, neither can I say that I have not the right to give the name of straights to the sides of non-Euclidean triangles because they are not in

235

conformity with the eternal idea of straight which I have by intuition.  I grant, indeed, that I have the intuitive idea of the side of the Euclidean triangle, but I have equally the intuitive idea of the side of the non-Euclidean triangle.  Why should I have the right to apply the name of straight to the first of these ideas and not to the second?  Wherein does this syllable form an integrant part of this intuitive idea?  Evidently when we say that the Euclidean straight is a *true* straight and that the non-Euclidean straight is not a true straight, we simply mean that the first intuitive idea corresponds to a *more noteworthy* object than the second.  But how do we decide that this object is more noteworthy?  This question I have investigated in 'Science and Hypothesis.'

It is here that we saw experience come in.  If the Euclidean straight is more noteworthy than the non-Euclidean straight, it is so chiefly because it differs little from certain noteworthy natural objects from which the non-Euclidean straight differs greatly.  But, it will be said, the definition of the non-Euclidean straight is artificial; if we for a moment adopt it, we shall see that two circles of different radius both receive the name of non-Euclidean straights, while of two circles of the same radius one can satisfy the definition without the other being able to satisfy it, and then if we transport one of these so-called straights without deforming it, it will cease to be a straight.  But by what right do we consider as equal these two figures which the Euclidean geometers call two circles with the same radius?  It is because by transporting one of them without deforming it we can make it coincide with the other.  And why do we say this transportation is effected without deformation?  It is impossible to give a good reason for it.  Among all the motions conceivable, there are some of which the Euclidean geometers say that they are not accompanied by deformation; but there are others of which the non-Euclidean geometers would say that they are not accompanied by deformation.  In the first, called Euclidean motions, the Euclidean straights remain Euclidean straights and the non-Euclidean straights do not remain non-Euclidean straights; in the motions of the second sort, or non-Euclidean motions, the non-Euclidean straights remain non-Euclidean straights and the Euclidean straights do not remain Euclidean straights. It has, therefore, not been demonstrated that it was unreasonable to call straights the sides of non-Euclidean triangles; it has only been shown that that would be unreasonable if one continued to call the Euclidean motions motions without deformation; but it has at the same time been shown that it would be just as unreasonable to call straights the sides of Euclidean triangles if the non-Euclidean motions were called motions without deformation.

Now when we say that the Euclidean motions are the *true* motions without deformation, what do we mean? We simply mean that they are *more noteworthy* than the others. And why are they more noteworthy? It is because certain noteworthy natural bodies, the solid bodies, undergo motions almost similar.

And then when we ask: Can one imagine non-Euclidean space? that means: Can we imagine a world where there would be noteworthy natural objects affecting almost the form of non-Euclidean straights, and noteworthy natural bodies frequently undergoing motions almost similar to the non-Euclidean motions? I have shown in 'Science and Hypothesis' that to this question we must answer yes.

It has often been observed that if all the bodies in the universe were dilated simultaneously and in the same proportion, we should have no means of perceiving it, since all our measuring instruments would grow at the same time as the objects themselves which they serve to measure. The world, after this dilatation, would continue on its course without anything apprising us of so considerable an event. In other words, two worlds similar to one another (understanding the word similitude in the sense of Euclid, Book VI.) would be absolutely indistinguishable. But more; worlds will be indistinguishable not only if they are equal or similar, that is, if we can pass from one to the other by changing the axes of coordinates, or by changing the scale to which lengths are referred; but they will still be indistinguishable if we can pass from one to the other by any 'point-transformation' whatever. I will explain my meaning. I suppose that to each point of one corresponds one point of the other and only one, and inversely; and besides that the coordinates of a point are continuous functions, *otherwise altogether arbitrary,* of the corresponding point.  I suppose besides that to each object of the first world corresponds in the second an object of the same nature placed precisely at the corresponding point. I suppose finally that this correspondence fulfilled at the initial instant is maintained indefinitely.  We should have no means of distinguishing these two worlds one from the other.   The relativity of space is not ordinarily understood in so broad a sense; it is thus, however, that it would be proper to understand it.

If one of these universes is our Euclidean world, what its inhabitants will call straight will be our Euclidean straight; but what the inhabitants of the second world will call straight will be a curve which will have the same properties in relation to the world they inhabit and in relation to the motions that they will call motions without deformation.  Their geometry will, therefore, be Euclidean geometry, but their straight will not be our Euclidean straight.  It will be its transform by the point-transformation which carries over from our world to theirs.   The straights of these men will not be our straights, but they will have among themselves the same relations as our straights to one another.  It is in this sense I say their geometry will be ours. If then we wish after all to proclaim that they deceive themselves, that their straight is not the true straight, if we still are unwilling to admit that such an affirmation has no meaning, at least we must confess that these people have no means whatever of recognizing their error.

### 2. *Qualitative Geometry*

All that is relatively easy to understand, and I have already so often repeated it that I think it needless to expatiate further on the matter.   Euclidean space is not a form imposed upon our sensibility, since we can imagine non-Euclidean space; but the two spaces, Euclidean and non-Euclidean, have a common basis, that amorphous continuum of which I spoke in the beginning. From this continuum we can get either Euclidean space or Lobachevskian space, just as we can, by tracing upon it a proper graduation, transform an ungraduated thermometer into a Fahrenheit or a Réaumur thermometer.

And then comes a question: Is not this amorphous continuum, that our analysis has allowed to survive, a form imposed upon our sensibility? If so, we should have enlarged the prison in which this sensibility is confined, but it would always be a prison.

This continuum has a certain number of properties, exempt from all idea of measurement. The study of these properties is the object of a science which has been cultivated by many great geometers and in particular by Riemann and Betti and which has received the name of analysis situs. In this science abstraction is made of every quantitative idea and, for example, if we ascertain that on a line the point $B$ is between the points $A$ and $C$, we shall be content with this ascertainment and shall not trouble to know whether the line $ABC$ is straight or curved, nor whether the length $AB$ is equal to the length $BC$, or whether it is twice as great.

The theorems of analysis situs have, therefore, this peculiarity, that they would remain true if the figures were copied by an inexpert draftsman who should grossly change all the proportions and replace the straights by lines more or less sinuous. In mathematical terms, they are not altered by any 'point-transformation' whatsoever. It has often been said that metric geometry was quantitative, while projective geometry was purely qualitative. That is not altogether true. The straight is still distinguished from other lines by properties which remain quantitative in some respects. The real qualitative geometry is, therefore, analysis situs.

The same questions which came up apropos of the truths of Euclidean geometry, come up anew apropos of the theorems of analysis situs. Are they obtainable by deductive reasoning? Are they disguised conventions? Are they experimental verities? Are they the characteristics of a form imposed either upon our sensibility or upon our understanding?

I wish simply to observe that the last two solutions exclude each other. We can not admit at the same time that it is impossible to imagine space of four dimensions and that experience proves to us that space has three dimensions. The experimenter puts to nature a question: Is it this or that? and he can not put it without imagining the two terms of the alternative. If it were impossible to imagine one of these terms, it would be futile and besides impossible to consult experience. There is no need of observation to know that the hand of a watch is not marking the hour 15 on the dial, because we know beforehand that there are only 12, and we could not look at the mark 15 to see if the hand is there, because this mark does not exist.

Note likewise that in analysis situs the empiricists are disembarrassed of one of the gravest objections that can be leveled against them, of that which renders absolutely vain in advance all their efforts to apply their thesis to the verities of Euclidean geometry. These verities are rigorous and all experimentation can only be approximate. In analysis situs approximate experiments may suffice to give a rigorous theorem and, for instance, if it is seen that space can not have either two or less than two dimensions, nor four or more than four, we are certain that it has exactly three, since it could not have two and a half or three and a half.

Of all the theorems of analysis situs, the most important is that which is expressed in saying that space has three dimensions. This it is that we are about to consider, and we shall put the question in these terms: When we say that space has three dimensions, what do we mean?

### 3. *The Physical Continuum of Several Dimensions*

I have explained in 'Science and Hypothesis' whence we derive the notion of physical continuity and how that of mathematical continuity has arisen from it. It happens that we are capable of distinguishing two impressions one from the other, while each is indistinguishable from a third. Thus we can readily distinguish a weight of 12 grams from a weight of 10 grams, while a weight of 11 grams could be distinguished from neither the one nor the other. Such a statement, translated into symbols, may be written:

$$A = B, \quad B = C, \quad A < C.$$

This would be the formula of the physical continuum, as crude experience gives it to us, whence arises an intolerable contradiction that has been obviated by the introduction of the mathematical continuum. This is a scale of which the steps (commensurable or incommensurable numbers) are infinite in number but are exterior to one another, instead of encroaching on one another as do the elements of the physical continuum, in conformity with the preceding formula.

The physical continuum is, so to speak, a nebula not resolved; the most perfect instruments could not attain to its resolution. Doubtless if we measured the weights with a good balance instead of judging them by the hand, we could distinguish the weight of 11 grams from those of 10 and 12 grams, and our formula would become:

$$A < B, \quad B < C, \quad A < C.$$

But we should always find between $A$ and $B$ and between $B$ and $C$ new elements $D$ and $E$, such that

$$A = D, \quad D = B, \quad A < B; \quad B = E, \quad E = C, \quad B < C,$$

and the difficulty would only have receded and the nebula would always remain unresolved; the mind alone can resolve it and the mathematical continuum it is which is the nebula resolved into stars.

Yet up to this point we have not introduced the notion of the number of dimensions. What is meant when we say that a mathematical continuum or that a physical continuum has two or three dimensions?

First we must introduce the notion of cut, studying first physical continua. We have seen what characterizes the physical continuum. Each of the elements of this continuum consists of a manifold of impressions; and it may happen either that an element can not be discriminated from another element of the same continuum, if this new element corresponds to a manifold of impressions not sufficiently different, or, on the contrary, that the discrimination is possible; finally it may happen that two elements indistinguishable from a third may, nevertheless, be distinguished one from the other.

That postulated, if $A$ and $B$ are two distinguishable elements of a continuum $C$, a series of elements may be found, $E_1, E_2, \cdots, E_n$, all belonging to this same continuum $C$ and such that each of

17

them is indistinguishable from the preceding, that $E_1$ is indistinguishable from $A$, and $E_n$ indistinguishable from $B$. Therefore we can go from $A$ to $B$ by a continuous route and without quitting $C$. If this condition is fulfilled for any two elements $A$ and $B$ of the continuum $C$, we may say that this continuum $C$ is all in one piece. Now let us distinguish certain of the elements of $C$ which may either be all distinguishable from one another, or themselves form one or several continua. The assemblage of the elements thus chosen arbitrarily among all those of $C$ will form what I shall call the *cut* or the *cuts*.

Take on $C$ any two elements $A$ and $B$. Either we can also find a series of elements $E_1$, $E_2$, $\cdots$, $E_n$, such: (1) that they all belong to $C$; (2) that each of them is indistinguishable from the following, $E_1$ indistinguishable from $A$ and $E_n$ from B; (3) *and besides that none of the elements E is indistinguishable from any element of the cut.* Or else, on the contrary, in each of the series $E_1$, $E_2$, $\cdots$, $E_n$ satisfying the first two conditions, there will be an element $E$ indistinguishable from one of the elements of the cut. In the first case we can go from $A$ to $B$ by a continuous route without quitting $C$ and *without meeting the cuts;* in the second case that is impossible.

If then for any two elements $A$ and $B$ of the continuum $C$, it is always the first case which presents itself, we shall say that $C$ remains all in one piece despite the cuts.

Thus, if we choose the cuts in a certain way, otherwise arbitrary, it may happen either that the continuum remains all in one piece or that it does not remain all in one piece; in this latter hypothesis we shall then say that it is *divided* by the cuts.

It will be noticed that all these definitions are constructed in setting out solely from this very simple fact, that two manifolds of impressions sometimes can be discriminated, sometimes can not be. That postulated, if, to *divide* a continuum, it suffices to consider as cuts a certain number of elements all distinguishable from one another, we say that this continuum *is of one dimension;* if, on the contrary, to divide a continuum, it is necessary to consider as cuts a system of elements themselves forming one or several continua, we shall say that this continuum is *of several dimensions.*

If to divide a continuum $C$, cuts forming one or several continua of one dimension suffice, we shall say that $C$ is a continuum *of two dimensions;* if cuts suffice which form one or several continua of two dimensions at most, we shall say that $C$ is a continuum *of three dimensions;* and so on.

To justify this definition it is proper to see whether it is in this way that geometers introduce the notion of three dimensions at the beginning of their works. Now, what do we see? Usually they begin by defining surfaces as the boundaries of solids or pieces of space, lines as the boundaries of surfaces, points as the boundaries of lines, and they affirm that the same procedure can not be pushed further.

This is just the idea given above: to divide space, cuts that are called surfaces are necessary; to divide surfaces, cuts that are called lines are necessary; to divide lines, cuts that are called points are necessary; we can go no further, the point can not be divided, so the point is not a continuum. Then lines which can be divided by cuts which are not continua will be continua of one dimension; surfaces which can be divided by continuous cuts of one dimension will be continua of two dimensions; finally, space which can be divided by continuous cuts of two dimensions will be a continuum of three dimensions.

Thus the definition I have just given does not differ essentially from the usual definitions; I have only endeavored to give it a form applicable not to the mathematical continuum, but to the physical continuum, which alone is susceptible of representation, and yet to retain all its precision. Moreover, we see that this definition applies not alone to space; that in all which falls under our senses we find the characteristics of the physical continuum, which would allow of the same classification; that it would be easy to find there examples of continua of four, of five, dimensions, in the sense of the preceding definition; such examples occur of themselves to the mind.

I should explain finally, if I had the time, that this science, of which I spoke above and to which Riemann gave the name of analysis situs, teaches us to make distinctions among continua of the same number of dimensions and that the classification of these continua rests also on the consideration of cuts.

From this notion has arisen that of the mathematical continuum of several dimensions in the same way that the physical continuum of one dimension engendered the mathematical continuum of one dimension. The formula

$$A > C, \quad A = B, \quad B = C,$$

which summed up the data of crude experience, implied an intolerable contradiction. To get free from it, it was necessary to introduce a new notion while still respecting the essential characteristics of the physical continuum of several dimensions. The mathematical continuum of one dimension admitted of a scale whose divisions, infinite in number, corresponded to the different values, commensurable or not, of one same magnitude. To have the mathematical continuum of $n$ dimensions, it will suffice to take $n$ like scales whose divisions correspond to different values of $n$ independent magnitudes called coordinates. We thus shall have an image of the physical continuum of $n$ dimensions, and this image will be as faithful as it can be after the determination not to allow the contradiction of which I spoke above.

#### 4. *The Notion of Point*

It seems now that the question we put to ourselves at the start is answered. When we say that space has three dimensions, it will be said, we mean that the manifold of points of space satisfies the definition we have just given of the physical continuum of three dimensions. To be content with that would be to suppose that we know what is the manifold of points of space, or even one point of space.

Now that is not as simple as one might think. Every one believes he knows what a point is, and it is just because we know it too well that we think there is no need of defining it. Surely we can not be required to know how to define it, because in going back from definition to definition a time must come when we must stop. But at what moment should we stop?

We shall stop first when we reach an object which falls under our senses or that we can represent to ourselves; definition then will become useless; we do not define the sheep to a child; we say to him: *See* the sheep.

So, then, we should ask ourselves if it is possible to represent to ourselves a point of space. Those who answer yes do not reflect that they represent to themselves in reality a white spot made with the chalk on a blackboard or a black spot made with a pen on white paper, and that they can represent to themselves only an object or rather the impressions that this object made on their senses.

When they try to represent to themselves a point, they represent the impressions that very little objects made them feel. It is needless to add that two different objects, though both very little, may produce extremely different impressions, but I shall not dwell on this difficulty, which would still require some discussion.

But it is not a question of that; it does not suffice to represent *one* point, it is necessary to represent *a certain* point and to have the means of distinguishing it from an *other* point. And in fact, that we may be able to apply to a continuum the rule I have above expounded and by which one may recognize the number of its dimensions, we must rely upon the fact that two elements of this continuum sometimes can and sometimes can not be distinguished. It is necessary therefore that we should in certain cases know how to represent to ourselves *a specific* element and to distinguish it from an *other* element.

The question is to know whether the point that I represented to myself an hour ago is the same as this that I now represent to myself, or whether it is a different point. In other words, how do we know whether the point occupied by the object $A$ at the instant $\alpha$ is the same as the point occupied by the object $B$ at the instant $\beta$, or still better, what this means?

I am seated in my room; an object is placed on my table; during a second I do not move, no one touches the object. I am tempted to say that the point $A$ which this object occupied at the beginning of this second is identical with the point $B$ which it occupies at its end. Not at all; from the point $A$ to the point $B$ is 30 kilometers, because the object has been carried along in the motion of the earth. We can not know whether an object, be it large or small, has not changed its absolute position in space, and not only can we not affirm it, but this affirmation has no meaning and in any case can not correspond to any representation.

But then we may ask ourselves if the relative position of an object with regard to other objects has changed or not, and first whether the relative position of this object with regard to our body has changed. If the impressions this object makes upon us have not changed, we shall be inclined to judge that neither has this relative position changed; if they have changed, we shall judge that this object has changed either in state or in relative position. It remains to decide which of the *two*. I have explained in 'Science and Hypothesis' how we have been led to distinguish the changes of position. Moreover, I shall return to that further on. We come to know, therefore, whether the relative position of an object with regard to our body has or has not remained the same.

If now we see that two objects have retained their relative position with regard to our body, we conclude that the relative position of these two objects with regard to one another has not changed; but we reach this conclusion only by indirect reasoning. The only thing that we know directly is the relative position of the objects with regard to our body. *A fortiori* it is only by indirect reasoning that we think we know (and, moreover, this belief is delusive) whether the absolute position of the object has changed.

In a word, the system of coordinate axes to which we naturally refer all exterior objects is a system of axes invariably bound to our body, and carried around with us.

It is impossible to represent to oneself absolute space; when I try to represent to myself simultaneously objects and myself in motion in absolute space, in reality I represent to myself my own self montionless and seeing move around me different objects and a man that is exterior to me, but that I convene to call me.

Will the difficulty be solved if we agree to refer everything to these axes bound to our body? Shall we know then what is a point thus defined by its relative position with regard to ourselves? Many persons will answer yes and will say that they 'localize' exterior objects.

What does this mean? To localize an object simply means to represent to oneself the movements that would be necessary to reach it. I will explain myself. It is not a question of representing the movements themselves in space, but solely of representing to oneself the muscular sensations which accompany these movements and which do not presuppose the preexistence of the notion of space.

If we suppose two different objects which successively occupy the same relative position with regard to ourselves, the impressions that these two objects make upon us will be very different; if we localize them at the same point, this is simply because it is necessary to make the same movements to reach them; apart from that, one can not just see what they could have in common.

But, given an object, we can conceive many different series of movements which equally enable us to reach it. If then we represent to ourselves a point by representing to ourselves the series of muscular sensations which accompany the movements which enable us to reach this point, there will be many ways entirely different of representing to oneself the same point. If one is not satisfied with this solution, but wishes, for instance, to bring in the visual sensations along with the muscular sensations, there will be one or two more ways of representing to oneself this same point and the difficulty will only be increased. In any case the following question comes up: Why do we think that all these representations so different from one another still represent the same point?

Another remark: I have just said that it is to our own body that we naturally refer exterior objects; that we carry about everywhere with us a system of axes to which we refer all the points of space, and that this system of axes seems to be invariably bound to our body. It should be noticed that rigorously we could not speak of axes invariably bound to the body unless the different parts of this body were themselves invariably bound to one another. As this is not the case, we ought, before referring exterior objects to these fictitious axes, to suppose our body brought back to the initial attitude.

## 5. *The Notion of Displacement*

I have shown in 'Science and Hypothesis' the preponderant rôle played by the movements of our body in the genesis of the notion of space. For a being completely immovable there would be neither space nor geometry; in vain would exterior objects be displaced about him, the variations which these displacements would make in his impressions would not be attributed by this being to changes of position, but to simple changes of state; this being would have no means of distinguishing these two sorts of changes, and this distinction, fundamental for us, would have no meaning for him.

The movements that we impress upon our members have as effect the varying of the impressions produced on our senses by external objects; other causes may likewise make them vary; but we are led to distinguish the changes produced by our own motions and we easily discriminate them for two reasons: (1) because they are voluntary; (2) because they are accompanied by muscular sensations.

So we naturally divide the changes that our impressions may undergo into two categories to which perhaps I have given an inappropriate designation: (1) the internal changes, which are voluntary and accompanied by muscular sensations; (2) the external changes, having the opposite characteristics.

We then observe that among the external changes are some which can be corrected, thanks to an internal change which brings everything back to the primitive state; others can not be corrected in this way (it is thus that, when an exterior object is displaced, we may then by changing our own position replace ourselves as regards this object in the same relative position as before, so as to reestablish the original aggregate of impressions; if this object was not displaced, but changed its state, that is impossible). Thence comes a new distinction among external changes: those which may be so corrected we call changes of position; and the others, changes of state.

Think, for example, of a sphere with one hemisphere blue and the other red; it first presents to us the blue hemisphere, then it so revolves as to present the red hemisphere. Now think of a spherical vase containing a blue liquid which becomes red in consequence of a chemical reaction. In both cases the sensation of red has replaced that of blue; our senses have experienced the same impressions which have succeeded each other in the same order, and yet these two changes are regarded by us as very different; the first is a displacement, the second a change of state. Why? Because in the first case it is sufficient for me to go around the sphere to place myself opposite the blue hemisphere and reestablish the original blue sensation.

Still more; if the two hemispheres, in place of being red and blue, had been yellow and green, how should I have interpreted the revolution of the sphere? Before, the red succeeded the blue, now the green succeeds the yellow; and yet I say that the two spheres have undergone the same revolution, that each has turned about its axis; yet I can not say that the green is to yellow as the red is to blue; how then am I led to decide that the two spheres have undergone the *same* displacement? Evidently because, in one case as in the other, I am able to reestablish the original sensation by going around the sphere, by making the same movements, and I know that I have made the same movements because I have felt the same muscular sensations; to know it, I do not need, therefore, to know geometry in advance and to represent to myself the movements of my body in geometric space.

Another example: An object is displaced before my eye; its image was first formed at the center of the retina; then it is formed at the border; the old sensation was carried to me by a nerve fiber ending at the center of the retina; the new sensation is carried to me by *another* nerve fiber starting from the border of the retina; these two sensations are qualitatively different; otherwise, how could I distinguish them?

Why then am I led to decide that these two sensations, qualitatively different, represent the same image, which has been displaced? It is because I *can follow the object with the eye* and by a displacement of the eye, voluntary and accompanied by muscular sensations, bring back the image to the center of the retina and reestablish the primitive sensation.

I suppose that the image of a red object has gone from the center $A$ to the border $B$ of the retina, then that the image of a blue object goes in its turn from the center $A$ to the border $B$ of the retina; I shall decide that these two objects have undergone the *same* displacement. Why? Because in both cases I shall have been able to reestablish the primitive sensation, and that to do it I shall have had to execute the *same* movement of the eye, and I shall know that my eye has executed the same movement because I shall have felt the *same* muscular sensations.

If I could not move my eye, should I have any reason to suppose that the sensation of red at the center of the retina is to the sensation of red at the border of the retina as that of blue at the center is to that of blue at the border? I should only have four sensations qualitatively different, and if I were asked if they are connected by the proportion I have just stated, the question would seem to me ridiculous, just as if I were asked if there is an analogous proportion between an auditory sensation, a tactile sensation and an olfactory sensation.

Let us now consider the internal changes, that is, those which are produced by the voluntary movements of our body and which are accompanied by muscular changes. They give rise to the two following observations, analogous to those we have just made on the subject of external changes.

1. I may suppose that my body has moved from one point to another, but that the same *attitude* is retained; all the parts of the body have therefore retained or resumed the same *relative* situation, although their absolute situation in space may have varied. I may suppose that not only has the position of my body changed, but that its attitude is no longer the same, that, for instance, my arms which before were folded are now stretched out.

I should therefore distinguish the simple changes of position without change of attitude, and the changes of attitude. Both would appear to me under form of muscular sensations. How then am I led to distinguish them? It is that the first may serve to correct an external change, and that the others can not, or at least can only give an imperfect correction.

This fact I proceed to explain as I would explain it to some one who already knew geometry, but it need not thence be concluded that it is necessary already to know geometry to make this distinction; before knowing geometry I ascertain the fact (experimentally, so to speak), without being able to explain it. But merely to make the distinction between the two kinds of change, I do not need to *explain* the fact, it suffices me *to ascertain* it.

However that may be, the explanation is easy. Suppose that

an exterior object is displaced; if we wish the different parts of our body to resume with regard to this object their initial relative position, it is necessary that these different parts should have resumed likewise their initial relative position with regard to one another. Only the internal changes which satisfy this latter condition will be capable of correcting the external change produced by the displacement of that object. If, therefore, the relative position of my eye with regard to my finger has changed, I shall still be able to replace the eye in its initial relative situation with regard to the object and reestablish thus the primitive visual sensations, but then the relative position of the finger with regard to the object will have changed and the tactile sensations will not be reestablished.

2. We ascertain likewise that the same external change may be corrected by two internal changes corresponding to different muscular sensations. Here again I can ascertain this without knowing geometry; and I have no need of anything else; but I proceed to give the explanation of the fact, employing geometrical language. To go from the position $A$ to the position $B$ I may take several routes. To the first of these routes will correspond a series $S$ of muscular sensations; to a second route will correspond another series $S''$, of muscular sensations which generally will be completely different, since other muscles will be used.

How am I led to regard these two series $S$ and $S''$ as corresponding to the same displacement $AB$? It is because these two series are capable of correcting the same external change. Apart from that, they have nothing in common.

Let us now consider two external changes: $\alpha$ and $\beta$, which shall be, for instance, the rotation of a sphere half blue, half red, and that of a sphere half yellow, half green; these two changes have nothing in common, since the one is for us the passing of blue into red and the other the passing of yellow into green. Consider, on the other hand, two series of internal changes $S$ and $S''$; like the others, they will have nothing in common. And yet I say that $\alpha$ and $\beta$ correspond to the same displacement, and that $S$ and $S''$ correspond also to the same displacement. Why? Simply because $S$ can correct $\alpha$ as well as $\beta$ and because $\alpha$ can be corrected by $S''$ as well as by $S$. And then a question suggests itself:

If I have ascertained that $S$ corrects $\alpha$ and $\beta$ and that $S'''$ corrects $\alpha$, am I certain that $S''$ likewise corrects $\beta$? Experiment alone can teach us whether this law is verified. If it were not verified, at least approximately, there would be no geometry, there would be no space, because we should have no more interest in classifying the internal and external changes as I have just done, and, for instance, in distinguishing changes of state from changes of position.

It is interesting to see what has been the rôle of experience in all this. It has shown me that a certain law is approximately verified. It has not told me *how* space is, and that it satisfies the condition in question. I knew, in fact, before all experience, that space satisfied this condition or that it would not be; nor have I any right to say that experience told me that geometry is possible; I very well see that geometry is possible, since it does not imply contradiction; experience only tells me that geometry is useful.

### 6. *Visual Space*

Although motor impressions have had, as I have just explained, an altogether preponderant influence in the genesis of the notion of space, which never would have taken birth without them, it will not be without interest to examine also the rôle of visual impressions and to investigate how many dimensions 'visual space' has, and for that purpose to apply to these impressions the definition of § 3.

A first difficulty presents itself: consider a red color sensation affecting a certain point of the retina; and on the other hand a blue color sensation affecting the same point of the retina. It is necessary that we have some means of recognizing that these two sensations, qualitatively different, have something in common. Now, according to the considerations expounded in the preceding paragraph, we have been able to recognize this only by the movements of the eye and the observations to which they have given rise. If the eye were immovable, or if we were unconscious of its movements, we should not have been able to recognize that these two sensations, of different quality, had something in common; we should not have been able to disengage from them what gives them a geometric character. The visual sensations, without the muscular sensations, would have nothing geometric, so that it may be said there is no pure visual space.

To do away with this difficulty, consider only sensations of the same nature, red sensations, for instance, differing one from another only as regards the point of the retina that they affect. It is clear that I have no reason for making such an arbitrary choice among all the possible visual sensations, for the purpose of uniting in the same class all the sensations of the same color, whatever may be the point of the retina affected. I should never have dreamt of it, had I not before learned, by the means we have just seen, to distinguish changes of state from changes of position, that is, if my eye were immovable. Two sensations of the same color affecting two different parts of the retina would have appeared to me as qualitatively distinct, just as two sensations of different color.

In restricting myself to red sensations, I therefore impose upon myself an artificial limitation and I neglect systematically one whole side of the question; but it is only by this artifice that I am able to analyze visual space without mingling any motor sensation.

Imagine a line traced on the retina and dividing in two its surface; and set apart the red sensations affecting a point of this line, or those differing from them too little to be distinguished from them. The aggregate of these sensations will form a sort of cut that I shall call $C$, and it is clear that this cut suffices to divide the manifold of possible red sensations, and that if I take two red sensations affecting two points situated on one side and the other of the line, I can not pass from one of these sensations to the other in a continuous way without passing at a certain moment through a sensation belonging to the cut.

If, therefore, the cut has $n$ dimensions, the total manifold of my red sensations, or if you wish, the whole visual space, will have $n + 1$.

Now, I distinguish the red sensations affecting a point of the cut $C$. The assemblage of these sensations will form a new cut $C'$. It is clear that this will divide the cut $C$, always giving to the word divide the same meaning.

If, therefore, the cut $C'$ has $n$ dimensions, the cut $C$ will have $n + 1$ and the whole of visual space $n + 2$.

If all the red sensations affecting the same point of the retina were regarded as identical, the cut $C'$ reducing to a single element would have 0 dimensions, and visual space would have 2.

And yet most often it is said that the eye gives us the sense of a third dimension, and enables us in a certain measure to recognize the distance of objects. When we seek to analyze this feeling, we ascertain that it reduces either to the consciousness of the convergence of the eyes, or to that of the effort of accommodation which the ciliary muscle makes to focus the image.

Two red sensations affecting the same point of the retina will therefore be regarded as identical only if they are accompanied by the same sensation of convergence and also by the same sensation of effort of accommodation or at least by sensations of convergence and accommodation so slightly different as to be indistinguishable.

On this account the cut $C'$ is itself a continuum and the cut $C$ has more than one dimension.

But it happens precisely that experience teaches us that when two visual sensations are accompanied by the same sensation of convergence, they are likewise accompanied by the same sensation of accommodation. If then we form a new cut $C''$ with all those of the sensations of the cut $C'$, which are accompanied by a certain sensation of convergence, in accordance with the preceding law they will all be indistinguishable and may be regarded as identical. Therefore $C''$ will not be a continuum and will have 0 dimension; and as $C''$ divides $C'$ it will thence result that $C'$ has one, $C$ two and *the whole visual space three dimensions.*

But would it be the same if experience had taught us the contrary and if a certain sensation of convergence were not always accompanied by the same sensation of accommodation? In this case two sensations affecting the same point of the retina and accompanied by the same sense of convergence, two sensations which consequently would both appertain to the cut $C''$, could nevertheless be distinguished since they would be accompanied by two different sensations of accommodation. Therefore $C''$ would be in its turn a continuum and would have one dimension (at least) ; then $C'$ would have two, $C$ three and *the whole visual space would have four dimensions.*

Will it then be said that it is experience which teaches us that space has three dimensions, since it is in setting out from an experimental law that we have come to attribute three to it? But we have therein performed, so to speak, only an experiment in physiology ; and as also it would suffice to fit over the eyes glasses of suitable construction to put an end to the accord between the feelings of convergence and of accommodation, are we to say that putting on spectacles is enough to make space have four dimensions and that the optician who constructed them has given one more dimension to space? Evidently not ; all we can say is that experience has taught us that it is convenient to attribute three dimensions to space.

But visual space is only one part of space, and in even the notion of this space there is something artificial, as I have explained at the beginning. The real space is motor space and this it is that we shall examine in the following chapter.

# CHAPTER IV

## SPACE AND ITS THREE DIMENSIONS

### 1. *The Group of Displacements*

LET us sum up briefly the results obtained. We proposed to investigate what was meant in saying that space has three dimensions and we have asked first what is a physical continuum and when it may be said to have $n$ dimensions. If we consider different systems of impressions and compare them with one another, we often recognize that two of these systems of impressions are indistinguishable (which is ordinarily expressed in saying that they are too close to one another, and that our senses are too crude, for us to distinguish them) and we ascertain besides that two of these systems can sometimes be discriminated from one another though indistinguishable from a third system. In that case we say the manifold of these systems of impressions forms a physical continuum $C$. And each of these systems is called an *element* of the continuum $C$.

How many dimensions has this continuum? Take first two elements $A$ and $B$ of $C$, and suppose there exists a series $\Sigma$ of elements, all belonging to the continuum $C$, of such a sort that $A$ and $B$ are the two extreme terms of this series and that each term of the series is indistinguishable from the preceding. If such a series $\Sigma$ can be found, we say that $A$ and $B$ are joined to one another; and if any two elements of $C$ are joined to one another, we say that $C$ is all of one piece.

Now take on the continuum $C$ a certain number of elements in a way altogether arbitrary. The aggregate of these elements will be called a *cut*. Among the various series $\Sigma$ which join $A$ to $B$, we shall distinguish those of which an element is indistinguishable from one of the elements of the cut (we shall say that these are they which *cut* the cut) and those of which *all* the elements are distinguishable from all those of the cut. If *all* the series $\Sigma$ which join $A$ to $B$ cut the cut, we shall say that $A$ and $B$ are *separated* by the cut, and that the cut *divides C.* If we can not find on *C* two elements which are separated by the cut, we shall say that the cut *does not divide C.*

These definitions laid down, if the continuum *C* can be divided by cuts which do not themselves form a continuum, this continuum *C* has only one dimension; in the contrary case it has several. If a cut forming a continuum of 1 dimension suffices to divide *C, C* will have 2 dimensions; if a cut forming a continuum of 2 dimensions suffices, *C* will have 3 dimensions, etc. Thanks to these definitions, we can always recognize how many dimensions any physical continuum has. It only remains to find a physical continuum which is, so to speak, equivalent to space, of such a sort that to every point of space corresponds an element of this continuum, and that to points of space very near one another correspond indistinguishable elements. Space will have then as many dimensions as this continuum.

The intermediation of this physical continuum, capable of representation, is indispensable; because we can not represent space to ourselves, and that for a multitude of reasons. Space is a mathematical continuum, it is infinite, and we can represent to ourselves only physical continua and finite objects. The different elements of space, which we call points, are all alike, and, to apply our definition, it is necessary that we know how to distinguish the elements from one another, at least if they are not too close. Finally absolute space is nonsense, and it is necessary for us to begin by referring space to a system of axes invariably bound to our body (which we must always suppose put back in the initial attitude).

Then I have sought to form with our visual sensations a physical continuum equivalent to space; that certainly is easy and this example is particularly appropriate for the discussion of the number of dimensions; this discussion has enabled us to see in what measure it is allowable to say that 'visual space' has three dimensions. Only this solution is incomplete and artificial. I have explained why, and it is not on visual space, but on motor space that it is necessary to bring our efforts to bear. I have then recalled what is the origin of the distinction we make between

18

changes of position and changes of state. Among the changes
which occur in our impressions, we distinguish, first the *internal*
changes, voluntary and accompanied by muscular sensations, and
the *external* changes, having opposite characteristics. We ascertain that it may happen that an external change may be *corrected*
by an internal change which reestablishes the primitive sensations. The external changes, capable of being corrected by an
internal change are called *changes of position,* those not capable
of it are called *changes of state.* The internal changes capable
of correcting an external change are called *displacements of the
whole body;* the others are called *changes of attitude.*

Now let $\alpha$ and $\beta$ be two external changes, $\alpha'$ and $\beta'$ two internal
changes. Suppose that $\alpha$ may be corrected either by $\alpha'$ or by $\beta'$,
and that $\alpha'$ can correct either $\alpha$ or $\beta$; experience tells us then that
$\beta'$ can likewise correct $\beta$. In this case we say that $\alpha$ and $\beta$ correspond to the *same* displacement and also that $\alpha'$ and $\beta'$ correspond to the *same* displacement. That postulated, we can
imagine a physical continuum which we shall call *the continuum
or group of displacements* and which we shall define in the following manner. The elements of this continuum shall be the internal changes capable of correcting an external change. Two of
these internal changes $\alpha'$ and $\beta'$ shall be regarded as indistinguishable: (1) if they are so naturally, that is, if they are
too close to one another; (2) if $\alpha'$ is capable of correcting
the same external change as a third internal change naturally indistinguishable from $\beta'$. In this second case, they will
be, so to speak, indistinguishable by convention, I mean by agreeing to disregard circumstances which might distinguish them.

Our continuum is now entirely defined, since we know its elements and have fixed under what conditions they may be regarded as indistinguishable. We thus have all that is necessary
to apply our definition and determine how many dimensions this
continuum has. We shall recognize that it has *six.* The continuum of displacements is, therefore, not equivalent to space,
since the number of dimensions is not the same; it is only related
to space. Now how do we know that this continuum of displacements has six dimensions? We know it *by experience.*

It would be easy to describe the experiments by which we

could arrive at this result. It would be seen that in this continuum cuts can be made which divide it and which are continua; that these cuts themselves can be divided by other cuts of the second order which yet are continua, and that this would stop only after cuts of the sixth order which would no longer be continua. From our definitions that would mean that the group of displacements has six dimensions.

That would be easy, I have said, but that would be rather long; and would it not be a little superficial? This group of displacements, we have seen, is related to space, and space could be deduced from it, but it is not equivalent to space, since it has not the same number of dimensions; and when we shall have shown how the notion of this continuum can be formed and how that of space may be deduced from it, it might always be asked why space of three dimensions is much more familiar to us than this continuum of six dimensions, and consequently doubted whether it was by this detour that the notion of space was formed in the human mind.

## 2. *Identity of Two Points*

What is a point? How do we know whether two points of space are identical or different? Or, in other words, when I say: The object $A$ occupied at the instant $\alpha$ the point which the object $B$ occupies at the instant $\beta$, what does that mean?

Such is the problem we set ourselves in the preceding chapter, §4. As I have explained it, it is not a question of comparing the positions of the objects $A$ and $B$ in absolute space; the question then would manifestly have no meaning. It is a question of comparing the positions of these two objects with regard to axes invariably bound to my body, supposing always this body replaced in the same attitude.

I suppose that between the instants $\alpha$ and $\beta$ I have moved neither my body nor my eye, as I know from my muscular sense. Nor have I moved either my head, my arm or my hand. I ascertain that at the instant $\alpha$ impressions that I attributed to the object $A$ were transmitted to me, some by one of the fibers of my optic nerve, the others by one of the sensitive tactile nerves of my finger; I ascertain that at the instant $\beta$ other impressions which I attribute to the object $B$ are transmitted to me, some by this same fiber of the optic nerve, the others by this same tactile nerve.

Here I must pause for an explanation; how am I told that this impression which I attribute to $A$, and that which I attribute to $B$, impressions which are qualitatively different, are transmitted to me by the same nerve? Must we suppose, to take for example the visual sensations, that $A$ produces two simultaneous sensations, a sensation purely luminous $a$ and a colored sensation $a'$, that $B$ produces in the same way simultaneously a luminous sensation $b$ and a colored sensation $b'$, that if these different sensations are transmitted to me by the same retinal fiber, $a$ is identical with $b$, but that in general the colored sensations $a'$ and $b'$ produced by different bodies are different? In that case it would be the identity of the sensation $a$ which accompanies $a'$ with the sensation $b$ which accompanies $b'$, which would tell that all these sensations are transmitted to me by the same fiber.

However it may be with this hypothesis and although I am led to prefer to it others considerably more complicated, it is certain that we are told in some way that there is something in common between these sensations $a + a'$ and $b + b'$, without which we should have no means of recognizing that the object $B$ has taken the place of the object $A$.

Therefore I do not further insist and I recall the hypothesis I have just made: I suppose that I have ascertained that the impressions which I attribute to $B$ are transmitted to me at the instant $\beta$ by the same fibers, optic as well as tactile, which, at the instant $\alpha$, had transmitted to me the impressions that I attributed to $A$. If it is so, we shall not hesitate to declare that the point occupied by $B$ at the instant $\beta$ is identical with the point occupied by $A$ at the instant $\alpha$.

I have just enunciated two conditions for these points being identical; one is relative to sight, the other to touch. Let us consider them separately. The first is necessary, but is not sufficient. The second is at once necessary and sufficient. A person knowing geometry could easily explain this in the following manner: Let $O$ be the point of the retina where is formed at the instant $\alpha$ the image of the body $A$; let $M$ be the point of space occupied at the instant $\alpha$ by this body $A$; let $M'$ be the point of space occupied at the instant $\beta$ by the body $B$. For this body $B$ to form its image in $O$, it is not necessary that the points $M$ and $M'$ coincide; since vision acts at a distance, it suffices for the three points $O$ $M$ $M'$ to be in a straight line. This condition that the two objects form their image on $O$ is therefore necessary, but not sufficient for the points $M$ and $M'$ to coincide. Let now $P$ be the point occupied by my finger and where it remains, since it does not budge. As touch does not act at a distance, if the body $A$ touches my finger at the instant $\alpha$, it is because $M$ and $P$ coincide; if $B$ touches my finger at the instant $\beta$, it is because $M'$ and $P$ coincide. Therefore $M$ and $M'$ coincide. Thus this condition that if $A$ touches my finger at the instant $\alpha$, $B$ touches it at the instant $\beta$, is at once necessary and sufficient for $M$ and $M'$ to coincide.

But we who, as yet, do not know geometry can not reason thus; all that we can do is to ascertain experimentally that the first condition relative to sight may be fulfilled without the second, which is relative to touch, but that the second can not be fulfilled without the first.

Suppose experience had taught us the contrary, as might well be; this hypothesis contains nothing absurd. Suppose, therefore, that we had ascertained experimentally that the condition relative to touch may be fulfilled without that of sight being fulfilled and that, on the contrary, that of sight can not be fulfilled without that of touch being also. It is clear that if this were so we should conclude that it is touch which may be exercised at a distance, and that sight does not operate at a distance.

But this is not all; up to this time I have supposed that to determine the place of an object I have made use only of my eye and a single finger; but I could just as well have employed other means, for example, all my other fingers.

I suppose that my first finger receives at the instant $\alpha$ a tactile impression which I attribute to the object $A$. I make a series of movements, corresponding to a series $S$ of muscular sensations. After these movements, at the instant $\alpha'$, my *second* finger receives a tactile impression that I attribute likewise to $A$. Afterward, at the instant $\beta$, without my having budged, as my muscular sense tells me, this same second finger transmits to me anew a tactile impression which I attribute this time to the object $B$; I then make a series of movements, corresponding to a series $S'$ of muscular sensations. I know that this series $S'$ is the inverse of the series $S$ and corresponds to contrary movements. I know this because many previous experiences have shown me that if I made successively the two series of movements corresponding to $S$ and to $S'$, the primitive impressions would be reestablished, in other words, that the two series mutually compensate. That settled, should I expect that at the instant $\beta'$, when the second series of movements is ended, my *first finger* would feel a tactile impression attributable to the object $B$?

To answer this question, those already knowing geometry would reason as follows: There are chances that the object $A$ has not budged, between the instants $\alpha$ and $\alpha'$, nor the object $B$ between the instants $\beta$ and $\beta'$; assume this. At the instant $\alpha$, the object $A$ occupied a certain point $M$ of space. Now at this instant it touched my first finger, and *as touch does not operate at a distance,* my first finger was likewise at the point $M$. I afterward made the series $S$ of movements and at the end of this series, at the instant $\alpha'$, I ascertained that the object $A$ touched my second finger. I thence conclude that this second finger was then at $M$, that is, that the movements $S$ had the result of bringing the second finger to the place of the first. At the instant $\beta$ the object $B$ has come in contact with my second finger: as I have not budged, this second finger has remained at $M$; therefore the object $B$ has come to $M$; by hypothesis it does not budge up to the instant $\beta'$. But between the instants $\beta$ and $\beta'$ I have made the movements $S'$; as these movements are the inverse of the movements $S$, they must have for effect bringing the first finger in the place of the second. At the instant $\beta'$ this first finger will, therefore, be at $M$; and as the object $B$ is likewise at $M$, this object $B$ will touch my first finger. To the question put, the answer should therefore be yes.

We who do not yet know geometry can not reason thus; but we ascertain that this anticipation is ordinarily realized; and we can always explain the exceptions by saying that the object $A$ has moved between the instants $\alpha$ and $\alpha'$, or the object $B$ between the instants $\beta$ and $\beta'$.

But could not experience have given a contrary result? Would this contrary result have been absurd in itself? Evidently not. What should we have done then if experience had given this contrary result? Would all geometry thus have become impossible? Not the least in the world. We should have contented ourselves with concluding that *touch can operate at a distance.*

When I say, touch does not operate at a distance, but sight operates at a distance, this assertion has only one meaning, which is as follows: To recognize whether $B$ occupies at the instant $\beta$ the point occupied by $A$ at the instant $\alpha$, I can use a multitude of different criteria. In one my eye intervenes, in another my first finger, in another my second finger, etc. Well, it is sufficient for the criterion relative to one of my fingers to be satisfied in order that all the others should be satisfied, but it is not sufficient that the criterion relative to the eye should be. This is the sense of my assertion, I content myself with affirming an experimental fact which is ordinarily verified.

At the end of the preceding chapter we analyzed visual space; we saw that to engender this space it is necessary to bring in the retinal sensations, the sensation of convergence and the sensation of accommodation; that if these last two were not always in accord, visual space would have four dimensions in place of three; we also saw that if we brought in only the retinal sensations, we should obtain 'simple visual space,' of only two dimensions. On the other hand, consider tactile space, limiting ourselves to the sensations of a single finger, that is in sum to the assemblage of positions this finger can occupy. This tactile space that we shall analyze in the following section and which consequently I ask permission not to consider further for the moment, this tactile space, I say, has three dimensions. Why has space properly so called as many dimensions as tactile space and more than simple visual space? It is because touch does not operate at a distance, while vision does operate at a distance. These two assertions have the same meaning and we have just seen what this is.

Now I return to a point over which I passed rapidly in order not to interrupt the discussion. How do we know that the impressions made on our retina by $A$ at the instant $\alpha$ and $B$ at the instant $\beta$ are transmitted by the same retinal fiber, although these impressions are qualitatively different? I have suggested a simple hypothesis, while adding that other hypotheses, decidedly more complex, would seem to me more probably true. Here then are these hypotheses, of which I have already said a word. How do we know that the impressions produced by the red object $A$ at the instant $\alpha$, and by the blue object $B$ at the instant $\beta$, if these two objects have been imaged on the same point of the retina, have something in common? The simple hypothesis above made may be rejected and we may suppose that these two impressions, qualitatively different, are transmitted by two different though contiguous nervous fibers. What means have I then of knowing that these fibers are contiguous? It is probable that we should have none, if the eye were immovable. It is the movements of the eye which have told us that there is the same relation between the sensation of blue at the point $A$ and the sensation of blue at the point $B$ of the retina as between the sensation of red at the point $A$ and the sensation of red at the point $B$. They have shown us, in fact, that the same movements, corresponding to the same muscular sensations, carry us from the first to the second, or from the third to the fourth. I do not emphasize these considerations, which belong, as one sees, to the question of local signs raised by Lotze.

### 3. *Tactile Space*

Thus I know how to recognize the identity of two points, the point occupied by $A$ at the instant $\alpha$ and the point occupied by $B$ at the instant $\beta$, but only *on one condition*, namely, that I have not budged between the instants $\alpha$ and $\beta$. That does not suffice for our object. Suppose, therefore, that I have moved in any manner in the interval between these two instants, how shall I know whether the point occupied by $A$ at the instant $\alpha$ is identical with the point occupied by $B$ at the instant $\beta$? I suppose that at the instant $\alpha$, the object $A$ was in contact with my first finger and that in the same way, at the instant $\beta$, the object $B$ touches this first finger; but at the same time, my muscular sense has told me that in the interval my body has moved. I have considered above two series of muscular sensations $S$ and $S'$, and

I have said it sometimes happens that we are led to consider two such series $S$ and $S'$ as inverse one of the other, because we have often observed that when these two series succeed one another our primitive impressions are reestablished.

If then my muscular sense tells me that I have moved between the two instants $\alpha$ and $\beta$, but so as to feel successively the two series of muscular sensations $S$ and $S'$ that I consider inverses, I shall still conclude, just as if I had not budged, that the points occupied by $A$ at the instant $\alpha$ and by $B$ at the instant $\beta$ are identical, if I ascertain that my first finger touches $A$ at the instant $\alpha$, and $B$ at the instant $\beta$.

This solution is not yet completely satisfactory, as one will see. Let us see, in fact, how many dimensions it would make us attribute to space. I wish to compare the two points occupied by $A$ and $B$ at the instants $\alpha$ and $\beta$, or (what amounts to the same thing since I suppose that my finger touches $A$ at the instant $\alpha$ and $B$ at the instant $\beta$) I wish to compare the two points occupied by my finger at the two instants $\alpha$ and $\beta$. The sole means I use for this comparison is the series $\Sigma$ of muscular sensations which have accompanied the movements of my body between these two instants. The different imaginable series $\Sigma$ form evidently a physical continuum of which the number of dimensions is very great. Let us agree, as I have done, not to consider as distinct the two series $\Sigma$ and $\Sigma + S + S'$, when $S$ and $S'$ are inverses one of the other in the sense above given to this word; in spite of this agreement, the aggregate of distinct series $\Sigma$ will still form a physical continuum and the number of dimensions will be less but still very great.

To each of these series $\Sigma$ corresponds a point of space; to two series $\Sigma$ and $\Sigma'$ thus correspond two points $M$ and $M'$. The means we have hitherto used enable us to recognize that $M$ and $M'$ are not distinct in two cases: (1) if $\Sigma$ is identical with $\Sigma'$; (2) if $\Sigma' = \Sigma + S + S'$, $S$ and $S'$ being inverses one of the other. If in all the other cases we should regard $M$ and $M'$ as distinct, the manifold of points would have as many dimensions as the aggregate of distinct series $\Sigma$, that is, much more than three.

For those who already know geometry, the following explanation would be easily comprehensible. Among the imaginable series of muscular sensations, there are those which correspond to series of movements where the finger does not budge. I say that if one does not consider as distinct the series $\Sigma$ and $\Sigma + \sigma$, where the series $\sigma$ corresponds to movements where the finger does not budge, the aggregate of series will constitute a continuum of three dimensions, but that if one regards as distinct two series $\Sigma$ and $\Sigma'$ unless $\Sigma' = \Sigma + S + S'$, $S$ and $S'$ being inverses, the aggregate of series will constitute a continuum of more than three dimensions.

In fact, let there be in space a surface $A$, on this surface a line $B$, on this line a point $M$. Let $C_0$ be the aggregate of all series $\Sigma$. Let $C_1$ be the aggregate of all the series $\Sigma$, such that at the end of corresponding movements the finger is found upon the surface $A$, and $C_2$ or $C_3$ the aggregate of series $\Sigma$ such that at the end the finger is found on $B$, or at $M$. It is clear, first that $C_1$ will constitute a cut which will divide $C_0$, that $C_2$ will be a cut which will divide $C_1$, and $C_3$ a cut which will divide $C_2$. Thence it results, in accordance with our definitions, that if $C_3$ is a continuum of $n$ dimensions, $C_0$ will be a physical continuum of $n + 3$ dimensions.

Therefore, let $\Sigma$ and $\Sigma' = \Sigma + \sigma$ be two series forming part of $C_3$; for both, at the end of the movements, the finger is found at $M$; thence results that at the beginning and at the end of the series $\sigma$, the finger is at the same point $M$. This series $\sigma$ is therefore one of those which correspond to movements where the finger does not budge. If $\Sigma$ and $\Sigma + \sigma$ are not regarded as distinct, all the series of $C_3$ blend into one; therefore $C_3$ will have 0 dimension, and $C_0$ will have 3, as I wished to prove. If, on the contrary, I do not regard $\Sigma$ and $\Sigma + \sigma$ as blending (unless $\sigma = S + S'$, $S$ and $S'$ being inverses), it is clear that $C_3$ will contain a great number of series of distinct sensations; because, without the finger budging, the body may take a multitude of different attitudes. Then $C_3$ will form a continuum and $C_0$ will have more than three dimensions, and this also I wished to prove.

We who do not yet know geometry can not reason in this way; we can only verify. But then a question arises; how, before knowing geometry, have we been led to distinguish from the others these series $\sigma$ where the finger does not budge? It is, in fact, only after having made this distinction that we could be led to regard $\Sigma$ and $\Sigma + \sigma$ as identical, and it is on this condition alone, as we have just seen, that we can arrive at space of three dimensions.

We are led to distinguish the series $\sigma$, because it often happens that when we have executed the movements which correspond to these series $\sigma$ of muscular sensations, the tactile sensations which are transmitted to us by the nerve of the finger that we have called the first finger, persist and are not altered by these movements. Experience alone tells us that and it alone could tell us.

If we have distinguished the series of muscular sensations $S + S'$ formed by the union of two inverse series, it is because they preserve the totality of our impressions; if now we distinguish the series $\sigma$, it is because they preserve *certain* of our impressions. (When I say that a series of muscular sensations $S$ 'preserves' one of our impressions $A$, I mean that we ascertain that if we feel the impression $A$, then the muscular sensations $S$, we *still* feel the impression $A$ *after* these sensations $S$.)

I have said above it often happens that the series $\sigma$ do not alter the tactile impressions felt by our first finger; I said *often*, I did not say *always*. This it is that we express in our ordinary language by saying that the tactile impressions would not be altered if the finger has not moved, *on the condition* that *neither has* the object $A$, which was in contact with this finger, moved. Before knowing geometry, we could not give this explanation; all we could do is to ascertain that the impression often persists, but not always.

But that the impression often continues is enough to make the series $\sigma$ appear remarkable to us, to lead us to put in the same class the series $\Sigma$ and $\Sigma + \sigma$, and hence not regard them as distinct. Under these conditions we have seen that they will engender a physical continuum of three dimensions.

Behold then a space of three dimensions engendered by my first finger. Each of my fingers will create one like it. It remains to consider how we are led to regard them as identical with visual space, as identical with geometric space.

But one reflection before going further; according to the foregoing, we know the points of space, or more generally the final situation of our body, only by the series of muscular sensations revealing to us the movements which have carried us from a certain initial situation to this final situation. But it is clear that this final situation will depend, on the one hand, upon these movements and, *on the other hand, upon the initial situation* from which we set out. Now these movements are revealed to us by our muscular sensations; but nothing tells us the initial situation; nothing can distinguish it for us from all the other possible situations. This puts well in evidence the essential relativity of space.

#### 4. *Identity of the Different Spaces*

We are therefore led to compare the two continua $C$ and $C'$ engendered, for instance, one by my first finger $D$, the other by my second finger $D'$. These two physical continua both have three dimensions. To each element of the continuum $C$, or, if you prefer, to each point of the first tactile space, corresponds a series of muscular sensations $\Sigma$, which carry me from a certain initial situation to a certain final situation.[1] Moreover, the same point of this first space will correspond to $\Sigma$ and to $\Sigma + \sigma$, if $\sigma$ is a series of which we know that it does not make the finger $D$ move.

Similarly to each element of the continuum $C'$, or to each point of the second tactile space, corresponds a series of sensations $\Sigma'$, and the same point will correspond to $\Sigma'$ and to $\Sigma' + \sigma'$, if $\sigma'$ is a series which does not make the finger $D'$ move.

What makes us distinguish the various series designated $\sigma$ from those called $\sigma'$ is that the first do not alter the tactile impressions felt by the finger $D$ and the second preserve those the finger $D'$ feels.

Now see what we ascertain: in the beginning my finger $D'$ feels a sensation $A'$; I make movements which produce muscular sensations $S$; my finger $D$ feels the impression $A$; I make movements which produce a series of sensations $\sigma$; my finger $D$ continues to feel the impression $A$, since this is the characteristic

[1] In place of saying that we refer space to axes rigidly bound to our body, perhaps it would be better to say, in conformity to what precedes, that we refer it to axes rigidly bound to the initial situation of our body.

property of the series $\sigma$; I then make movements which produce the series $S'$ of muscular sensations, *inverse* to $S$ in the sense above given to this word. I ascertain then that my finger $D'$ feels anew the impression $A'$. (It is of course understood that $S$ has been suitably chosen.)

This means that the series $S + \sigma + S'$, preserving the tactile impressions of the finger $D'$, is one of the series I have called $\sigma'$. Inversely, if one takes any series $\sigma'$, $S' + \sigma' + S$ will be one of the series that we call $\sigma$.

Thus if $S$ is suitably chosen, $S + \sigma + S'$ will be a series $\sigma'$, and by making $\sigma$ vary in all possible ways, we shall obtain all the possible series $\sigma'$.

Not yet knowing geometry, we limit ourselves to verifying all that, but here is how those who know geometry would explain the fact. In the beginning my finger $D'$ is at the point $M$, in contact with the object $a$, which makes it feel the impression $A'$. I make the movements corresponding to the series $S$; I have said that this series should be suitably chosen, I should so make this choice that these movements carry the finger $D$ to the point originally occupied by the finger $D'$, that is, to the point $M$; this finger $D$ will thus be in contact with the object $a$, which will make it feel the impression $A$.

I then make the movements corresponding to the series $\sigma$; in these movements, by hypothesis, the position of the finger $D$ does not change, this finger therefore remains in contact with the object $a$ and continues to feel the impression $A$. Finally I make the movements corresponding to the series $S'$. As $S'$ is inverse to $S$, these movements carry the finger $D'$ to the point previously occupied by the finger $D$, that is, to the point $M$. If, as may be supposed, the object $a$ has not budged, this finger $D'$ will be in contact with this object and will feel anew the impression $A'$. . . . *Q. E. D.*

Let us see the consequences. I consider a series of muscular sensations $\Sigma$. To this series will correspond a point $M$ of the first tactile space. Now take again the two series $S$ and $S'$, inverses of one another, of which we have just spoken. To the series $S + \Sigma + S'$ will correspond a point $N$ of the second tactile space, since to any series of muscular sensations corresponds, as we have said, a point, whether in the first space or in the second.

I am going to consider the two points $N$ and $M$, thus defined, as corresponding. What authorizes me so to do? For this correspondence to be admissible, it is necessary that if two points $M$ and $M'$, corresponding in the first space to two series $\Sigma$ and $\Sigma'$, are identical, so also are the two corresponding points of the second space $N$ and $N'$, that is the two points which correspond to the two series $S + \Sigma + S'$ and $S + \Sigma' + S'$. Now we shall see that this condition is fulfilled.

First a remark. As $S$ and $S'$ are inverses of one another, we shall have $S + S' = 0$, and consequently $S + S' + \Sigma = \Sigma + S + S' = \Sigma$, or again $\Sigma + S + S' + \Sigma' = \Sigma + \Sigma'$; but it does not follow that we have $S + \Sigma + S' = \Sigma$; because, though we have used the addition sign to represent the succession of our sensations, it is clear that the order of this succession is not indifferent: we can not, therefore, as in ordinary addition, invert the order of the terms; to use abridged language, our operations are associative, but not commutative.

That fixed, in order that $\Sigma$ and $\Sigma'$ should correspond to the same point $M = M'$ of the first space, it is necessary and sufficient for us to have $\Sigma' = \Sigma + \sigma$. We shall then have: $S + \Sigma' + S' = S + \Sigma + \sigma + S' = S + \Sigma + S' + S + \sigma + S'$.

But we have just ascertained that $S + \sigma + S'$ was one of the series $\sigma'$. We shall therefore have: $S + \Sigma' + S' = S + \Sigma + S' + \sigma'$, which means that the series $S + \Sigma' + S'$ and $S + \Sigma + S'$ correspond to the same point $N = N'$ of the second space. Q. E. D.

Our two spaces therefore correspond point for point; they can be 'transformed' one into the other; they are isomorphic. How are we led to conclude thence that they are identical?

Consider the two series $\sigma$ and $S + \sigma + S' = \sigma'$. I have said that often, but not always, the series $\sigma$ preserves the tactile impression $A$ felt by the finger $D$; and similarly it often happens, but not always, that the series $\sigma'$ preserves the tactile impression $A'$ felt by the finger $D'$. Now I ascertain that it happens *very often* (that is, much more often than what I have just called 'often') that when the series $\sigma$ has preserved the impression $A$ of the finger *D,* the series $\sigma'$ preserves at the same time the impression $A'$ of the finger $D'$; and, inversely, that if the first impression is altered, the second is likewise. That happens *very often,* but not always.

We interpret this experimental fact by saying that the unknown object *a* which gives the impression *A* to the finger *D* is identical with the unknown object $a'$ which gives the impression $A'$ to the finger $D'$. And in fact when the first object moves, which the disappearance of the impression *A* tells us, the second likewise moves, since the impression $A'$ disappears likewise. When the first object remains motionless, the second remains motionless. If these two objects are identical, as the first is at the point *M* of the first space and the second at the point *N* of the second space, these two points are identical. This is how we are led to regard these two spaces as identical; or better, this is what we mean when we say that they are identical.

What we have just said of the identity of the two tactile spaces makes unnecessary our discussing the question of the identity of tactile space and visual space, which could be treated in the same way.

### 5. *Space and Empiricism*

It seems that I am about to be led to conclusions in conformity with empiristic ideas. I have, in fact, sought to put in evidence the rôle of experience and to analyze the experimental facts which intervene in the genesis of space of three dimensions. But whatever may be the importance of these facts, there is one thing we must not forget and to which besides I have more than once called attention. These experimental facts are often verified but not always. That evidently does not mean that space has often three dimensions, but not always.

I know well that it is easy to save oneself and that, if the facts do not verify, it will be easily explained by saying that the exterior objects have moved. If experience succeeds, we say that it teaches us about space; if it does not succeed, we hie to exterior objects which we accuse of having moved; in other words, if it does not succeed, it is given a fillip.

These fillips are legitimate; I do not refuse to admit them; but they suffice to tell us that the properties of space are not experimental truths, properly so called. If we had wished to verify other laws, we could have succeeded also, by giving other analogous fillips. Should we not always have been able to justify these fillips by the same reasons? One could at most have said to us: 'Your fillips are doubtless legitimate, but you abuse them; why move the exterior objects so often?'

To sum up, experience does not prove to us that space has three dimensions; it only proves to us that it is convenient to attribute three to it, because thus the number of fillips is reduced to a minimum.

I will add that experience brings us into contact only with representative space, which is a physical continuum, never with geometric space, which is a mathematical continuum. At the very most it would appear to tell us that it is convenient to give to geometric space three dimensions, so that it may have as many as representative space.

The empiric question may be put under another form. Is it impossible to conceive physical phenomena, the mechanical phenomena, for example, otherwise than in space of three dimensions? We should thus have an objective experimental proof, so to speak, independent of our physiology, of our modes of representation.

But it is not so; I shall not here discuss the question completely, I shall confine myself to recalling the striking example given us by the mechanics of Hertz. You know that the great physicist did not believe in the existence of forces, properly so called; he supposed that visible material points are subjected to certain invisible bonds which join them to other invisible points and that it is the effect of these invisible bonds that we attribute to forces.

But that is only a part of his ideas. Suppose a system formed of $n$ material points, visible or not; that will give in all $3n$ coordinates; let us regard them as the coordinates of a *single* point in space of $3n$ dimensions. This single point would be constrained to remain upon a surface (of any number of dimensions $< 3n$) in virtue of the bonds of which we have just spoken; to go on this surface from one point to another, it would always take the shortest way; this would be the single principle which would sum up all mechanics.

Whatever should be thought of this hypothesis, whether we be allured by its simplicity, or repelled by its artificial character, the simple fact that Hertz was able to conceive it, and to regard it as more convenient than our habitual hypotheses, suffices to prove that our ordinary ideas, and, in particular, the three dimensions of space, are in no wise imposed upon mechanics with an invincible force.

### 6. *Mind and Space*

Experience, therefore, has played only a single rôle, it has served as occasion. But this rôle was none the less very important; and I have thought it necessary to give it prominence. This rôle would have been useless if there existed an *a priori* form imposing itself upon our sensitivity, and which was space of three dimensions.

Does this form exist, or, if you choose, can we represent to ourselves space of more than three dimensions? And first what does this question mean? In the true sense of the word, it is clear that we can not represent to ourselves space of four, nor space of three, dimensions; we can not first represent them to ourselves empty, and no more can we represent to ourselves an object either in space of four, or in space of three, dimensions: (1) Because these spaces are both infinite and we can not represent to ourselves a figure *in* space, that is, the part *in* the whole, without representing the whole, and that is impossible, because it is infinite; (2) because these spaces are both mathematical continua, and we can represent to ourselves only the physical continuum; (3) because these spaces are both homogeneous, and the frames in which we enclose our sensations, being limited, can not be homogeneous.

Thus the question put can only be understood in one way; is it possible to imagine that, the results of the experiences related above having been different, we might have been led to attribute to space more than three dimensions; to imagine, for instance, that the sensation of accommodation might not be constantly in accord with the sensation of convergence of the eyes;

19

or indeed that the experiences of which we have spoken in § 2, and of which we express the result by saying ' that touch does not operate at a distance,' might have led us to an inverse conclusion.

And then yes evidently that is possible; from the moment one imagines an experience, one imagines just there by the two contrary results it may give. That is possible, but that is difficult, because we have to overcome a multitude of associations of ideas, which are the fruit of a long personal experience and of the still longer experience of the race. Is it these associations (or at least those of them that we have inherited from our ancestors), which constitute this *a priori* form of which it is said that we have pure intuition? Then I do not see why one should declare it refractory to analysis and should deny me the right of investigating its origin.

When it is said that our sensations are 'extended' only one thing can be meant, that is that they are always associated with the idea of certain muscular sensations, corresponding to the movements which enable us to reach the object which causes them, which enable us, in other words, to defend ourselves against it. And it is just because this association is useful for the defense of the organism, that it is so old in the history of the species and that it seems to us indestructible. Nevertheless, it is only an association and we can conceive that it may be broken; so that we may not say that sensation can not enter consciousness without entering in space, but that in fact it does not enter consciousness without entering in space, which means, without being entangled in this association.

No more can I understand one's saying that the idea of time is logically subsequent to space, since we can represent it to ourselves only under the form of a straight line; as well say that time is logically subsequent to the cultivation of the prairies, since it is usually represented armed with a scythe. That one can not represent to himself simultaneously the different parts of time, goes without saying, since the essential character of these parts is precisely not to be simultaneous. That does not mean that we have not the intuition of time. So far as that goes, no more should we have that of space, because neither can we represent it, in the proper sense of the word, for the reasons I have mentioned. What we represent to ourselves under the name of straight is a crude image which as ill resembles the geometric straight as it does time itself.

Why has it been said that every attempt to give a fourth dimension to space always carries this one back to one of the other three? It is easy to understand. Consider our muscular sensations and the 'series' they may form. In consequence of numerous experiences, the ideas of these series are associated together in a very complex woof, our series are *classed*. Allow me, for convenience of language, to express my thought in a way altogether crude and even inexact by saying that our series of muscular sensations are classed in three classes corresponding to the three dimensions of space. Of course this classification is much more complicated than that, but that will suffice to make my reasoning understood. If I wish to imagine a fourth dimension, I shall suppose another series of muscular sensations, making part of a fourth class. But as *all* my muscular sensations have already been classed in one of the three preexistent classes, I can only represent to myself a series belonging to one of these three classes, so that my fourth dimension is carried back to one of the other three.

What does that prove? This: that it would have been necessary first to destroy the old classification and replace it by a new one in which the series of muscular sensations should have been distributed into four classes. The difficulty would have disappeared.

It is presented sometimes under a more striking form. Suppose I am enclosed in a chamber between the six impassable boundaries formed by the four walls, the floor and the ceiling; it will be impossible for me to get out and to imagine my getting out. Pardon, can you not imagine that the door opens, or that two of these walls separate? But of course, you answer, one must suppose that these walls remain immovable. Yes, but it is evident that I have the right to move; and then the walls that we suppose absolutely at rest will be in motion with regard to me. Yes, but such a relative motion can not be arbitrary; when objects are at rest, their relative motion with regard to any axes is that of a rigid solid; now, the apparent motions that you imagine are not in conformity with the laws of motion of a rigid solid. Yes, but it is experience which has taught us the laws of motion of a rigid solid; nothing would prevent our *imagining* them different. To sum up, for me to imagine that I get out of my prison, I have only to imagine that the walls seem to open, when I move.

I believe, therefore, that if by space is understood a mathematical continuum of three dimensions, were it otherwise amorphous, it is the mind which constructs it, but it does not construct it out of nothing; it needs materials and models. These materials, like these models, preexist within it. But there is not a single model which is imposed upon it; it has *choice;* it may choose, for instance, between space of four and space of three dimensions. What then is the rôle of experience? It gives the indications following which the choice is made.

Another thing: whence does space get its quantitative character? It comes from the rôle which the series of muscular sensations play in its genesis. These are series which may *repeat themselves,* and it is from their repetition that number comes; it is because they can repeat themselves indefinitely that space is infinite. And finally we have seen, at the end of section 3, that it is also because of this that space is relative. So it is repetition which has given to space its essential characteristics; now, repetition supposes time; this is enough to tell that time is logically anterior to space.

### 7. *Rôle of the Semicircular Canals*

I have not hitherto spoken of the rôle of certain organs to which the physiologists attribute with reason a capital importance, I mean the semicircular canals. Numerous experiments have sufficiently shown that these canals are necessary to our sense of orientation; but the physiologists are not entirely in accord; two opposing theories have been proposed, that of Mach-Delage and that of M. de Cyon.

M. de Cyon is a physiologist who has made his name illustrious by important discoveries on the innervation of the heart; I can not, however, agree with his ideas on the question before us. Not being a physiologist, I hesitate to criticize the experiments he has directed against the adverse theory of Mach-Delage; it seems to me, however, that they are not convincing, because in many of them the *total* pressure was made to vary in one of the canals, while, physiologically, what varies is the *difference* between the pressures on the two extremities of the canal; in others the organs were subjected to profound lesions, which must alter their functions

Besides, this is not important; the experiments, if they were irreproachable, might be convincing against the old theory. They would not be convincing *for* the new theory. In fact, if I have rightly understood the theory, my explaining it will be enough for one to understand that it is impossible to conceive of an experiment confirming it.

The three pairs of canals would have as sole function to tell us that space has three dimensions. Japanese mice have only two pairs of canals; they believe, it would seem, that space has only two dimensions, and they manifest this opinion in the strangest way; they put themselves in a circle, and, so ordered, they spin rapidly around. The lampreys, having only one pair of canals, believe that space has only one dimension, but their manifestations are less turbulent.

It is evident that such a theory is inadmissible. The sense-organs are designed to tell us of *changes* which happen in the exterior world. We could not understand why the Creator should have given us organs destined to cry without cease: Remember that space has three dimensions, since the number of these three dimensions is not subject to change.

We must, therefore, come back to the theory of Mach-Delage. What the nerves of the canals can tell us is the difference of pressure on the two extremities of the same canal, and thereby: (1) the direction of the vertical with regard to three axes rigidly bound to the head; (2) the three components of the acceleration of translation of the center of gravity of the head; (3) the centrifugal forces developed by the rotation of the head; (4) the acceleration of the motion of rotation of the head.

It follows from the experiments of M. Delage that it is this last indication which is much the most important; doubtless because the nerves are less sensible to the difference of pressure itself than to the brusque variations of this difference. The first three indications may thus be neglected.

Knowing the acceleration of the motion of rotation of the head at each instant, we deduce from it, by an unconscious integration, the final orientation of the head, referred to a certain initial orientation taken as origin. The circular canals contribute, therefore, to inform us of the movements that we have executed, and that on the same ground as the muscular sensations. When, therefore, above we speak of the series $S$ or of the series $\Sigma$, we should say, not that these were series of muscular sensations alone, but that they were series at the same time of muscular sensations and of sensations due to the semicircular canals. Apart from this addition, we should have nothing to change in what precedes.

In the series $S$ and $\Sigma$, these sensations of the semicircular canals evidently hold a very important place. Yet alone they would not suffice, because they can tell us only of the movements of the head; they tell us nothing of the relative movements of the body or of the members in regard to the head. And more, it seems that they tell us only of the rotations of the head and not of the translations it may undergo.

# PART II

## THE PHYSICAL SCIENCES

### CHAPTER V

#### ANALYSIS AND PHYSICS

#### I

You have doubtless often been asked of what good is mathematics and whether these delicate constructions entirely mind-made are not artificial and born of our caprice.

Among those who put this question I should make a distinction; practical people ask of us only the means of money-making. These merit no reply; rather would it be proper to ask of them what is the good of accumulating so much wealth and whether, to get time to acquire it, we are to neglect art and science, which alone give us souls capable of enjoying it, 'and for life's sake to sacrifice all reasons for living.'

Besides, a science made solely in view of applications is impossible; truths are fecund only if bound together. If we devote ourselves solely to those truths whence we expect an immediate result, the intermediary links are wanting and there will no longer be a chain.

The men most disdainful of theory get from it, without suspecting it, their daily bread; deprived of this food, progress would quickly cease, and we should soon congeal into the immobility of old China.

But enough of uncompromising practicians! Besides these, there are those who are only interested in nature and who ask us if we can enable them to know it better.

To answer these, we have only to show them the two monuments already rough-hewn, Celestial Mechanics and Mathematical Physics.

They would doubtless concede that these structures are well worth the trouble they have cost us. But this is not enough. Mathematics has a triple aim. It must furnish an instrument for the study of nature. But that is not all: it has a philosophic aim and, I dare maintain, an esthetic aim. It must aid the philosopher to fathom the notions of number, of space, of time. And above all, its adepts find therein delights analogous to those given by painting and music. They admire the delicate harmony of numbers and forms; they marvel when a new discovery opens to them an unexpected perspective; and has not the joy they thus feel the esthetic character, even though the senses take no part therein? Only a privileged few are called to enjoy it fully, it is true, but is not this the case for all the noblest arts?

This is why I do not hesitate to say that mathematics deserves to be cultivated for its own sake, and the theories inapplicable to physics as well as the others. Even if the physical aim and the esthetic aim were not united, we ought not to sacrifice either.

But more: these two aims are inseparable and the best means of attaining one is to aim at the other, or at least never to lose sight of it. This is what I am about to try to demonstrate in setting forth the nature of the relations between the pure science and its applications.

The mathematician should not be for the physicist a mere purveyor of formulas; there should be between them a more intimate collaboration. Mathematical physics and pure analysis are not merely adjacent powers, maintaining good neighborly relations; they mutually interpenetrate and their spirit is the same. This will be better understood when I have shown what physics gets from mathematics and what mathematics, in return, borrows from physics.

## II

The physicist can not ask of the analyst to reveal to him a new truth; the latter could at most only aid him to foresee it. It is a long time since one still dreamt of forestalling experiment, or of constructing the entire world on certain premature hypotheses. Since all those constructions in which one yet took a naïve delight it is an age, to-day only their ruins remain.

All laws are therefore deduced from experiment; but to enunciate them, a special language is needful; ordinary language is too poor, it is besides too vague, to express relations so delicate, so rich, and so precise.

This therefore is one reason why the physicist can not do without mathematics; it furnishes him the only language he can speak. And a well-made language is no indifferent thing; not to go beyond physics, the unknown man who invented the word *heat* devoted many generations to error. Heat has been treated as a substance, simply because it was designated by a substantive, and it has been thought indestructible.

On the other hand, he who invented the word *electricity* had the unmerited good fortune to implicitly endow physics with a *new* law, that of the conservation of electricity, which, by a pure chance, has been found exact, at least until now.

Well, to continue the simile, the writers who embellish a language, who treat it as an object of art, make of it at the same time a more supple instrument, more apt for rendering shades of thought.

We understand, then, how the analyst, who pursues a purely esthetic aim, helps create, just by that, a language more fit to satisfy the physicist.

But this is not all: law springs from experiment, but not immediately. Experiment is individual, the law deduced from it is general; experiment is only approximate, the law is precise, or at least pretends to be. Experiment is made under conditions always complex, the enunciation of the law eliminates these complications. This is what is called 'correcting the systematic errors.'

In a word, to get the law from experiment, it is necessary to generalize; this is a necessity imposed upon the most circumspect observer. But how generalize? Every particular truth may evidently be extended in an infinity of ways. Among these thousand routes opening before us, it is necessary to make a choice, at least provisional; in this choice, what shall guide us?

It can only be analogy. But how vague is this word! Primitive man knew only crude analogies, those which strike the senses, those of colors or of sounds. He never would have dreamt of likening light to radiant heat.

What has taught us to know the true, profound analogies, those the eyes do not see but reason divines?

It is the mathematical spirit, which disdains matter to cling only to pure form. This it is which has taught us to give the same name to things differing only in material, to call by the same name, for instance, the multiplication of quaternions and that of whole numbers.

If quaternions, of which I have just spoken, had not been so promptly utilized by the English physicists, many persons would doubtless see in them only a useless fancy, and yet, in teaching us to liken what appearances separate, they would have already rendered us more apt to penetrate the secrets of nature.

Such are the services the physicist should expect of analysis; but for this science to be able to render them, it must be cultivated in the broadest fashion without immediate expectation of utility—the mathematician must have worked as artist.

What we ask of him is to help us to see, to discern our way in the labyrinth which opens before us. Now, he sees best who stands highest. Examples abound, and I limit myself to the most striking.

The first will show us how to change the language suffices to reveal generalizations not before suspected.

When Newton's law has been substituted for Kepler's we still know only elliptic motion. Now, in so far as concerns this motion, the two laws differ only in form; we pass from one to the other by a simple differentiation. And yet from Newton's law may be deduced by an immediate generalization all the effects of perturbations and the whole of celestial mechanics. If, on the other hand, Kepler's enunciation had been retained, no one would ever have regarded the orbits of the perturbed plants, those complicated curves of which no one has ever written the equation, as the natural generalizations of the ellipse. The progress of observations would only have served to create belief in chaos.

The second example is equally deserving of consideration.

When Maxwell began his work, the laws of electro-dynamics admitted up to his time accounted for all the known facts. It was not a new experiment which came to invalidate them. But in looking at them under a new bias, Maxwell saw that the equations became more symmetrical when a term was added, and besides, this term was too small to produce effects appreciable with the old methods.

You know that Maxwell's *a priori* views awaited for twenty years an experimental confirmation; or, if you prefer, Maxwell was twenty years ahead of experiment. How was this triumph obtained?

It was because Maxwell was profoundly steeped in the sense of mathematical symmetry; would he have been so, if others before him had not studied this symmetry for its own beauty?

It was because Maxwell was accustomed to 'think in vectors,' and yet it was through the theory of imaginaries (neomonics) that vectors were introduced into analysis. And those who invented imaginaries hardly suspected the advantage which would be obtained from them for the study of the real world, of this the name given them is proof sufficient.

In a word, Maxwell was perhaps not an able analyst, but this ability would have been for him only a useless and bothersome baggage. On the other hand, he had in the highest degree the intimate sense of mathematical analogies. Therefore it is that he made good mathematical physics.

Maxwell's example teaches us still another thing.

How should the equations of mathematical physics be treated? Should we simply deduce all the consequences, and regard them as intangible realities? Far from it; what they should teach us above all is what can and what should be changed. It is thus that we get from them something useful.

The third example goes to show us how we may perceive mathematical analogies between phenomena which have physically no relation either apparent or real, so that the laws of one of these phenomena aid us to divine those of the other.

The very same equation, that of Laplace, is met in the theory of Newtonian attraction, in that of the motion of liquids, in that of the electric potential, in that of magnetism, in that of the propagation of heat and in still many others. What is the result? These theories seem images copied one from the other; they are mutually illuminating, borrowing their language from each other; ask electricians if they do not felicitate themselves on having invented the phrase flow of force, suggested by hydrodynamics and the theory of heat.

Thus mathematical analogies not only may make us foresee physical analogies, but besides do not not cease to be useful when these latter fail.

To sum up, the aim of mathematical physics is not only to facilitate for the physicist the numerical calculation of certain constants or the integration of certain differential equations. It is besides, it is above all, to reveal to him the hidden harmony of things in making him see them in a new way.

Of all the parts of analysis, the most elevated, the purest, so to speak, will be the most fruitful in the hands of those who know how to use them.

### III

Let us now see what analysis owes to physics.

It would be necessary to have completely forgotten the history of science not to remember that the desire to understand nature has had on the development of mathematics the most constant and happiest influence.

In the first place the physicist sets us problems whose solution he expects of us. But in proposing them to us, he has largely paid us in advance for the service we shall render him, if we solve them.

If I may be allowed to continue my comparison with the fine arts, the pure mathematician who should forget the existence of the exterior world would be like a painter who knew how to harmoniously combine colors and forms, but who lacked models. His creative power would soon be exhausted.

The combinations which numbers and symbols may form are an infinite multitude. In this multitude how shall we choose those which are worthy to fix our attention? Shall we let ourselves be guided solely by our caprice? This caprice, which itself would besides soon tire, would doubtless carry us very far apart and we should quickly cease to understand each other.

But this is only the smaller side of the question. Physics will doubtless prevent our straying, but it will also preserve us from a danger much more formidable; it will prevent our ceaselessly going around in the same circle.

History proves that physics has not only forced us to choose among problems which came in a crowd; it has imposed upon us such as we should without it never have dreamed of. However varied may be the imagination of man, nature is still a thousand times richer. To follow her we must take ways we have neglected, and these paths lead us often to summits whence we discover new countries. What could be more useful!

It is with mathematical symbols as with physical realities; it is in comparing the different aspects of things that we are able to comprehend their inner harmony, which alone is beautiful and consequently worthy of our efforts.

The first example I shall cite is so old we are tempted to forget it; it is nevertheless the most important of all.

The sole natural object of mathematical thought is the whole number. It is the external world which has imposed the continuum upon us, which we doubtless have invented, but which it has forced us to invent. Without it there would be no infinitesimal analysis; all mathematical science would reduce itself to arithmetic or to the theory of substitutions.

On the contrary, we have devoted to the study of the continuum almost all our time and all our strength. Who will regret it; who will think that this time and this strength have been wasted? Analysis unfolds before us infinite perspectives that arithmetic never suspects; it shows us at a glance a majestic assemblage whose array is simple and symmetric; on the contrary, in the theory of numbers, where reigns the unforeseen, the view is, so to speak, arrested at every step.

Doubtless it will be said that outside of the whole number there is no rigor, and consequently no mathematical truth; that the whole number hides everywhere, and that we must strive to render transparent the screens which cloak it, even if to do so we must resign ourselves to interminable repetitions. Let us not be such purists and let us be grateful to the continuum, which, if *all* springs from the whole number, was alone capable of making *so much* proceed therefrom.

Need I also recall that M. Hermite obtained a surprising advantage from the introduction of continuous variables into the theory of numbers? Thus the whole number's own domain is itself invaded, and this invasion has established order where disorder reigned.

See what we owe to the continuum and consequently to physical nature.

Fourier's series is a precious instrument of which analysis makes continual use, it is by this means that it has been able to represent discontinuous functions; Fourier invented it to solve a problem of physics relative to the propagation of heat. If this problem had not come up naturally, we should never have dared to give discontinuity its rights; we should still long have regarded continuous functions as the only true functions.

The notion of function has been thereby considerably extended and has received from some logician-analysts an unforeseen development. These analysts have thus adventured into regions where reigns the purest abstraction and have gone as far away as possible from the real world. Yet it is a problem of physics which has furnished them the occasion.

After Fourier's series, other analogous series have entered the domain of analysis; they have entered by the same door; they have been imagined in view of applications.

The theory of partial differential equations of the second order has an analogous history. It has been developed chiefly by and for physics. But it may take many forms, because such an equation does not suffice to determine the unknown function, it is necessary to adjoin to it complementary conditions which are called conditions at the limits; whence many different problems.

If the analysts had abandoned themselves to their natural tendencies, they would never have known but one, that which Madame Kovalevski has treated in her celebrated memoir. But there are a multitude of others which they would have ignored. Each of the theories of physics, that of electricity, that of heat, presents us these equations under a new aspect. It may, therefore, be said that without these theories we should not know partial differential equations.

It is needless to multiply examples. I have given enough to be able to conclude: when physicists ask of us the solution of a problem, it is not a duty-service they impose upon us, it is on the contrary we who owe them thanks.

## IV

But this is not all; physics not only gives us the occasion to solve problems; it aids us to find the means thereto, and that in two ways. It makes us foresee the solution; it suggests arguments to us.

I have spoken above of Laplace's equation which is met in a multitude of diverse physical theories. It is found again in geometry, in the theory of conformal representation and in pure analysis, in that of imaginaries.

In this way, in the study of functions of complex variables, the analyst, alongside of the geometric image, which is his usual instrument, finds many physical images which he may make use of with the same success. Thanks to these images, he can see at a glance what pure deduction would show him only successively. He masses thus the separate elements of the solution, and by a sort of intuition divines before being able to demonstrate.

To divine before demonstrating! Need I recall that thus have been made all the important discoveries? How many are the truths that physical analogies permit us to present and that we are not in condition to establish by rigorous reasoning!

For example, mathematical physics introduces a great number of developments in series. No one doubts that these developments converge; but the mathematical certitude is lacking. These are so many conquests assured for the investigators who shall come after us.

On the other hand, physics furnishes us not alone solutions; it furnishes us besides, in a certain measure, arguments. It will suffice to recall how Felix Klein, in a question relative to Riemann surfaces, has had recourse to the properties of electric currents.

It is true, the arguments of this species are not rigorous, in the sense the analyst attaches to this word. And here a question arises: How can a demonstration not sufficiently rigorous for the analyst suffice for the physicist? It seems there can not be two rigors, that rigor is or is not, and that, where it is not there can not be deduction.

This apparent paradox will be better understood by recalling under what conditions number is applied to natural phenomena. Whence come in general the difficulties encountered in seeking rigor? We strike them almost always in seeking to establish that some quantity tends to some limit, or that some function is continuous, or that it has a derivative.

Now the numbers the physicist measures by experiment are never known except approximately; and besides, any function always differs as little as you choose from a discontinuous function, and at the same time it differs as little as you choose from a continuous function. The physicist may, therefore, at will suppose that the function studied is continuous, or that it is discontinuous; that it has or has not a derivative; and may do so without fear of ever being contradicted, either by present experience or by any future experiment. We see that with such liberty he makes sport of difficulties which stop the analyst. He may always reason as if all the functions which occur in his calculations were entire polynomials.

Thus the sketch which suffices for physics is not the deduction which analysis requires. It does not follow thence that one can not aid in finding the other. So many physical sketches have already been transformed into rigorous demonstrations that to-day this transformation is easy. There would be plenty of examples did I not fear in citing them to tire the reader.

I hope I have said enough to show that pure analysis and mathematical physics may serve one another without making any sacrifice one to the other, and that each of these two sciences should rejoice in all which elevates its associate.

# CHAPTER VI

## ASTRONOMY

GOVERNMENTS and parliaments must find that astronomy is one of the sciences which cost most dear: the least instrument costs hundreds of thousands of dollars, the least observatory costs millions; each eclipse carries with it supplementary appropriations. And all that for stars which are so far away, which are complete strangers to our electoral contests, and in all probability will never take any part in them. It must be that our politicians have retained a remnant of idealism, a vague instinct for what is grand; truly, I think they have been calumniated; they should be encouraged and shown that this instinct does not deceive them, that they are not dupes of that idealism.

We might indeed speak to them of navigation, of which no one can underestimate the importance, and which has need of astronomy. But this would be to take the question by its smaller side.

Astronomy is useful because it raises us above ourselves; it is useful because it is grand; that is what we should say. It shows us how small is man's body, how great his mind, since his intelligence can embrace the whole of this dazzling immensity, where his body is only an obscure point, and enjoy its silent harmony. Thus we attain the consciousness of our power, and this is something which can not cost too dear, since this consciousness makes us mightier.

But what I should wish before all to show is, to what point astronomy has facilitated the work of the other sciences, more directly useful, since it has given us a soul capable of comprehending nature.

Think how diminished humanity would be if, under heavens constantly overclouded, as Jupiter's must be, it had forever remained ignorant of the stars. Do you think that in such a world we should be what we are? I know well that under this somber vault we should have been deprived of the light of the

sun, necessary to organisms like those which inhabit the earth.
But if you please, we shall assume that these clouds are phos-
phorescent and emit a soft and constant light. Since we are
making hypotheses, another will cost no more. Well! I repeat
my question: Do you think that in such a world we should be
what we are?

The stars send us not only that visible and gross light which
strikes our bodily eyes, but from them also comes to us a light far
more subtle, which illuminates our minds and whose  effects I
shall try to show you. You know what man was on the earth
some thousands of years ago, and what he is to-day. Isolated
amidst a nature where everything was a mystery to him, terrified
at each unexpected manifestation of incomprehensible forces, he
was incapable of seeing in the conduct of the universe anything
but caprice; he attributed all phenomena to the action of a multitude of little genii, fantastic and exacting, and to act on the
world he sought to conciliate them by means analogous to those
employed to gain the good graces of a minister or a deputy.
Even his failures did not enlighten him, any more than to-day
a beggar refused is discouraged to the point of ceasing to beg.

To-day we no longer beg of nature; we command her, because
we have discovered certain of her secrets and shall discover
others each day. We command her in the name of laws she can
not challenge, because they are hers; these laws we do not madly
ask her to change, we are the first to submit to them. Nature
can only be governed by obeying her.

What a change must our souls have undergone to pass from the
one state to the other! Does any one believe that, without the
lessons of the stars, under the heavens perpetually overclouded
that I have just supposed, they would have changed so quickly?
Would the metamorphosis have been possible, or at least would it
not have been much slower?

And first of all, astronomy it is which taught that there are
laws. The Chaldeans, who were the first to observe the heavens
with some attention, saw that this multitude of luminous points
is not a confused crowd wandering at random, but rather a disciplined army. Doubtless the rules of this discipline escaped them,
but the harmonious spectacle of the starry night sufficed to give

them the impression of regularity, and that was in itself already a great thing. Besides, these rules were discerned by Hipparchus, Ptolemy, Copernicus, Kepler, one after another, and finally, it is needless to recall that Newton it was who enunciated the oldest, the most precise, the most simple, the most general of all natural laws.

And then, taught by this example, we have seen our little terrestrial world better and, under the apparent disorder, there also we have found again the harmony that the study of the heavens had revealed to us. It also is regular, it also obeys immutable laws, but they are more complicated, in apparent conflict one with another, and an eye untrained by other sights would have seen there only chaos and the reign of chance or caprice. If we had not known the stars, some bold spirits might perhaps have sought to foresee physical phenomena; but their failures would have been frequent, and they would have excited only the derision of the vulgar; do we not see, that even in our day the meteorologists sometimes deceive themselves, and that certain persons are inclined to laugh at them.

How often would the physicists, disheartened by so many checks, have fallen into discouragement, if they had not had, to sustain their confidence, the brilliant example of the success of the astronomers! This success showed them that nature obeys laws; it only remained to know what laws; for that they only needed patience, and they had the right to demand that the sceptics should give them credit.

This is not all: astronomy has not only taught us that there are laws, but that from these laws there is no escape, that with them there is no possible compromise. How much time should we have needed to comprehend that fact, if we had known only the terrestrial world, where each elemental force would always seem to us in conflict with other forces? Astronomy has taught us that the laws are infinitely precise, and that if those we enunciate are approximative, it is because we do not know them well. Aristotle, the most scientific mind of antiquity, still accorded a part to accident, to chance, and seemed to think that the laws of nature, at least here below, determine only the large features of phenomena. How much has the ever-increasing precision of astronomical predictions contributed to correct such an error, which would have rendered nature unintelligible!

But are these laws not local, varying in different places, like those which men make; does not that which is truth in one corner of the universe, on our globe, for instance, or in our little solar system, become error a little farther away? And then could it not be asked whether laws depending on space do not also depend upon time, whether they are not simple habitudes, transitory, therefore, and ephemeral? Again it is astronomy that answers this question. Consider the double stars; all describe conics; thus, as far as the telescope carries, it does not reach the limits of the domain which obeys Newton's law.

Even the simplicity of this law is a lesson for us; how many complicated phenomena are contained in the two lines of its enunciation; persons who do not understand celestial mechanics may form some idea of it at least from the size of the treatises devoted to this science; and then it may be hoped that the complication of physical phenomena likewise hides from us some simple cause still unknown.

It is therefore astronomy which has shown us what are the general characteristics of natural laws; but among these characteristics there is one, the most subtle and the most important of all, which I shall ask leave to stress.

How was the order of the universe understood by the ancients; for instance, by Pythagoras, Plato or Aristotle? It was either an immutable type fixed once for all, or an ideal to which the world sought to approach. Kepler himself still thought thus when, for instance, he sought whether the distances of the planets from the sun had not some relation to the five regular polyhedrons. This idea contained nothing absurd, but it was sterile, since nature is not so made. Newton has shown us that a law is only a necessary relation between the present state of the world and its immediately subsequent state. All the other laws since discovered are nothing else; they are in sum, differential equations; but it is astronomy which furnished the first model for them, without which we should doubtless long have erred.

Astronomy has also taught us to set at naught appearances.

The day Copernicus proved that what was thought the most stable was in motion, that what was thought moving was fixed, he showed us how deceptive could be the infantile reasonings which spring directly from the immediate data of our senses. True, his ideas did not easily triumph, but since this triumph there is no longer a prejudice so inveterate that we can not shake it off. How can we estimate the value of the new weapon thus won?

The ancients thought everything was made for man, and this illusion must be very tenacious, since it must ever be combated. Yet it is necessary to divest oneself of it; or else one will be only an eternal myope, incapable of seeing the truth. To comprehend nature one must be able to get out of self, so to speak, and to contemplate her from many different points of view; otherwise we never shall know more than one side. Now, to get out of self is what he who refers everything to himself can not do. Who delivered us from this illusion? It was those who showed us that the earth is only one of the smallest planets of the solar system, and that the solar system itself is only an imperceptible point in the infinite spaces of the stellar universe.

At the same time astronomy taught us not to be afraid of big numbers. This was needful, not only for knowing the heavens, but to know the earth itself; and was not so easy as it seems to us to-day. Let us try to go back and picture to ourselves what a Greek would have thought if told that red light vibrates four hundred millions of millions of times per second. Without any doubt, such an assertion would have appeared to him pure madness, and he never would have lowered himself to test it. To-day a hypothesis will no longer appear absurd to us because it obliges us to imagine objects much larger or smaller than those our senses are capable of showing us, and we no longer comprehend those scruples which arrested our predecessors and prevented them from discovering certain truths simply because they were afraid of them. But why? It is because we have seen the heavens enlarging and enlarging without cease; because we know that the sun is 150 millions of kilometers from the earth and that the distances of the nearest stars are hundreds of thousands of times greater yet. Habituated to the contemplation of the infinitely great, we have become apt to comprehend the infinitely small. Thanks to the education it has received, our imagination, like the eagle's eye that the sun does not dazzle, can look truth in the face.

Was I wrong in saying that it is astronomy which has made us a soul capable of comprehending nature; that under heavens always overcast and starless, the earth itself would have been for us eternally unintelligible; that we should there have seen only caprice and disorder; and that, not knowing the world, we should never have been able to subdue it? What science could have been more useful? And in thus speaking I put myself at the point of view of those who only value practical applications. Certainly, this point of view is not mine; as for me, on the contrary, if I admire the conquests of industry, it is above all because if they free us from material cares, they will one day give to all the leisure to contemplate nature. I do not say: Science is useful, because it teaches us to construct machines. I say: Machines are useful, because in working for us, they will some day leave us more time to make science. But finally it is worth remarking that between the two points of view there is no antagonism, and that man having pursued a disinterested aim, all else has been added unto him.

Auguste Comte has said somewhere, that it would be idle to seek to know the composition of the sun, since this knowledge would be of no use to sociology. How could he be so short-sighted? Have we not just seen that it is by astronomy that, to speak his language, humanity has passed from the theological to the positive state? He found an explanation for that because it had happened. But how has he not understood that what remained to do was not less considerable and would be not less profitable? Physical astronomy, which he seems to condemn, has already begun to bear fruit, and it will give us much more, for it only dates from yesterday.

First was discovered the nature of the sun, what the founder of positivism wished to deny us, and there bodies were found which exist on the earth, but had here remained undiscovered; for example, helium, that gas almost as light as hydrogen. That already contradicted Comte. But to the spectroscope we owe a lesson precious in a quite different way; in the most distant stars, it shows us the same substances. It might have been asked whether the terrestrial elements were not due to some chance which had brought together more tenuous atoms to construct of them the more complex edifice that the chemists call atom; whether, in other regions of the universe, other fortuitous meetings had not engendered edifices entirely different. Now we know that this is not so, that the laws of our chemistry are the general laws of nature, and that they owe nothing to the chance which caused us to be born on the earth.

But, it will be said, astronomy has given to the other sciences all it can give them, and now that the heavens have procured for us the instruments which enable us to study terrestrial nature, they could without danger veil themselves forever. After what we have just said, is there still need to answer this objection? One could have reasoned the same in Ptolemy's time; then also men thought they knew everything, and they still had almost everything to learn.

The stars are majestic laboratories, gigantic crucibles, such as no chemist could dream. There reign temperatures impossible for us to realize. Their only defect is being a little far away; but the telescope will soon bring them near to us, and then we shall see how matter acts there. What good fortune for the physicist and the chemist!

Matter will there exhibit itself to us under a thousand different states, from those rarefied gases which seem to form the nebulæ and which are luminous with I know not what glimmering of mysterious origin, even to the incandescent stars and to the planets so near and yet so different.

Perchance even, the stars will some day teach us something about life; that seems an insensate dream and I do not at all see how it can be realized; but, a hundred years ago, would not the chemistry of the stars have also appeared a mad dream?

But limiting our views to horizons less distant, there still will remain to us promises less contingent and yet sufficiently seductive. If the past has given us much, we may rest assured that the future will give us still more.

In sum, it is incredible how useful belief in astrology has been to humanity. If Kepler and Tycho Brahe made a living, it was because they sold to naïve kings predictions founded on the conjunctions of the stars. If these princes had not been so credulous, we should perhaps still believe that nature obeys caprice, and we should still wallow in ignorance.

## CHAPTER VII

### The History of Mathematical Physics

*The Past and the Future of Physics.*—What is the present state of mathematical physics? What are the problems it is led to set itself? What is its future? Is its orientation about to be modified?

Ten years hence will the aim and the methods of this science appear to our immediate successors in the same light as to ourselves; or, on the contrary, are we about to witness a profound transformation? Such are the questions we are forced to raise in entering to-day upon our investigation.

If it is easy to propound them: to answer is difficult. If we felt tempted to risk a prediction, we should easily resist this temptation, by thinking of all the stupidities the most eminent savants of a hundred years ago would have uttered, if some one had asked them what the science of the nineteenth century would be. They would have thought themselves bold in their predictions, and after the event, how very timid we should have found them. Do not, therefore, expect of me any prophecy.

But if, like all prudent physicians, I shun giving a prognosis, yet I can not dispense with a little diagnostic; well, yes, there are indications of a serious crisis, as if we might expect an approaching transformation. Still, be not too anxious: we are sure the patient will not die of it, and we may even hope that this crisis will be salutary, for the history of the past seems to guarantee us this. This crisis, in fact, is not the first, and to understand it, it is important to recall those which have preceded. Pardon then a brief historical sketch.

*The Physics of Central Forces.*—Mathematical physics, as we know, was born of celestial mechanics, which gave birth to it at the end of the eighteenth century, at the moment when it itself attained its complete development. During its first years especially, the infant strikingly resembled its mother.

The astronomic universe is formed of masses, very great, no doubt, but separated by intervals so immense that they appear to us only as material points. These points attract each other inversely as the square of the distance, and this attraction is the sole force which influences their movements. But if our senses were sufficiently keen to show us all the details of the bodies which the physicist studies, the spectacle thus disclosed would scarcely differ from the one the astronomer contemplates. There also we should see material points, separated from one another by intervals, enormous in comparison with their dimensions, and describing orbits according to regular laws. These infinitesimal stars are the atoms. Like the stars proper, they attract or repel each other, and this attraction or this repulsion, following the straight line which joins them, depends only on the distance. The law according to which this force varies as function of the distance is perhaps not the law of Newton, but it is an analogous law; in place of the exponent —2, we have probably a different exponent, and it is from this change of exponent that arises all the diversity of physical phenomena, the variety of qualities and of sensations, all the world, colored and sonorous, which surrounds us; in a word, all nature.

Such is the primitive conception in all its purity. It only remains to seek in the different cases what value should be given to this exponent in order to explain all the facts. It is on this model that Laplace, for example, constructed his beautiful theory of capillarity; he regards it only as a particular case of attraction, or, as he says, of universal gravitation, and no one is astonished to find it in the middle of one of the five volumes of the 'Mécanique céleste.' More recently Briot believes he penetrated the final secret of optics in demonstrating that the atoms of ether attract each other in the inverse ratio of the sixth power of the distance; and Maxwell himself, does he not say somewhere that the atoms of gases repel each other in the inverse ratio of the fifth power of the distance? We have the exponent — 6, or — 5, in place of the exponent — 2, but it is always an exponent.

Among the theories of this epoch, one alone is an exception, that of Fourier; in it are indeed atoms acting at a distance one upon the other; they mutually transmit heat, but they do not attract, they never budge. From this point of view, Fourier's theory must have appeared to the eyes of his contemporaries, to those of Fourier himself, as imperfect and provisional.

This conception was not without grandeur; it was seductive, and many among us have not finally renounced it; they know that one will attain the ultimate elements of things only by patiently disentangling the complicated skein that our senses give us; that it is necessary to advance step by step, neglecting no intermediary; that our fathers were wrong in wishing to skip stations; but they believe that when one shall have arrived at these ultimate elements, there again will be found the majestic simplicity of celestial mechanics.

Neither has this conception been useless; it has rendered us an inestimable service, since it has contributed to make precise the fundamental notion of the physical law.

I will explain myself; how did the ancients understand law? It was for them an internal harmony, static, so to say, and immutable; or else it was like a model that nature tried to imitate. For us a law is something quite different; it is a constant relation between the phenomenon of to-day and that of to-morrow; in a word, it is a differential equation.

Behold the ideal form of physical law; well, it is Newton's law which first clothed it forth. If then one has acclimated this form in physics, it is precisely by copying as far as possible this law of Newton, that is by imitating celestial mechanics. This is, moreover, the idea I have tried to bring out in Chapter VI.

*The Physics of the Principles.*—Nevertheless, a day arrived when the conception of central forces no longer appeared sufficient, and this is the first of those crises of which I just now spoke.

What was done then? The attempt to penetrate into the detail of the structure of the universe, to isolate the pieces of this vast mechanism, to analyze one by one the forces which put them in motion, was abandoned, and we were content to take as guides certain general principles the express object of which is to spare us this minute study. How so? Suppose we have before us any machine; the initial wheel work and the final wheel work alone are visible, but the transmission, the intermediary machinery by which the movement is communicated from one to the other, is hidden in the interior and escapes our view; we do not know whether the communication is made by gearing or by belts, by connecting-rods or by other contrivances. Do we say that it is impossible for us to understand anything about this machine so long as we are not permitted to take it to pieces? You know well we do not, and that the principle of the conservation of energy suffices to determine for us the most interesting point. We easily ascertain that the final wheel turns ten times less quickly than the initial wheel, since these two wheels are visible; we are able thence to conclude that a couple applied to the one will be balanced by a couple ten times greater applied to the other. For that there is no need to penetrate the mechanism of this equilibrium and to know how the forces compensate each other in the interior of the machine; it suffices to be assured that this compensation can not fail to occur.

Well, in regard to the universe, the principle of the conservation of energy is able to render us the same service. The universe is also a machine, much more complicated than all those of industry, of which almost all the parts are profoundly hidden from us; but in observing the motion of those that we can see, we are able, by the aid of this principle, to draw conclusions which remain true whatever may be the details of the invisible mechanism which animates them.

The principle of the conservation of energy, or Mayer's principle, is certainly the most important, but it is not the only one; there are others from which we can derive the same advantage. These are:

Carnot's principle, or the principle of the degradation of energy.

Newton's principle, or the principle of the equality of action and reaction.

The principle of relativity, according to which the laws of physical phenomena must be the same for a stationary observer as for an observer carried along in a uniform motion of translation; so that we have not and can not have any means of discerning whether or not we are carried along in such a motion.

The principle of the conservation of mass, or Lavoisier's principle.

I will add the principle of least action.

The application of these five or six general principles to the different physical phenomena is sufficient for our learning of them all that we could reasonably hope to know of them. The most remarkable example of this new mathematical physics is, beyond question, Maxwell's electromagnetic theory of light.

We know nothing as to what the ether is, how its molecules are disposed, whether they attract or repel each other; but we know that this medium transmits at the same time the optical perturbations and the electrical perturbations; we know that this transmission must take place in conformity with the general principles of mechanics, and that suffices us for the establishment of the equations of the electromagnetic field.

These principles are results of experiments boldly generalized; but they seem to derive from their very generality a high degree of certainty. In fact, the more general they are, the more frequent are the opportunities to check them, and the verifications multiplying, taking the most varied, the most unexpected forms, end by no longer leaving place for doubt.

*Utility of the Old Physics.*—Such is the second phase of the history of mathematical physics and we have not yet emerged from it. Shall we say that the first has been useless? that during fifty years science went the wrong way, and that there is nothing left but to forget so many accumulated efforts that a vicious conception condemned in advance to failure? Not the least in the world. Do you think the second phase could have come into existence without the first? The hypothesis of central forces contained all the principles; it involved them as necessary consequences; it involved both the conservation of energy and that of masses, and the equality of action and reaction, and the law of least action, which appeared, it is true, not as experimental truths, but as theorems; the enunciation of which had at the same time something more precise and less general than under their present form.

It is the mathematical physics of our fathers which has familiarized us little by little with these various principles; which has habituated us to recognize them under the different vestments in which they disguise themselves. They have been compared with the data of experience, it has been seen how it was necessary to modify their enunciation to adapt them to these data; thereby they have been extended and consolidated. Thus they came to be regarded as experimental truths; the conception of central forces became then a useless support, or rather an embarrassment, since it made the principles partake of its hypothetical character.

The frames then have not broken, because they are elastic; but they have enlarged; our fathers, who established them, did not labor in vain, and we recognize in the science of to-day the general traits of the sketch which they traced.

# CHAPTER VIII

## The Present Crisis of Mathematical Physics

*The New Crisis.*—Are we now about to enter upon a third period? Are we on the eve of a second crisis? These principles on which we have built all, are they about to crumble away in their turn? This has been for some time a pertinent question.

When I speak thus, you no doubt think of radium, that grand revolutionist of the present time, and in fact I shall come back to it presently; but there is something else. It is not alone the conservation of energy which is in question; all the other principles are equally in danger, as we shall see in passing them successively in review.

*Carnot's Principle.*—Let us commence with the principle of Carnot. This is the only one which does not present itself as an immediate consequence of the hypothesis of central forces; more than that, it seems, if not to directly contradict that hypothesis, at least not to be reconciled with it without a certain effort. If physical phenomena were due exclusively to the movements of atoms whose mutual attraction depended only on the distance, it seems that all these phenomena should be reversible; if all the initial velocities were reversed, these atoms, always subjected to the same forces, ought to go over their trajectories in the contrary sense, just as the earth would describe in the retrograde sense this same elliptic orbit which it describes in the direct sense, if the initial conditions of its motion had been reversed. On this account, if a physical phenomenon is possible, the inverse phenomenon should be equally so, and one should be able to reascend the course of time. Now, it is not so in nature, and this is precisely what the principle of Carnot teaches us; heat can pass from the warm body to the cold body; it is impossible afterward to make it take the inverse route and to reestablish differences of temperature which have been effaced. Motion can be wholly dissipated and transformed into heat by friction; the contrary transformation can never be made except partially.

We have striven to reconcile this apparent contradiction. If the world tends toward uniformity, this is not because its ultimate parts, at first unlike, tend to become less and less different; it is because, shifting at random, they end by blending. For an eye which should distinguish all the elements, the variety would remain always as great; each grain of this dust preserves its originality and does not model itself on its neighbors; but as the blend becomes more and more intimate, our gross senses perceive only the uniformity. This is why for example, temperatures tend to a level, without the possibility of going backwards.

A drop of wine falls into a glass of water; whatever may be the law of the internal motion of the liquid, we shall soon see it colored of a uniform rosy tint, and however much from this moment one may shake it afterwards, the wine and the water do not seem capable of again separating. Here we have the type of the irreversible physical phenomenon: to hide a grain of barley in a heap of wheat, this is easy; afterwards to find it again and get it out, this is practically impossible. All this Maxwell and Boltzmann have explained; but the one who has seen it most clearly, in a book too little read because it is a little difficult to read, is Gibbs, in his 'Elementary Principles of Statistical Mechanics.'

For those who take this point of view, Carnot's principle is only an imperfect principle, a sort of concession to the infirmity of our senses; it is because our eyes are too gross that we do not distinguish the elements of the blend; it is because our hands are too gross that we can not force them to separate; the imaginary demon of Maxwell, who is able to sort the molecules one by one, could well constrain the world to return backward. Can it return of itself? That is not impossible; that is only infinitely improbable. The chances are that we should wait a long time for the concourse of circumstances which would permit a retrogradation; but sooner or later they will occur, after years whose number it would take millions of figures to write. These reservations, however, all remained theoretic; they were not very disquieting, and Carnot's principle retained all its practical value. But here the scene changes. The biologist, armed with his microscope, long ago noticed in his preparations irregular movements of little particles in suspension; this is the Brownian movement. He first thought this was a vital phenomenon, but soon he saw that the inanimate bodies danced with no less ardor than the others; then he turned the matter over to the physicists. Unhappily, the physicists remained long uninterested in this question; one concentrates the light to illuminate the microscopic preparation, thought they; with light goes heat; thence inequalities of temperature and in the liquid interior currents which produce the movements referred to. It occurred to M. Gouy to look more closely, and he saw, or thought he saw, that this explanation is untenable, that the movements become brisker as the particles are smaller, but that they are not influenced by the mode of illumination. If then these movements never cease, or rather are reborn without cease, without borrowing anything from an external source of energy, what ought we to believe? To be sure, we should not on this account renounce our belief in the conservation of energy, but we see under our eyes now motion transformed into heat by friction, now inversely heat changed into motion, and that without loss since the movement lasts forever. This is the contrary of Carnot's principle. If this be so, to see the world return backward, we no longer have need of the infinitely keen eye of Maxwell's demon; our microscope suffices. Bodies too large, those, for example, which are a tenth of a millimeter, are hit from all sides by moving atoms, but they do not budge, because these shocks are very numerous and the law of chance makes them compensate each other; but the smaller particles receive too few shocks for this compensation to take place with certainty and are incessantly knocked about. And behold already one of our principles in peril.

*The Principle of Relativity.*—Let us pass to the principle of relativity: this not only is confirmed by daily experience, not only is it a necessary consequence of the hypothesis of central forces, but it is irresistibly imposed upon our good sense, and yet it also is assailed. Consider two electrified bodies; though they seem to us at rest, they are both carried along by the motion of the earth; an electric charge in motion, Rowland has taught us, is equivalent to a current; these two charged bodies are, therefore, equivalent to two parallel currents of the same

21

sense and these two currents should attract each other. In measuring this attraction, we shall measure the velocity of the earth; not its velocity in relation to the sun or the fixed stars, but its absolute velocity.

I well know what will be said: It is not its absolute velocity that is measured, it is its velocity in relation to the ether. How unsatisfactory that is! Is it not evident that from the principle so understood we could no longer infer anything? It could no longer tell us anything just because it would no longer fear any contradiction. If we succeed in measuring anything, we shall always be free to say that this is not the absolute velocity, and if it is not the velocity in relation to the ether, it might always be the velocity in relation to some new unknown fluid with which we might fill space.

Indeed, experiment has taken upon itself to ruin this interpretation of the principle of relativity; all attempts to measure the velocity of the earth in relation to the ether have led to negative results. This time experimental physics has been more faithful to the principle than mathematical physics; the theorists, to put in accord their other general views, would not have spared it; but experiment has been stubborn in confirming it. The means have been varied; finally Michelson pushed precision to its last limits; nothing came of it. It is precisely to explain this obstinacy that the mathematicians are forced to-day to employ all their ingenuity.

Their task was not easy, and if Lorentz has got through it, it is only by accumulating hypotheses.

The most ingenious idea was that of local time. Imagine two observers who wish to adjust their timepieces by optical signals; they exchange signals, but as they know that the transmission of light is not instantaneous, they are careful to cross them. When station B perceives the signal from station A, its clock should not mark the same hour as that of station A at the moment of sending the signal, but this hour augmented by a constant representing the duration of the transmission. Suppose, for example, that station A sends its signal when its clock marks the hour $O$, and that station B perceives it when its clock marks the hour $t$. The clocks are adjusted if the slowness equal to $t$ represents the duration of the transmission, and to verify it, station B sends in its turn a signal when its clock marks $O$; then station A should perceive it when its clock marks $t$. The timepieces are then adjusted.

And in fact they mark the same hour at the same physical instant, but on the one condition, that the two stations are fixed. Otherwise the duration of the transmission will not be the same in the two senses, since the station A, for example, moves forward to meet the optical perturbation emanating from B, whereas the station B flees before the perturbation emanating from A. The watches adjusted in that way will not mark, therefore, the true time; they will mark what may be called the *local time,* so that one of them will be slow of the other. It matters little, since we have no means of perceiving it. All the phenomena which happen at A, for example, will be late, but all will be equally so, and the observer will not perceive it, since his watch is slow; so, as the principle of relativity requires, he will have no means of knowing whether he is at rest or in absolute motion.

Unhappily, that does not suffice, and complementary hypotheses are necessary; it is necessary to admit that bodies in motion undergo a uniform contraction in the sense of the motion. One of the diameters of the earth, for example, is shrunk by one two-hundred-millionth in consequence of our planet's motion, while the other diameter retains its normal length. Thus the last little differences are compensated. And then, there is still the hypothesis about forces. Forces, whatever be their origin, gravity as well as elasticity, would be reduced in a certain proportion in a world animated by a uniform translation; or, rather, this would happen for the components perpendicular to the translation; the components parallel would not change. Resume, then, our example of two electrified bodies; these bodies repel each other, but at the same time if all is carried along in a uniform translation, they are equivalent to two parallel currents of the same sense which attract each other. This electrodynamic attraction diminishes, therefore, the electrostatic repulsion, and the total repulsion is feebler than if the two bodies were at rest. But since to measure this repulsion we must balance it by another force, and all these other forces are reduced in the same proportion, we perceive nothing. Thus all seems arranged, but are all the doubts dissipated? What would happen if one could communicate by non-luminous signals whose velocity of propagation differed from that of light? If, after having adjusted the watches by the optical procedure, we wished to verify the adjustment by the aid of these new signals, we should observe discrepancies which would render evident the common translation of the two stations. And are such signals inconceivable, if we admit with Laplace that universal gravitation is transmitted a million times more rapidly than light?

Thus, the principle of relativity has been valiantly defended in these latter times, but the very energy of the defense proves how serious was the attack.

*Newton's Principle.*—Let us speak now of the principle of Newton, on the equality of action and reaction. This is intimately bound up with the preceding, and it seems indeed that the fall of the one would involve that of the other. Thus we must not be astonished to find here the same difficulties.

Electrical phenomena, according to the theory of Lorentz, are due to the displacements of little charged particles, called electrons, immersed in the medium we call ether. The movements of these electrons produce perturbations in the neighboring ether; these perturbations propagate themselves in every direction with the velocity of light, and in turn other electrons, originally at rest, are made to vibrate when the perturbation reaches the parts of the ether which touch them. The electrons, therefore, act on one another, but this action is not direct, it is accomplished through the ether as intermediary. Under these conditions can there be compensation between action and reaction, at least for an observer who should take account only of the movements of matter, that is, of the electrons, and who should be ignorant of those of the ether that he could not see? Evidently not. Even if the compensation should be exact, it could not be simultaneous. The perturbation is propagated with a finite velocity; it, therefore, reaches the second electron only when the first has long ago entered upon its rest. This second electron, therefore, will undergo, after a delay, the action of the first, but will certainly not at that moment react upon it, since around this first electron nothing any longer budges.

The analysis of the facts permits us to be still more precise. Imagine, for example, a Hertzian oscillator, like those used in wireless telegraphy; it sends out energy in every direction; but we can provide it with a parabolic mirror, as Hertz did with his smallest oscillators, so as to send all the energy produced in a single direction. What happens then according to the theory? The apparatus recoils, as if it were a cannon and the projected energy a ball; and that is contrary to the principle of Newton, since our projectile here has no mass, it is not matter, it is energy. The case is still the same, moreover, with a beacon light provided with a reflector, since light is nothing but a perturbation of the electromagnetic field. This beacon light should recoil as if the light it sends out were a projectile. What is the force that should produce this recoil? It is what is called the Maxwell-Bartholi pressure. It is very minute, and it has been difficult to put it in evidence even with the most sensitive radiometers; but it suffices that it exists.

If all the energy issuing from our oscillator falls on a receiver, this will act as if it had received a mechanical shock, which will represent in a sense the compensation of the oscillator's recoil; the reaction will be equal to the action, but it will not be simultaneous; the receiver will move on, but not at the moment when the oscillator recoils. If the energy propagates itself indefinitely without encountering a receiver, the compensation will never occur.

Shall we say that the space which separates the oscillator from the receiver and which the perturbation must pass over in going from the one to the other is not void, that it is full not only of ether, but of air, or even in the interplanetary spaces of some fluid subtile but still ponderable; that this matter undergoes the shock like the receiver at the moment when the energy reaches it, and recoils in its turn when the perturbation quits it? That would save Newton's principle, but that is not true. If energy in its diffusion remained always attached to some material substratum, then matter in motion would carry along light with it, and Fizeau has demonstrated that it does nothing of the sort, at least for air. Michelson and Morley have since confirmed this. It might be supposed also that the movements of matter proper are exactly compensated by those of the ether; but that would lead us to the same reflections as before now. The principle so understood will explain everything, since, whatever might be the visible movements, we always could imagine hypothetical movements which compensate them. But if it is able to explain everything, this is because it does not enable us to foresee anything; it does not enable us to decide between the different possible hypotheses, since it explains everything beforehand. It therefore becomes useless.

And then the suppositions that it would be necessary to make on the movements of the ether are not very satisfactory. If the electric charges double, it would be natural to imagine that the velocities of the diverse atoms of ether double also; but, for the compensation, it would be necessary that the mean velocity of the ether quadruple.

This is why I have long thought that these consequences of theory, contrary to Newton's principle, would end some day by being abandoned, and yet the recent experiments on the movements of the electrons issuing from radium seem rather to confirm them.

*Lavoisier's Principle.*—I arrive at the principle of Lavoisier on the conservation of mass. Certainly, this is one not to be touched without unsettling all mechanics. And now certain persons think that it seems true to us only because in mechanics merely moderate velocities are considered, but that it would cease to be true for bodies animated by velocities comparable to that of light. Now these velocities are believed at present to have been realized; the cathode rays and those of radium may be formed of very minute particles or of electrons which are displaced with velocities smaller no doubt than that of light, but which might be its one tenth or one third.

These rays can be deflected, whether by an electric field, or by a magnetic field, and we are able, by comparing these deflections, to measure at the same time the velocity of the electrons and their mass (or rather the relation of their mass to their charge). But when it was seen that these velocities approached that of light, it was decided that a correction was necessary. These molecules, being electrified, can not be displaced without agitating the ether; to put them in motion it is necessary to over-come a double inertia, that of the molecule itself and that of the ether. The total or apparent mass that one measures is com-posed, therefore, of two parts: the real or mechanical mass of the molecule and the electrodynamic mass representing the inertia of the ether.

The calculations of Abraham and the experiments of Kauf-mann have then shown that the mechanical mass, properly so called, is null, and that the mass of the electrons, or, at least, of the negative electrons, is of exclusively electrodynamic origin. This is what forces us to change the definition of mass; we can not any longer distinguish mechanical mass and electrodynamic mass, since then the first would vanish; there is no mass other than electrodynamic inertia. But in this case the mass can no longer be constant; it augments with the velocity, and it even depends on the direction, and a body animated by a notable velocity will not oppose the same inertia to the forces which tend to deflect it from its route, as to those which tend to accelerate or to retard its progress.

There is still a resource; the ultimate elements of bodies are electrons, some charged negatively, the others charged positively. The negative electrons have no mass, this is understood; but the positive electrons, from the little we know of them, seem much greater. Perhaps they have, besides their electrodynamic mass, a true mechanical mass. The real mass of a body would, then, be the sum of the mechanical masses of its positive electrons, the negative electrons not counting; mass so defined might still be constant.

Alas! this resource also evades us. Recall what we have said of the principle of relativity and of the efforts made to save it. And it is not merely a principle which it is a question of saving, it is the indubitable results of the experiments of Michelson.

Well, as was above seen, Lorentz, to account for these results, was obliged to suppose that all forces, whatever their origin, were reduced in the same proportion in a medium animated by a uniform translation; this is not sufficient; it is not enough that this take place for the real forces, it must also be the same for the forces of inertia; it is therefore necessary, he says, that *the masses of all the particles be influenced by a translation to the same degree as the electromagnetic masses of the electrons.*

So the mechanical masses must vary in accordance with the same laws as the electrodynamic masses; they can not, therefore, be constant.

Need I point out that the fall of Lavoisier's principle involves that of Newton's? This latter signifies that the center of gravity of an isolated system moves in a straight line; but if there is no longer a constant mass, there is no longer a center of gravity, we no longer know even what this is. This is why I said above that the experiments on the cathode rays appeared to justify the doubts of Lorentz concerning Newton's principle.

From all these results, if they were confirmed, would arise an entirely new mechanics, which would be, above all, characterized by this fact, that no velocity could surpass that of light,[1] any more than any temperature can fall below absolute zero.

No more for an observer, carried along himself in a translation he does not suspect, could any apparent velocity surpass that of light; and this would be then a contradiction, if we did not recall that this observer would not use the same clocks as a fixed observer, but, indeed, clocks marking 'local time.'

Here we are then facing a question I content myself with stating. If there is no longer any mass, what becomes of Newton's law? Mass has two aspects: it is at the same time a coefficient of inertia and an attracting mass entering as factor into Newtonian attraction. If the coefficient of inertia is not constant, can the attracting mass be? That is the question.

*Mayer's Principle.*—At least, the principle of the conservation of energy yet remained to us, and this seemed more solid. Shall I recall to you how it was in its turn thrown into discredit? This event has made more noise than the preceding, and it is in all the memoirs. From the first works of Becquerel, and, above all, when the Curies had discovered radium, it was seen that every radioactive body was an inexhaustible source of radiation. Its activity seemed to subsist without alteration throughout the months and the years. This was in itself a strain on the prin-

---

[1] Because bodies would oppose an increasing inertia to the causes which would tend to accelerate their motion; and this inertia would become infinite when one approached the velocity of light.

ciples; these radiations were in fact energy, and from the same morsel of radium this issued and forever issued. But these quantities of energy were too slight to be measured; at least that was the belief and we were not much disquieted.

The scene changed when Curie bethought himself to put radium in a calorimeter; it was then seen that the quantity of heat incessantly created was very notable.

The explanations proposed were numerous; but in such case we can not say, the more the better. In so far as no one of them has prevailed over the others, we can not be sure there is a good one among them. Since some time, however, one of these explanations seems to be getting the upper hand and we may reasonably hope that we hold the key to the mystery.

Sir W. Ramsay has striven to show that radium is in process of transformation, that it contains a store of energy enormous but not inexhaustible. The transformation of radium then would produce a million times more heat than all known transformations; radium would wear itself out in 1,250 years; this is quite short, and you see that we are at least certain to have this point settled some hundreds of years from now. While waiting, our doubts remain.

# CHAPTER IX

## THE FUTURE OF MATHEMATICAL PHYSICS

*The Principles and Experiment.*—In the midst of so much ruin, what remains standing? The principle of least action is hitherto intact, and Larmor appears to believe that it will long survive the others; in reality, it is still more vague and more general.

In presence of this general collapse of the principles, what attitude will mathematical physics take? And first, before too much excitement, it is proper to ask if all that is really true. All these derogations to the principles are encountered only among infinitesimals; the microscope is necessary to see the Brownian movement; electrons are very light; radium is very rare, and one never has more than some milligrams of it at a time. And, then, it may be asked whether, besides the infinitesimal seen, there was not another infinitesimal unseen counterpoise to the first.

So there is an interlocutory question, and, as it seems, only experiment can solve it. We shall, therefore, only have to hand over the matter to the experimenters, and, while waiting for them to finally decide the debate, not to preoccupy ourselves with these disquieting problems, and to tranquilly continue our work as if the principles were still uncontested. Certes, we have much to do without leaving the domain where they may be applied in all security; we have enough to employ our activity during this period of doubts.

*The Rôle of the Analyst.*—And as to these doubts, is it indeed true that we can do nothing to disembarrass science of them? It must indeed be said, it is not alone experimental physics that has given birth to them; mathematical physics has well contributed. It is the experimenters who have seen radium throw out energy, but it is the theorists who have put in evidence all the difficulties raised by the propagation of light across a medium in motion; but for these it is probable we should not have become conscious of them. Well, then, if they have done their best to put us into this embarrassment, it is proper also that they help us to get out of it.

They must subject to critical examination all these new views I have just outlined before you, and abandon the principles only after having made a loyal effort to save them. What can they do in this sense? That is what I will try to explain.

It is a question before all of endeavoring to obtain a more satisfactory theory of the electrodynamics of bodies in motion. It is there especially, as I have sufficiently shown above, that difficulties accumulate. It is useless to heap up hypotheses, we can not satisfy all the principles at once; so far, one has succeeded in safeguarding some only on condition of sacrificing the others; but all hope of obtaining better results is not yet lost. Let us take, then, the theory of Lorentz, turn it in all senses, modify it little by little, and perhaps everything will arrange itself.

Thus in place of supposing that bodies in motion undergo a contraction in the sense of the motion, and that this contraction is the same whatever be the nature of these bodies and the forces to which they are otherwise subjected, could we not make a more simple and natural hypothesis? We might imagine, for example, that it is the ether which is modified when it is in relative motion in reference to the material medium which penetrates it, that, when it is thus modified, it no longer transmits perturbations with the same velocity in every direction. It might transmit more rapidly those which are propagated parallel to the motion of the medium, whether in the same sense or in the opposite sense, and less rapidly those which are propagated perpendicularly. The wave surfaces would no longer be spheres, but ellipsoids, and we could dispense with that extraordinary contraction of all bodies.

I cite this only as an example, since the modifications that might be essayed would be evidently susceptible of infinite variation.

*Aberration and Astronomy.*—It is possible also that astronomy may some day furnish us data on this point; she it was in the main who raised the question in making us acquainted with the phenomenon of the aberration of light. If we make crudely the theory of aberration, we reach a very curious result. The apparent positions of the stars differ from their real positions because of the earth's motion, and as this motion is variable, these apparent positions vary. The real position we can not ascertain, but we can observe the variations of the apparent position. The observations of the aberration show us, therefore, not the earth's motion, but the variations of this motion; they can not, therefore, give us information about the absolute motion of the earth.

At least this is true in first approximation, but the case would be no longer the same if we could appreciate the thousandths of a second. Then it would be seen that the amplitude of the oscillation depends not alone on the variation of the motion, a variation which is well known, since it is the motion of our globe on its elliptic orbit, but on the mean value of this motion, so that the constant of aberration would not be quite the same for all the stars, and the differences would tell us the absolute motion of the earth in space.

This, then, would be, under another form, the ruin of the principle of relativity. We are far, it is true, from appreciating the thousandth of a second, but, after all, say some, the earth's total absolute velocity is perhaps much greater than its relative velocity with respect to the sun. If, for example, it were 300 kilometers per second in place of 30, this would suffice to make the phenomenon observable.

I believe that in reasoning thus one admits a too simple theory of aberration. Michelson has shown us, I have told you, that the physical procedures are powerless to put in evidence absolute motion; I am persuaded that the same will be true of the astronomic procedures, however far precision be carried.

However that may be, the data astronomy will furnish us in this regard will some day be precious to the physicist. Meanwhile, I believe that the theorists, recalling the experience of Michelson, may anticipate a negative result, and that they would accomplish a useful work in constructing a theory of aberration which would explain this in advance.

*Electrons and Spectra.*—This dynamics of electrons can be approached from many sides, but among the ways leading thither is one which has been somewhat neglected, and yet this is one of those which promise us the most surprises. It is movements of electrons which produce the lines of the emission spectra; this is proved by the Zeeman effect; in an incandescent body what vibrates is sensitive to the magnet, therefore electrified. This is a very important first point, but no one has gone farther. Why are the lines of the spectrum distributed in accordance with a regular law? These laws have been studied by the experimenters in their least details; they are very precise and comparatively simple. A first study of these distributions recalls the harmonics encountered in acoustics; but the difference is great. Not only are the numbers of vibrations not the successive multiples of a single number, but we do not even find anything analogous to the roots of those transcendental equations to which we are led by so many problems of mathematical physics: that of the vibrations of an elastic body of any form, that of the Hertzian oscillations in a generator of any form, the problem of Fourier for the cooling of a solid body.

The laws are simpler, but they are of wholly other nature, and to cite only one of these differences, for the harmonics of high order, the number of vibrations tends toward a finite limit, instead of increasing indefinitely.

That has not yet been accounted for, and I believe that there we have one of the most important secrets of nature. A Japanese physicist, M. Nagaoka, has recently proposed an explanation; according to him, atoms are composed of a large positive electron surrounded by a ring formed of a great number of very small negative electrons. Such is the planet Saturn with its rings. This is a very interesting attempt, but not yet wholly satisfactory; this attempt should be renewed. We will penetrate, so to speak, into the inmost recess of matter. And from the particular point of view which we to-day occupy, when we know why the vibrations of incandescent bodies differ thus from ordinary elastic vibrations, why the electrons do not behave like the matter which is familiar to us, we shall better comprehend the dynamics of electrons and it will be perhaps more easy for us to reconcile it with the principles.

*Conventions Preceding Experiment.*—Suppose, now, that all

these efforts fail, and, after all, I do not believe they will, what must be done? Will it be necessary to seek to mend the broken principles by giving what we French call a *coup de pouce?* That evidently is always possible, and I retract nothing of what I have said above.

Have you not written, you might say if you wished to seek a quarrel with me—have you not written that the principles, though of experimental origin, are now unassailable by experiment because they have become conventions? And now you have just told us that the most recent conquests of experiment put these principles in danger.

Well, formerly I was right and to-day I am not wrong. Formerly I was right, and what is now happening is a new proof of it. Take, for example, the calorimetric experiment of Curie on radium. Is it possible to reconcile it with the principle of the conservation of energy? This has been attempted in many ways. But there is among them one I should like you to notice; this is not the explanation which tends to-day to prevail, but it is one of those which have been proposed. It has been conjectured that radium was only an intermediary, that it only stored radiations of unknown nature which flashed through space in every direction, traversing all bodies, save radium, without being altered by this passage and without exercising any action upon them. Radium alone took from them a little of their energy and afterward gave it out to us in various forms.

What an advantageous explanation, and how convenient! First, it is unverifiable and thus irrefutable. Then again it will serve to account for any derogation whatever to Mayer's principle; it answers in advance not only the objection of Curie, but all the objections that future experimenters might accumulate. This new and unknown energy would serve for everything.

This is just what I said, and therewith we are shown that our principle is unassailable by experiment.

But then, what have we gained by this stroke? The principle is intact, but thenceforth of what use is it? It enabled us to foresee that in such or such circumstance we could count on such a total quantity of energy; it limited us; but now that this indefinite provision of new energy is placed at our disposal, we are no longer limited by anything; and, as I have written in 'Science and Hypothesis,' if a principle ceases to be fecund, experiment without contradicting it directly will nevertheless have condemned it.

*Future Mathematical Physics.*—This, therefore, is not what would have to be done; it would be necessary to rebuild anew. If we were reduced to this necessity, we could moreover console ourselves. It would not be necessary thence to conclude that science can weave only a Penelope's web, that it can raise only ephemeral structures, which it is soon forced to demolish from top to bottom with its own hands.

As I have said, we have already passed through a like crisis. I have shown you that in the second mathematical physics, that of the principles, we find traces of the first, that of central forces; it will be just the same if we must know a third. Just so with the animal that exuviates, that breaks its too narrow carapace and makes itself a fresh one; under the new envelope one will recognize the essential traits of the organism which have persisted.

We can not foresee in what way we are about to expand; perhaps it is the kinetic theory of gases which is about to undergo development and serve as model to the others. Then the facts which first appeared to us as simple thereafter would be merely resultants of a very great number of elementary facts which only the laws of chance would make cooperate for a common end. Physical law would then assume an entirely new aspect; it would no longer be solely a differential equation, it would take the character of a statistical law.

Perhaps, too, we shall have to construct an entirely new mechanics that we only succeed in catching a glimpse of, where, inertia increasing with the velocity, the velocity of light would become an impassable limit. The ordinary mechanics, more simple, would remain a first approximation, since it would be true for velocities not too great, so that the old dynamics would still be found under the new. We should not have to regret having believed in the principles, and even, since velocities too great for the old formulas would always be only exceptional, the surest way in practise would be still to act as if we continued to believe in them. They are so useful, it would be necessary to keep a place for them. To determine to exclude them altogether would be to deprive oneself of a precious weapon. I hasten to say in conclusion that we are not yet there, and as yet nothing proves that the principles will not come forth from out the fray victorious and intact.[1]

[1] These considerations on mathematical physics are borrowed from my St. Louis address.

# PART III

## THE OBJECTIVE VALUE
## OF SCIENCE

---

### CHAPTER X

#### Is Science Artificial?

##### 1. *The Philosophy of M. LeRoy*

THERE are many reasons for being sceptics; should we push this scepticism to the very end or stop on the way? To go to the end is the most tempting solution, the easiest, and that which many have adopted, despairing of saving anything from the shipwreck.

Among the writings inspired by this tendency it is proper to place in the first rank those of M. LeRoy. This thinker is not only a philosopher and a writer of the greatest merit, but he has acquired a deep knowledge of the exact and physical sciences, and even has shown rare powers of mathematical invention. Let us recapitulate in a few words his doctrine, which has given rise to numerous discussions.

Science consists only of conventions, and to this circumstance solely does it owe its apparent certitude; the facts of science and, *a fortiori,* its laws are the artificial work of the scientist; science therefore can teach us nothing of the truth; it can only serve us as rule of action.

Here we recognize the philosophic theory known under the name of nominalism; all is not false in this theory; its legitimate domain must be left it, but out of this it should not be allowed to go.

This is not all; M. LeRoy's doctrine is not only nominalistic; it has besides another characteristic which it doubtless owes to M. Bergson, it is anti-intellectualistic. According to M. LeRoy, the intellect deforms all its touches, and that is still more true of its necessary instrument 'discourse.' There is reality only in our fugitive and changing impressions, and even this reality, when touched, vanishes.

And yet M. LeRoy is not a sceptic; if he regards the intellect as incurably powerless, it is only to give more scope to other sources of knowledge, to the heart, for instance, to sentiment, to instinct or to faith.

However great my esteem for M. LeRoy's talent, whatever the ingenuity of this thesis, I can not wholly accept it. Certes, I am in accord on many points with M. LeRoy, and he has even cited, in support of his view, various passages of my writings which I am by no means disposed to reject. I think myself only the more bound to explain why I can not go with him all the way.

M. LeRoy often complains of being accused of scepticism. He could not help being, though this accusation is probably unjust. Are not appearances against him? Nominalist in doctrine, but realist at heart, he seems to escape absolute nominalism only by a desperate act of faith.

The fact is that anti-intellectualistic philosophy in rejecting analysis and 'discourse,' just by that condemns itself to being intransmissible; it is a philosophy essentially internal, or, at the very least, only its negations can be transmitted; what wonder then that for an external observer it takes the shape of scepticism?

Therein lies the weak point of this philosophy; if it strives to remain faithful to itself, its energy is spent in a negation and a cry of enthusiasm. Each author may repeat this negation and this cry, may vary their form, but without adding anything.

And yet, would it not be more logical in remaining silent? See, you have written long articles; for that, it was necessary to use words. And therein have you not been much more 'discursive' and consequently much farther from life and truth than the animal who simply lives without philosophizing? Would not this animal be the true philosopher?

However, because no painter has made a perfect portrait, should we conclude that the best painting is not to paint? When a zoologist dissects an animal, certainly he 'alters it.' Yes, in dissecting it, he condemns himself to never know all of it; but in not dissecting it, he would condemn himself to never know anything of it and consequently to never see anything of it.

Certes, in man are other forces besides his intellect; no one has ever been mad enough to deny that. The first comer makes these blind forces act or lets them act; the philosopher must *speak* of them; to speak of them, he must know of them the little that can be known, he should therefore *see* them act. How? With what eyes, if not with his intellect? Heart, instinct, may guide it, but not render it useless; they may direct the look, but not replace the eye. It may be granted that the heart is the workman, and the intellect only the instrument. Yet is it an instrument not to be done without, if not for action, at least for philosophizing? Therefore a philosopher really anti-intellectualistic is impossible. Perhaps we shall have to declare for the supremacy of action; always it is our intellect which will thus conclude; in allowing precedence to action it will thus retain the superiority of the thinking reed. This also is a supremacy not to be disdained.

Pardon these brief reflections and pardon also their brevity, scarcely skimming the question. The process of intellectualism is not the subject I wish to treat: I wish to speak of science, and about it there is no doubt; by definition, so to speak, it will be intellectualistic or it will not be at all. Precisely the question is, whether it will be.

## 2. *Science, Rule of Action*

For M. LeRoy, science is only a rule of action. We are powerless to know anything and yet we are launched, we must act, and at all hazards we have established rules. It is the aggregate of these rules that is called science.

It is thus that men, desirous of diversion, have instituted rules of play, like those of tric-trac for instance, which, better than science itself, could rely upon the proof by universal consent. It is thus likewise that, unable to choose, but forced to choose, we toss up a coin, head or tail to win.

The rule of tric-trac is indeed a rule of action like science, but does any one think the comparison just and not see the difference? The rules of the game are arbitrary conventions and the contrary convention might have been adopted, *which would have been none the less good.* On the contrary, science is a rule of action which is successful, generally at least, and I add, while the contrary rule would not have succeeded.

If I say, to make hydrogen cause an acid to act on zinc, I formulate a rule which succeeds; I could have said, make distilled water act on gold; that also would have been a rule, only it would not have succeeded. If, therefore, scientific 'recipes' have a value, as rule of action, it is because we know they succeed, generally at least. But to know this is to know something and then why tell us we can know nothing?

Science foresees, and it is because it foresees that it can be useful and serve as rule of action. I well know that its previsions are often contradicted by the event; that shows that science is imperfect, and if I add that it will always remain so, I am certain that this is a prevision which, at least, will never be contradicted. Always the scientist is less often mistaken than a prophet who should predict at random. Besides the progress though slow is continuous, so that scientists, though more and more bold, are less and less misled. This is little, but it is enough.

I well know that M. LeRoy has somewhere said that science was mistaken oftener than one thought, that comets sometimes played tricks on astronomers, that scientists, who apparently are men, did not willingly speak of their failures, and that, if they should speak of them, they would have to count more defeats than victories.

That day, M. LeRoy evidently overreached himself. If science did not succeed, it could not serve as rule of action; whence would it get its value? Because it is 'lived,' that is, because we love it and believe in it? The alchemists had recipes for making gold, they loved them and had faith in them, and yet our recipes are the good ones, although our faith be less lively, because they succeed.

There is no escape from this dilemma; either science does not enable us to foresee, and then it is valueless as rule of action; or else it enables us to foresee, in a fashion more or less imperfect, and then it is not without value as means of knowledge.

It should not even be said that action is the goal of science; should we condemn studies of the star Sirius, under pretext that we shall probably never exercise any influence on that star? To my eyes, on the contrary, it is the knowledge which is the end, and the action which is the means. If I felicitate myself on the industrial development, it is not alone because it furnishes a facile argument to the advocates of science; it is above all because it gives to the scientist faith in himself and also because it offers him an immense field of experience where he clashes against forces too colossal to be tampered with. Without this ballast, who knows whether he would not quit solid ground, seduced by the mirage of some scholastic novelty, or whether he would not despair, believing he had fashioned only a dream?

### 3. *The Crude Fact and the Scientific Fact*

What was most paradoxical in M. LeRoy's thesis was that affirmation that *the scientist creates the fact;* this was at the same time its essential point and it is one of those which have been most discussed.

Perhaps, says he (I well believe that this was a concession), it is not the scientist that creates the fact in the rough; it is at least he who creates the scientific fact.

This distinction between the fact in the rough and the scientific fact does not by itself appear to me illegitimate. But I complain first that the boundary has not been traced either exactly or precisely; and then that the author has seemed to suppose that the crude fact, not being scientific, is outside of science.

Finally, I can not admit that the scientist creates without restraint the scientific fact, since it is the crude fact which imposes it upon him.

The examples given by M. LeRoy have greatly astonished me. The first is taken from the notion of atom. The atom chosen as example of fact! I avow that this choice has so disconcerted me that I prefer to say nothing about it. I have evidently misunderstood the author's thought and I could not fruitfully discuss it.

The second case taken as example is that of an eclipse where the crude phenomenon is a play of light and shadow, but where the astronomer can not intervene without introducing two foreign elements, to wit, a clock and Newton's law.

Finally, M. LeRoy cites the rotation of the earth; it has been answered: but this is not a fact, and he has replied: it was one for Galileo, who affirmed it, as for the inquisitor, who denied it. It always remains that this is not a fact in the same sense as those just spoken of and that to give them the same name is to expose one's self to many confusions.

Here then are four degrees:

1°. It grows dark, says the clown.

2°. The eclipse happened at nine o'clock, says the astronomer.

3°. The eclipse happened at the time deducible from the tables constructed according to Newton's law, says he again.

4°. That results from the earth's turning around the sun, says Galileo finally.

Where then is the boundary between the fact in the rough and the scientific fact? To read M. LeRoy one would believe that it is between the first and the second stage, but who does not see that there is a greater distance from the second to the third, and still more from the third to the fourth.

Allow me to cite two examples which perhaps will enlighten us a little.

I observe the deviation of a galvanometer by the aid of a movable mirror which projects a luminous image or spot on a divided scale. The crude fact is this: I see the spot displace itself on the scale, and the scientific fact is this: a current passes in the circuit.

Or again: when I make an experiment I should subject the result to certain corrections, because I know I must have made errors. These errors are of two kinds, some are accidental and these I shall correct by taking the mean; the others are systematic and I shall be able to correct those only by a thorough study of their causes. The first result obtained is then the fact in the rough, while the scientific fact is the final result after the finished corrections.

Reflecting on this latter example, we are led to subdivide our second stage, and in place of saying:

2. The eclipse happened at nine o'clock, we shall say:

2a. The eclipse happened when my clock pointed to nine, and

2*b*. My clock being ten minutes slow, the eclipse happened at ten minutes past nine.

And this is not all: the first stage also should be subdivided, and not between these two subdivisions will be the least distance; it is necessary to distinguish between the impression of obscurity felt by one witnessing an eclipse, and the affirmation: It grows dark, which this impression extorts from him. In a sense it is the first which is the only true fact in the rough, and the second is already a sort of scientific fact.

Now then our scale has six stages, and even though there is no reason for halting at this figure, there we shall stop.

What strikes me at the start is this. At the first of our six stages, the fact, still completely in the rough, is, so to speak, individual, it is completely distinct from all other possible facts. From the second stage, already it is no longer the same. The enunciation of the fact would suit an infinity of other facts. So soon as language intervenes, I have at my command only a finite number of terms to express the shades, in number infinite, that my impressions might cover. When I say: It grows dark, that well expresses the impressions I feel in being present at an eclipse; but even in obscurity a multitude of shades could be imagined, and if, instead of that actually realized, had happened a slightly different shade, yet I should still have enunciated this *other* fact by saying: It grows dark.

Second remark: even at the second stage, the enunciation of a fact can only be *true or false*. This is not so of any proposition; if this proposition is the enunciation of a convention, it can not be said that this enunciation is *true,* in the proper sense of the word, since it could not be true apart from me and is true only because I wish it to be.

When, for instance, I say the unit for length is the meter, this is a decree that I promulgate, it is not something ascertained which forces itself upon me. It is the same, as I think I have elsewhere shown, when it is a question, for example, of Euclid's postulate.

When I am asked: Is it growing dark? I always know whether I ought to reply yes or no. Although an infinity of possible facts may be susceptible of this same enunciation, it grows dark,

I shall always know whether the fact realized belongs or does not belong among those which answer to this enunciation. Facts are classed in categories, and if I am asked whether the fact that I ascertain belongs or does not belong in such a category, I shall not hesitate.

Doubtless this classification is sufficiently arbitrary to leave a large part to man's freedom or caprice. In a word, this classification is a convention. *This convention being given,* if I am asked: Is such a fact true? I shall always know what to answer, and my reply will be imposed upon me by the witness of my senses.

If therefore, during an eclipse, it is asked: Is it growing dark? all the world will answer yes. Doubtless those speaking a language where bright was called dark, and dark bright, would answer no. But of what importance is that?

In the same way, in mathematics, *when I have laid down the definitions, and the postulates which are conventions,* a theorem henceforth can only be true or false. But to answer the question: Is this theorem true? it is no longer to the witness of my senses that I shall have recourse, but to reasoning.

A statement of fact is always verifiable, and for the verification we have recourse either to the witness of our senses, or to the memory of this witness. This is properly what characterizes a fact. If you put the question to me: Is such a fact true? I shall begin by asking you, if there is occasion, to state precisely the conventions, by asking you, in other words, what language you have spoken; then once settled on this point, I shall interrogate my senses and shall answer yes or no. But it will be my senses that will have made answer, it will not be *you* when you say to me: I have spoken to you in English or in French.

Is there something to change in all that when we pass to the following stages? When I observe a galvanometer, as I have just said, if I ask an ignorant visitor: Is the current passing? he looks at the wire to try to see something pass; but if I put the same question to my assistant who understands my language, he will know I mean: Does the spot move? and he will look at the scale.

What difference is there then between the statement of a fact in the rough and the statement of a scientific fact? The same difference as between the statement of the same crude fact in French and in German. The scientific statement is the translation of the crude statement into a language which is distinguished above all from the common German or French, because it is spoken by a very much smaller number of people.

Yet let us not go too fast. To measure a current I may use a very great number of types of galvanometers or besides an electrodynamometer. And then when I shall say there is running in this circuit a current of so many amperes, that will mean: if I adapt to this circuit such a galvanometer I shall see the spot come to the division $a$; but that will mean equally: if I adapt to this circuit such an electrodynamometer, I shall see the spot go to the division $b$. And that will mean still many other things, because the current can manifest itself not only by mechanical effects, but by effects chemical, thermal, luminous, etc.

Here then is one same statement which suits a very great number of facts absolutely different. Why? It is because I assume a law according to which, whenever such a mechanical effect shall happen, such a chemical effect will happen also. Previous experiments, very numerous, have never shown this law to fail, and then I have understood that I could express by the same statement two facts so invariably bound one to the other.

When I am asked: Is the current passing? I can understand that that means: Will such a mechanical effect happen? But I can understand also: Will such a chemical effect happen? I shall then verify either the existence of the mechanical effect, or that of the chemical effect; that will be indifferent, since in both cases the answer must be the same.

And if the law should one day be found false? If it was perceived that the concordance of the two effects, mechanical and chemical, is not constant? That day it would be necessary to change the scientific language to free it from a grave ambiguity.

And after that? Is it thought that ordinary language by aid of which are expressed the facts of daily life is exempt from ambiguity?

*Shall we thence conclude that the facts of daily life are the work of the grammarians?*

You ask me: Is there a current? I try whether the mechanical effect exists, I ascertain it and I answer: Yes, there is a current. You understand at once that that means that the mechanical effect exists, and that the chemical effect, that I have not investigated, exists likewise. Imagine now, supposing an impossibility, the law we believe true, not to be, and the chemical effect not to exist. Under this hypothesis there will be two distinct facts, the one directly observed and which is true, the other inferred and which is false. It may strictly be said that we have created the second. So that error is the part of man's personal collaboration in the creation of the scientific fact.

But if we can say that the fact in question is false, is this not just because it is not a free and arbitrary creation of our mind, a disguised convention, in which case it would be neither true nor false. And in fact it was verifiable; I had not made the verification, but I could have made it. If I answered amiss, it was because I chose to reply too quickly, without having asked nature, who alone knew the secret.

When, after an experiment, I correct the accidental and systematic errors to bring out the scientific fact, the case is the same; the scientific fact will never be anything but the crude fact translated into another language. When I shall say: It is such an hour, that will be a short way of saying: There is such a relation between the hour indicated by my clock, and the hour it marked at the moment of the passing of such a star and such another star across the meridian. And this convention of language once adopted, when I shall be asked: Is it such an hour? it will not depend upon me to answer yes or no.

Let us pass to the stage before the last: the eclipse happened at the hour given by the tables deduced from Newton's laws. This is still a convention of language which is perfectly clear for those who know celestial mechanics or simply for those who have the tables calculated by the astronomers. I am asked: Did the eclipse happen at the hour predicted? I look in the nautical almanac, I see that the eclipse was announced for nine o'clock and I understand that the question means: Did the eclipse happen at nine o'clock? There still we have nothing to change in our conclusions. *The scientific fact is only the crude fact translated into a convenient language.*

It is true that at the last stage things change. Does the earth rotate? Is this a verifiable fact? Could Galileo and the Grand Inquisitor, to settle the matter, appeal to the witness of their senses? On the contrary, they were in accord about the appearances, and whatever had been the accumulated experiences, they would have remained in accord with regard to the appearances without ever agreeing on their interpretation. It is just on that account that they were obliged to have recourse to procedures of discussion so unscientific.

This is why I think they did not disagree about a *fact:* we have not the right to give the same name to the rotation of the earth, which was the object of their discussion, and to the facts crude or scientific we have hitherto passed in review.

After what precedes, it seems superfluous to investigate whether the fact in the rough is outside of science, because there can neither be science without scientific fact, nor scientific fact without fact in the rough, since the first is only the translation of the second.

And then, has one the right to say that the scientist creates the scientific fact? First of all, he does not create it from nothing, since he makes it with the fact in the rough. Consequently he does not make it freely and *as he chooses*. However able the worker may be, his freedom is always limited by the properties of the raw material on which he works.

After all, what do you mean when you speak of this free creation of the scientific fact and when you take as example the astronomer who intervenes actively in the phenomenon of the eclipse by bringing his clock? Do you mean: The eclipse happened at nine o'clock; but if the astronomer had wished it to happen at ten, that depended only on him, he had only to advance his clock an hour?

But the astronomer, in perpetrating that bad joke, would evidently have been guilty of an equivocation. When he tells me: The eclipse happened at nine, I understand that nine is the hour deduced from the crude indication of the pendulum by the usual series of corrections. If he has given me solely that crude indication, or if he has made corrections contrary to the habitual rules, he has changed the language agreed upon without forewarning me. If, on the contrary, he took care to forewarn me, I have nothing to complain of, but then it is always the same fact expressed in another language.

In sum, *all the scientist creates in a fact is the language in which he enunciates it.* If he predicts a fact, he will employ this language, and for all those who can speak and understand it, his prediction is free from ambiguity. Moreover, this prediction once made, it evidently does not depend upon him whether it is fulfilled or not.

What then remains of M. LeRoy's thesis? This remains: the scientist intervenes actively in choosing the facts worth observing. An isolated fact has by itself no interest; it becomes interesting if one has reason to think that it may aid in the prediction of other facts; or better, if, having been predicted, its verification is the confirmation of a law. Who shall choose the facts which, corresponding to these conditions, are worthy the freedom of the city in science? This is the free activity of the scientist.

And that is not all. I have said that the scientific fact is the translation of a crude fact into a certain language; I should add that every scientific fact is formed of many crude facts. This is sufficiently shown by the examples cited above. For instance, for the hour of the eclipse my clock marked the hour $\alpha$ at the instant of the eclipse; it marked the hour $\beta$ at the moment of the last transit of the meridian of a certain star that we take as origin of right ascensions; it marked the hour $\gamma$ at the moment of the preceding transit of this same star. There are three distinct facts (still it will be noticed that each of them results itself from two simultaneous facts in the rough; but let us pass this over). In place of that I say: The eclipse happened at the hour $24 \ (\alpha-\beta)/(\beta-\gamma)$, and the three facts are combined in a single scientific fact. I have concluded that the three readings $\alpha$, $\beta$, $\gamma$ made on my clock at three different moments lacked interest and that the only thing interesting was the combination $(\alpha-\beta)/(\beta-\gamma)$ of the three. In this conclusion is found the free activity of my mind.

But I have thus used up my power; I can not make this combination $(\alpha-\beta)/(\beta-\gamma)$ have such a value and not such another, since I can not influence either the value of $\alpha$, or that of $\beta$, or that of $\gamma$, which are imposed upon me as crude facts.

In sum, facts are facts, and *if it happens that they satisfy a prediction, this is not an effect of our free activity.* There is no precise frontier between the fact in the rough and the scientific fact; it can only be said that such an enunciation of fact is *more crude* or, on the contrary, *more scientific* than such another.

#### 4. *'Nominalism' and 'the Universal Invariant'*

If from facts we pass to laws, it is clear that the part of the free activity of the scientist will become much greater. But did not M. LeRoy make it still too great? This is what we are about to examine.

Recall first the examples he has given. When I say: Phosphorus melts at 44°, I think I am enunciating a law; in reality it is just the definition of phosphorus; if one should discover a body which, possessing otherwise all the properties of phosphorus, did not melt at 44°, we should give it another name, that is all, and the law would remain true.

Just so when I say: Heavy bodies falling freely pass over spaces proportional to the squares of the times, I only give the definition of free fall. Whenever the condition shall not be fulfilled, I shall say that the fall is not free, so that the law will never be wrong. It is clear that if laws were reduced to that, they could not serve in prediction; then they would be good for nothing, either as means of knowledge or as principle of action.

When I say: Phosphorus melts at 44°, I mean by that: All bodies possessing such or such a property (to wit, all the properties of phosphorus, save fusing-point) fuse at 44°. So understood, my proposition is indeed a law, and this law may be useful to me, because if I meet a body possessing these properties I shall be able to predict that it will fuse at 44°.

Doubtless the law may be found to be false. Then we shall read in the treatises on chemistry: "There are two bodies which chemists long confounded under the name of phosphorus; these two bodies differ only by their points of fusion." That would evidently not be the first time for chemists to attain to the separation of two bodies they were at first not able to distinguish; such, for example, are neodymium and praseodymium, long confounded under the name of didymium.

I do not think the chemists much fear that a like mischance will ever happen to phosphorus. And if, to suppose the impossible, it should happen, the two bodies would probably not have *identically* the same density, *identically* the same specific heat, etc., so that after having determined with care the density, for instance, one could still foresee the fusion point.

It is, moreover, unimportant; it suffices to remark that there is a law, and that this law, true or false, does not reduce to a tautology.

Will it be said that if we do not know on the earth a body which does not fuse at 44° while having all the other properties of phosphorus, we can not know whether it does not exist on other planets? Doubtless that may be maintained, and it would then be inferred that the law in question, which may serve as a rule of action to us who inhabit the earth, has yet no general value from the point of view of knowledge, and owes its interest only to the chance which has placed us on this globe. This is possible, but, if it were so, the law would be valueless, not because it reduced to a convention, but because it would be false.

The same is true in what concerns the fall of bodies. It would do me no good to have given the name of free fall to falls which happen in conformity with Galileo's law, if I did not know that elsewhere, in such circumstances, the fall will be *probably* free or *approximately* free. That then is a law which may be true or false, but which does not reduce to a convention.

Suppose the astronomers discover that the stars do not exactly obey Newton's law. They will have the choice between two attitudes; they may say that gravitation does not vary exactly as the inverse of the square of the distance, or else they may say that gravitation is not the only force which acts on the stars and that there is in addition a different sort of force.

In the second case, Newton's law will be considered as the definition of gravitation. This will be the nominalist attitude. The choice between the two attitudes is free, and is made from considerations of convenience, though these considerations are most often so strong that there remains practically little of this freedom.

We can break up this proposition: (1) The stars obey Newton's law, into two others; (2) gravitation obeys Newton's law; (3) gravitation is the only force acting on the stars. In this case proposition (2) is no longer anything but a definition and is beyond the test of experiment; but then it will be on proposition (3) that this check can be exercised. This is indeed necessary, since the resulting proposition (1) predicts verifiable facts in the rough.

It is thanks to these artifices that by an unconscious nominalism the scientists have elevated above the laws what they call principles. When a law has received a sufficient confirmation from experiment, we may adopt two attitudes: either we may leave this law in the fray; it will then remain subjected to an incessant revision, which without any doubt will end by demonstrating that it is only approximative. Or else we may elevate it into a *principle* by adopting conventions such that the proposition may be certainly true. For that the procedure is always the same. The primitive law enunciated a relation between two facts in the rough, *A* and *B*; between these two crude facts is introduced an abstract intermediary *C*, more or less fictitious (such was in the preceding example the impalpable entity, gravitation). And then we have a relation between *A* and *C* that we may suppose rigorous and which is the *principle;* and another between *C* and *B* which remains a *law* subject to revision.

The principle, henceforth crystallized, so to speak, is no longer subject to the test of experiment. It is not true or false, it is convenient.

Great advantages have often been found in proceeding in that way, but it is clear that if *all* the laws had been transformed into principles *nothing* would be left of science. Every law may be broken up into a principle and a law, but thereby it is very clear that, however far this partition be pushed, there will always remain laws.

Nominalism has therefore limits, and this is what one might fail to recognize if one took to the very letter M. LeRoy's assertions.

A rapid review of the sciences will make us comprehend better what are these limits. The nominalist attitude is justified only when it is convenient; when is it so?

Experiment teaches us relations between bodies; this is the fact in the rough; these relations are extremely complicated. Instead of envisaging directly the relation of the body $A$ and the body $B$, we introduce between them an intermediary, which is space, and we envisage three distinct relations: that of the body $A$ with the figure $A'$ of space, that of the body $B$ with the figure $B'$ of space, that of the two figures $A'$ and $B'$ to each other. Why is this detour advantageous? Because the relation of $A$ and $B$ was complicated, but differed little from that of $A'$ and $B'$, which is simple; so that this complicated relation may be replaced by the simple relation between $A'$ and $B'$ and by two other relations which tell us that the differences between $A$ and $A'$, on the one hand, between $B$ and $B'$, on the other hand, are *very small*. For example, if $A$ and $B$ are two natural solid bodies which are displaced with slight deformation, we envisage two movable *rigid* figures $A'$ and $B'$. The laws of the relative displacement of these figures $A'$ and $B'$ will be very simple; they will be those of geometry. And we shall afterward add that the body $A$, which always differs very little from $A'$, dilates from the effect of heat and bends from the effect of elasticity. These dilatations and flexions, just because they are very small, will be for our mind relatively easy to study. Just imagine to what complexities of language it would have been necessary to be resigned if we had wished to comprehend in the same enunciation the displacement of the solid, its dilatation and its flexure?

The relation between $A$ and $B$ was a rough law, and was broken up; we now have two laws which express the relations of $A$ and $A'$, of $B$ and $B'$, and a principle which expresses that of $A'$ with $B'$ It is the aggregate of these principles that is called geometry.

Two other remarks. We have a relation between two bodies $A$ and $B$, which we have replaced by a relation between two figures $A'$ and $B'$; but this same relation between the same two figures $A'$ and $B'$ could just as well have replaced advantageously a relation between two other bodies $A''$ and $B''$, entirely different from $A$ and $B$. And that in many ways. If the principles and geometry had not been invented, after having studied the relation of $A$ and $B$, it would be necessary to begin again *ab ovo* the study of the relation of $A''$ and $B''$. That is why geometry is so precious. A geometrical relation can advantageously replace a relation which, considered in the rough state, should be regarded as mechanical, it can replace another which should be regarded as optical, etc.

Yet let no one say: But that proves geometry an experimental science; in separating its principles from laws whence they have been drawn, you artificially separate it itself from the sciences which have given birth to it. The other sciences have likewise principles, but that does not preclude our having to call them experimental.

It must be recognized that it would have been difficult not to make this separation that is pretended to be artificial. We know the rôle that the kinematics of solid bodies has played in the genesis of geometry; should it then be said that geometry is only a branch of experimental kinematics? But the laws of the rectilinear propagation of light have also contributed to the formation of its principles. Must geometry be regarded both as a branch of kinematics and as a branch of optics? I recall besides that our Euclidean space which is the proper object of geometry has been chosen, for reasons of convenience, from among a certain number of types which preexist in our mind and which are called groups.

If we pass to mechanics, we still see great principles whose origin is analogous, and, as their 'radius of action,' so to speak, is smaller, there is no longer reason to separate them from mechanics proper and to regard this science as deductive.

In physics, finally, the rôle of the principles is still more diminished. And in fact they are only introduced when it is of advantage. Now they are advantageous precisely because they are few, since each of them very nearly replaces a great number of laws. Therefore it is not of interest to multiply them. Besides an outcome is necessary, and for that it is needful to end by leaving abstraction to take hold of reality.

Such are the limits of nominalism, and they are narrow.

M. LeRoy has insisted, however, and he has put the question under another form.

Since the enunciation of our laws may vary with the conventions that we adopt, since these conventions may modify even the

23

natural relations of these laws, is there in the manifold of these laws something independent of these conventions and which may, so to speak, play the rôle of *universal invariant?* For instance, the fiction has been introduced of beings who, having been educated in a world different from ours, would have been led to create a non-Euclidean geometry. If these beings were afterward suddenly transported into our world, they would observe the same laws as we, but they would enunciate them in an entirely different way. In truth there would still be something in common between the two enunciations, but this is because these beings do not yet differ enough from us. Beings still more strange may be imagined, and the part common to the two systems of enunciations will shrink more and more. Will it thus shrink in convergence toward zero, or will there remain an irreducible residue which will then be the universal invariant sought?

The question calls for precise statement. Is it desired that this common part of the enunciations be expressible in words? It is clear, then, that there are not words common to all languages, and we can not pretend to construct I know not what universal invariant which should be understood both by us and by the fictitious non-Euclidean geometers of whom I have just spoken; no more than we can construct a phrase which can be understood both by Germans who do not understand French and by French who do not understand German. But we have fixed rules which permit us to translate the French enunciations into German, and inversely. It is for that that grammars and dictionaries have been made. There are also fixed rules for translating the Euclidean language into the non-Euclidean language, or, if there are not, they could be made.

And even if there were neither interpreter nor dictionary, if the Germans and the French, after having lived centuries in separate worlds, found themselves all at once in contact, do you think there would be nothing in common between the science of the German books and that of the French books? The French and the Germans would certainly end by understanding each other, as the American Indians ended by understanding the language of their conquerors after the arrival of the Spanish.

But, it will be said, doubtless the French would be capable of understanding the Germans even without having learned German, but this is because there remains between the French and the Germans something in common, since both are men. We should still attain to an understanding with our hypothetical non-Euclideans, though they be not men, because they would still retain something human. But in any case a minimum of humanity is necessary.

This is possible, but I shall observe first that this little humanness which would remain in the non-Euclideans would suffice not only to make possible the translation of *a little* of their language, but to make possible the translation of *all* their language.

Now, that there must be a minimum is what I concede; suppose there exists I know not what fluid which penetrates between the molecules of our matter, without having any action on it and without being subject to any action coming from it. Suppose beings sensible to the influence of this fluid and insensible to that of our matter. It is clear that the science of these beings would differ absolutely from ours and that it would be idle to seek an 'invariant' common to these two sciences. Or again, if these beings rejected our logic and did not admit, for instance, the principle of contradiction.

But truly I think it without interest to examine such hypotheses.

And then, if we do not push whimsicality so far, if we introduce only fictitious beings having senses analogous to ours and sensible to the same impressions, and moreover admitting the principles of our logic, we shall then be able to conclude that their language, however different from ours it may be, would always be capable of translation. Now the possibility of translation implies the existence of an invariant. To translate is precisely to disengage this invariant. Thus, to decipher a cryptogram is to seek what in this document remains invariant, when the letters are permuted.

What now is the nature of this invariant it is easy to understand, and a word will suffice us. The invariant laws are the relations between the crude facts, while the relations between the 'scientific facts' remain always dependent on certain conventions.

# CHAPTER XI

## 5. *Contingence and Determinism*

I DO not intend to treat here the question of the contingence of the laws of nature, which is evidently insoluble, and on which so much has already been written. I only wish to call attention to what different meanings have been given to this word, contingence, and how advantageous it would be to distinguish them.

If we look at any particular law, we may be certain in advance that it can only be approximate. It is, in fact, deduced from experimental verifications, and these verifications were and could be only approximate. We should always expect that more precise measurements will oblige us to add new terms to our formulas; this is what has happened, for instance, in the case of Mariotte's law.

Moreover the statement of any law is necessarily incomplete. This enunciation should comprise the enumeration of *all* the antecedents in virtue of which a given consequent can happen. I should first describe *all* the conditions of the experiment to be made and the law would then be stated: If all the conditions are fulfilled, the phenomenon will happen.

But we shall be sure of not having forgotten *any* of these conditions only when we shall have described the state of the entire universe at the instant $t$; all the parts of this universe may, in fact, exercise an influence more or less great on the phenomenon which must happen at the instant $t + dt$.

Now it is clear that such a description could not be found in the enunciation of the law; besides, if it were made, the law would become incapable of application; if one required so many conditions, there would be very little chance of their ever being all realized at any moment.

Then as one can never be certain of not having forgotten some essential condition, it can not be said: If such and such conditions are realized, such a phenomenon will occur; it can only be said: If such and such conditions are realized, it is probable that such a phenomenon will occur, very nearly.

Take the law of gravitation, which is the least imperfect of all known laws. It enables us to foresee the motions of the planets. When I use it, for instance, to calculate the orbit of Saturn, I neglect the action of the stars, and in doing so I am certain of not deceiving myself, because I know that these stars are too far away for their action to be sensible.

I announce, then, with a quasi-certitude that the coordinates of Saturn at such an hour will be comprised between such and such limits. Yet is that certitude absolute? Could there not exist in the universe some gigantic mass, much greater than that of all the known stars and whose action could make itself felt at great distances? That mass might be animated by a colossal velocity, and after having circulated from all time at such distances that its influence had remained hitherto insensible to us, it might come all at once to pass near us. Surely it would produce in our solar system enormous perturbations that we could not have foreseen. All that can be said is that such an event is wholly improbable, and then, instead of saying: Saturn will be near such a point of the heavens, we must limit ourselves to saying: Saturn will probably be near such a point of the heavens. Although this probability may be practically equivalent to certainty, it is only a probability.

For all these reasons, no particular law will ever be more than approximate and probable. Scientists have never failed to recognize this truth; only they believe, right or wrong, that every law may be replaced by another closer and more probable, that this new law will itself be only provisional, but that the same movement can continue indefinitely, so that science in progressing will possess laws more and more probable, that the approximation will end by differing as little as you choose from exactitude and the probability from certitude.

If the scientists who think thus are right, still could it be said that *the* laws of nature are contingent, even though *each* law, taken in particular, may be qualified as contingent? Or must one require, before concluding the contingence *of the* natural laws, that this progress have an end, that the scientist finish some day by being arrested in his search for a closer and closer approximation, and that, beyond a certain limit, he thereafter meet in nature only caprice?

In the conception of which I have just spoken (and which I shall call the scientific conception), every law is only a statement imperfect and provisional, but it must one day be replaced by another, a superior law, of which it is only a crude image. No place therefore remains for the intervention of a free will.

It seems to me that the kinetic theory of gases will furnish us a striking example.

You know that in this theory all the properties of gases are explained by a simple hypothesis; it is supposed that all the gaseous molecules move in every direction with great velocities and that they follow rectilineal paths which are disturbed only when one molecule passes very near the sides of the vessel or another molecule. The effects our crude senses enable us to observe are the mean effects, and in these means, the great deviations compensate, or at least it is very improbable that they do not compensate; so that the observable phenomena follow simple laws such as that of Mariotte or of Gay-Lussac. But this compensation of deviations is only probable. The molecules incessantly change place and in these continual displacements the figures they form pass successively through all possible combinations. Singly these combinations are very numerous; almost all are in conformity with Mariotte's law, only a few deviate from it. These also will happen, only it would be necessary to wait a long time for them. If a gas were observed during a sufficiently long time, it would certainly be finally seen to deviate, for a very short time, from Mariotte's law. How long would it be necessary to wait? If it were desired to calculate the probable number of years, it would be found that this number is so great that to write only the number of places of figures employed would still require half a score places of figures. No matter; enough that it may be done.

I do not care to discuss here the value of this theory. It is evident that if it be adopted, Mariotte's law will thereafter appear only as contingent, since a day will come when it will not be true. And yet, think you the partisans of the kinetic theory are adversaries of determinism? Far from it; they are the most ultra of mechanists. Their molecules follow rigid paths, from which they depart only under the influence of forces which vary with the distance, following a perfectly determinate law. There remains in their system not the smallest place either for freedom, or for an evolutionary factor, properly so-called, or for anything whatever that could be called contingence. I add, to avoid mistake, that neither is there any evolution of Mariotte's law itself; it ceases to be true after I know not how many centuries; but at the end of a fraction of a second it again becomes true and that for an incalculable number of centuries.

And since I have pronounced the word evolution, let us clear away another mistake. It is often said: Who knows whether the laws do not evolve and whether we shall not one day discover that they were not at the Carboniferous epoch what they are to-day? What are we to understand by that? What we think we know about the past state of our globe, we deduce from its present state. And how is this deduction made? It is by means of laws supposed known. The law, being a relation between the antecedent and the consequent, enables us equally well to deduce the consequent from the antecedent, that is, to foresee the future, and to deduce the antecedent from the consequent, that is, to conclude from the present to the past. The astronomer who knows the present situation of the stars can from it deduce their future situation by Newton's law, and this is what he does when he constructs ephemerides; and he can equally deduce from it their past situation. The calculations he thus can make can not teach him that Newton's law will cease to be true in the future, since this law is precisely his point of departure; not more can they tell him it was not true in the past. Still, in what concerns the future, his ephemerides can one day be tested and our descendants will perhaps recognize that they were false. But in what concerns the past, the geologic past which had no witnesses, the results of his calculation, like those of all speculations where we seek to deduce the past from the present, escape by their very nature every species of test. So that if the laws of nature were not the same in the Carboniferous age as at the present epoch, we shall never be able to know it, since we can know nothing of this age, only what we deduce from the hypothesis of the permanence of these laws.

Perhaps it will be said that this hypothesis might lead to contradictory results and that we shall be obliged to abandon it. Thus, in what concerns the origin of life, we may conclude that there have always been living beings, since the present world shows us always life springing from life; and we may also conclude that there have not always been, since the application of the existent laws of physics to the present state of our globe teaches us that there was a time when this globe was so warm that life on it was impossible. But contradictions of this sort can always be removed in two ways; it may be supposed that the actual laws of nature are not exactly what we have assumed; or else it may be supposed that the laws of nature actually are what we have assumed, but that it has not always been so.

It is evident that the actual laws will never be sufficiently well known for us not to be able to adopt the first of these two solutions and for us to be constrained to infer the evolution of natural laws.

On the other hand, suppose such an evolution; assume, if you wish, that humanity lasts sufficiently long for this evolution to have witnesses. The *same* antecedent shall produce, for instance, different consequents at the Carboniferous epoch and at the Quaternary. That evidently means that the antecedents are closely alike; if all the circumstances were identical, the Carboniferous epoch would be indistinguishable from the Quaternary. Evidently this is not what is supposed. What remains is that such antecedent, accompanied by such accessory circumstance, produces such consequent; and that the same antecedent, accompanied by such other accessory circumstance, produces such other consequent. Time does not enter into the affair.

The law, such as ill-informed science would have stated it, and which would have affirmed that this antecedent always produces this consequent, without taking account of the accessory circumstances, this law, which was only approximate and probable, must be replaced by another law more approximate and more probable, which brings in these accessory circumstances. We always come back, therefore, to that same process which we have analyzed above, and if humanity should discover something of this sort, it would not say that it is the laws which have evoluted, but the circumstances which have changed.

Here, therefore, are several different senses of the word contingence. M. LeRoy retains them all and he does not sufficiently distinguish them, but he introduces a new one. Experimental laws are only approximate, and if some appear to us as exact, it is because we have artificially transformed them into what I have above called a principle. We have made this transformation freely, and as the caprice which has determined us to make it is something eminently contingent, we have communicated this contingence to the law itself. It is in this sense that we have the right to say that determinism supposes freedom, since it is freely that we become determinists. Perhaps it will be found that this is to give large scope to nominalism and that the introduction of this new sense of the word contingence will not help much to solve all those questions which naturally arise and of which we have just been speaking.

I do not at all wish to investigate here the foundations of the principle of induction; I know very well that I should not succeed; it is as difficult to justify this principle as to get on without it. I only wish to show how scientists apply it and are forced to apply it.

When the same antecedent recurs, the same consequent must likewise recur; such is the ordinary statement. But reduced to these terms this principle could be of no use. For one to be able to say that the same antecedent recurred, it would be necessary for the circumstances *all* to be reproduced, since no one is absolutely indifferent, and for them to be *exactly* reproduced. And, as that will never happen, the principle can have no application.

We should therefore modify the enunciation and say: If an antecedent $A$ has once produced a consequent $B$, an antecedent $A'$, slightly different from $A$, will produce a consequent $B'$, slightly different from $B$. But how shall we recognize that the antecedents $A$ and $A'$ are 'slightly different'? If some one of the circumstances can be expressed by a number, and this number has in the two cases values very near together, the sense of the phrase ' slightly different ' is relatively clear; the principle then signifies that the consequent is a continuous function of the antecedent. And as a practical rule, we reach this conclusion that we have the right to interpolate. This is in fact what scientists do every day, and without interpolation all science would be impossible.

Yet observe one thing. The law sought may be represented by a curve. Experiment has taught us certain points of this curve. In virtue of the principle we have just stated, we believe these points may be connected by a continuous graph. We trace this graph with the eye. New experiments will furnish us new points of the curve. If these points are outside of the graph traced in advance, we shall have to modify our curve, but not to abandon our principle. Through any points, however numerous they may be, a continuous curve may always be passed. Doubtless, if this curve is too capricious, we shall be shocked (and we shall even suspect errors of experiment), but the principle will not be directly put at fault.

Furthermore, among the circumstances of a phenomenon, there are some that we regard as negligible, and we shall consider $A$ and $A'$ as slightly different if they differ only by these accessory circumstances. For instance, I have ascertained that hydrogen unites with oxygen under the influence of the electric spark, and I am certain that these two gases will unite anew, although the longitude of Jupiter may have changed considerably in the interval. We assume, for instance, that the state of distant bodies can have no sensible influence on terrestrial phenomena, and that seems in fact requisite, but there are cases where the choice of these practically indifferent circumstances admits of more arbitrariness or, if you choose, requires more tact.

One more remark: The principle of induction would be inapplicable if there did not exist in nature a great quantity of bodies like one another, or almost alike, and if we could not infer, for instance, from one bit of phosphorus to another bit of phosphorus.

If we reflect on these considerations, the problem of determinism and of contingence will appear to us in a new light.

Suppose we were able to embrace the series of all phenomena of the universe in the whole sequence of time. We could envisage what might be called the *sequences;* I mean relations between antecedent and consequent. I do not wish to speak of constant relations or laws, I envisage separately (individually, so to speak) the different sequences realized.

We should then recognize that among these sequences there are no two altogether alike. But, if the principle of induction, as we have just stated it, is true, there will be those almost alike and that can be classed alongside one another. In other words, it is possible to make a classification of sequences.

It is to the possibility and the legitimacy of such a classification that determinism, in the end, reduces. This is all that the preceding analysis leaves of it. Perhaps under this modest form it will seem less appalling to the moralist.

It will doubtless be said that this is to come back by a detour to M. LeRoy's conclusion which a moment ago we seemed to reject: we are determinists voluntarily. And in fact all classification supposes the active intervention of the classifier. I agree that this may be maintained, but it seems to me that this detour will not have been useless and will have contributed to enlighten us a little.

### 6. *Objectivity of Science*

I arrive at the question set by the title of this article: What is the objective value of science? And first what should we understand by objectivity?

What guarantees the objectivity of the world in which we live is that this world is common to us with other thinking beings. Through the communications that we have with other men, we receive from them ready-made reasonings; we know that these reasonings do not come from us and at the same time we recognize in them the work of reasonable beings like ourselves. And as these reasonings appear to fit the world of our sensations, we think we may infer that these reasonable beings have seen the same thing as we; thus it is we know we have not been dreaming.

Such, therefore, is the first condition of objectivity; what is objective must be common to many minds and consequently transmissible from one to the other, and as this transmission can only come about by that 'discourse' which inspires so much distrust in M. LeRoy, we are even forced to conclude: no discourse, no objectivity.

The sensations of others will be for us a world eternally closed. We have no means of verifying that the sensation I call red is the same as that which my neighbor calls red.

Suppose that a cherry and a red poppy produce on me the sensation $A$ and on him the sensation $B$ and that, on the contrary, a leaf produces on me the sensation $B$ and on him the sensation $A$. It is clear we shall never know anything about it; since I shall call red the sensation $A$ and green the sensation $B$, while he will call the first green and the second red. In compensation, what we shall be able to ascertain is that, for him as for me, the cherry and the red poppy produce the *same* sensation, since he gives the same name to the sensations he feels and I do the same.

Sensations are therefore intransmissible, or rather all that is pure quality in them is intransmissible and forever impenetrable. But it is not the same with relations between these sensations.

From this point of view, all that is objective is devoid of all quality and is only pure relation. Certes, I shall not go so far as to say that objectivity is only pure quantity (this would be to particularize too far the nature of the relations in question), but we understand how some one could have been carried away into saying that the world is only a differential equation.

With due reserve regarding this paradoxical proposition, we must nevertheless admit that nothing is objective which is not transmissible, and consequently that the relations between the sensations can alone have an objective value.

Perhaps it will be said that the esthetic emotion, which is common to all mankind, is proof that the qualities of our sensations are also the same for all men and hence are objective. But if we think about this, we shall see that the proof is not complete; what is proved is that this emotion is aroused in John as in James by the sensations to which James and John give the same name or by the corresponding combinations of these sensations; either because this emotion is associated in John with the sensation $A$, which John calls red, while parallelly it is associated in James with the sensation *B*, which James calls red; or better because this emotion is aroused, not by the qualities themselves of the sensations, but by the harmonious combination of their relations of which we undergo the unconscious impression.

Such a sensation is beautiful, not because it possesses such a quality, but because it occupies such a place in the woof of our associations of ideas, so that it can not be excited without putting in motion the 'receiver' which is at the other end of the thread and which corresponds to the artistic emotion.

Whether we take the moral, the esthetic or the scientific point of view, it is always the same thing. Nothing is objective except what is identical for all; now we can only speak of such an identity if a comparison is possible, and can be translated into a 'money of exchange' capable of transmission from one mind to another. Nothing, therefore, will have objective value except what is transmissible by 'discourse,' that is, intelligible.

But this is only one side of the question. An absolutely disordered aggregate could not have objective value since it would be unintelligible, but no more can a well-ordered assemblage have it, if it does not correspond to sensations really experienced. It seems to me superfluous to recall this condition, and I should not have dreamed of it, if it had not lately been maintained that physics is not an experimental science. Although this opinion has no chance of being adopted either by physicists or by philosophers, it is well to be warned so as not to let oneself slip over the declivity which would lead thither. Two conditions are therefore to be fulfilled, and if the first separates reality[1] from the dream, the second distinguishes it from the romance.

Now what is science? I have explained in the preceding article, it is before all a classification, a manner of bringing together facts which appearances separate, though they were bound together by some natural and hidden kinship. Science, in other words, is a system of relations. Now we have just said, it is in the relations alone that objectivity must be sought; it

---

[1] I here use the word real as a synonym of objective; I thus conform to common usage; perhaps I am wrong, our dreams are real, but they are not objective.

would be vain to seek it in beings considered as isolated from one another.

To say that science can not have objective value since it teaches us only relations, this is to reason backward, since, precisely, it is relations alone which can be regarded as objective.

External objects, for instance, for which the word *object* was invented, are really *objects* and not fleeting and fugitive appearances, because they are not only groups of sensations, but groups cemented by a constant bond. It is this bond, and this bond alone, which is the object in itself, and this bond is a relation.

Therefore, when we ask what is the objective value of science, that does not mean: Does science teach us the true nature of things? but it means: Does it teach us the true relations of things?

To the first question, no one would hesitate to reply, no; but I think we may go farther; not only science can not teach us the nature of things; but nothing is capable of teaching it to us, and if any god knew it, he could not find words to express it. Not only can we not divine the response, but if it were given to us we could understand nothing of it; I ask myself even whether we really understand the question.

When, therefore, a scientific theory pretends to teach us what heat is, or what is electricity, or life, it is condemned beforehand; all it can give us is only a crude image. It is, therefore, provisional and crumbling.

The first question being out of reason, the second remains. Can science teach us the true relations of things? What it joins together should that be put asunder, what it puts asunder should that be joined together?

To understand the meaning of this new question, it is needful to refer to what was said above on the conditions of objectivity. Have these relations an objective value? That means: Are these relations the same for all? Will they still be the same for those who shall come after us?

It is clear that they are not the same for the scientist and the ignorant person. But that is unimportant, because if the ignorant person does not see them all at once, the scientist may succeed in making him see them by a series of experiments and reasonings.

The thing essential is that there are points on which all those acquainted with the experiments made can reach accord.

The question is to know whether this accord will be durable and whether it will persist for our successors. It may be asked whether the unions that the science of to-day makes will be confirmed by the science of to-morrow. To affirm that it will be so we can not invoke any *a priori* reason; but this is a question of fact, and science has already lived long enough for us to be able to find out by asking its history whether the edifices it builds stand the test of time, or whether they are only ephemeral constructions.

Now what do we see? At the first blush it seems to us that the theories last only a day and that ruins upon ruins accumulate. To-day the theories are born, to-morrow they are the fashion, the day after to-morrow they are classic, the fourth day they are superannuated, and the fifth they are forgotten. But if we look more closely, we see that what thus succumb are the theories properly so called, those which pretend to teach us what things are. But there is in them something which usually survives. If one of them taught us a true relation, this relation is definitively acquired, and it will be found again under a new disguise in the other theories which will successively come to reign in place of the old.

Take only a single example: The theory of the undulations of the ether taught us that light is a motion; to-day fashion favors the electromagnetic theory which teaches us that light is a current. We do not consider whether we could reconcile them and say that light is a current, and that this current is a motion. As it is probable in any case that this motion would not be identical with that which the partisans of the old theory presume, we might think ourselves justified in saying that this old theory is dethroned. And yet something of it remains, since between the hypothetical currents which Maxwell supposes there are the same relations as between the hypothetical motions that Fresnel supposed. There is, therefore, something which remains over and this something is the essential. This it is which explains how we see the present physicists pass without any embarrassment from the language of Fresnel to that of Maxwell. Doubtless many connections that were believed well established have been abandoned, but the greatest number remain and it would seem must remain.

And for these, then, what is the measure of their objectivity? Well, it is precisely the same as for our belief in external objects. These latter are real in this, that the sensations they make us feel appear to us as united to each other by I know not what indestructible cement and not by the hazard of a day. In the same way science reveals to us between phenomena other bonds finer but not less solid; these are threads so slender that they long remained unperceived, but once noticed there remains no way of not seeing them; they are therefore not less real than those which give their reality to external objects; small matter that they are more recently known, since neither can perish before the other.

It may be said, for instance, that the ether is no less real than any external body; to say this body exists is to say there is between the color of this body, its taste, its smell, an intimate bond, solid and persistent; to say the ether exists is to say there is a natural kinship between all the optical phenomena, and neither of the two propositions has less value than the other.

And the scientific syntheses have in a sense even more reality than those of the ordinary senses, since they embrace more terms and tend to absorb in them the partial syntheses.

It will be said that science is only a classification and that a classification can not be true, but convenient. But it is true that it is convenient, it is true that it is so not only for me, but for all men; it is true that it will remain convenient for our descendants; it is true finally that this can not be by chance.

In sum, the sole objective reality consists in the relations of things whence results the universal harmony. Doubtless these relations, this harmony, could not be conceived outside of a mind which conceives them. But they are nevertheless objective because they are, will become, or will remain, common to all thinking beings.

This will permit us to revert to the question of the rotation of the earth which will give us at the same time a chance to make clear what precedes by an example.

### 7. *The Rotation of the Earth*

". . . Therefore," have I said in *Science and Hypothesis*, "this affirmation, the earth turns round, has no meaning . . . or rather these two propositions, the earth turns round, and, it is more convenient to suppose that the earth turns round, have one and the same meaning."

These words have given rise to the strangest interpretations. Some have thought they saw in them the rehabilitation of Ptolemy's system, and perhaps the justification of Galileo's condemnation.

Those who had read attentively the whole volume could not, however, delude themselves. This truth, the earth turns round, was put on the same footing as Euclid's postulate, for example. Was that to reject it? But better; in the same language it may very well be said: These two propositions, the external world exists, or, it is more convenient to suppose that it exists, have one and the same meaning. So the hypothesis of the rotation of the earth would have the same degree of certitude as the very existence of external objects.

But after what we have just explained in the fourth part, we may go farther. A physical theory, we have said, is by so much the more true as it puts in evidence more true relations. In the light of this new principle, let us examine the question which occupies us.

No, there is no absolute space; these two contradictory propositions: 'The earth turns round' and 'The earth does not turn round' are, therefore, neither of them more true than the other. To affirm one while denying the other, *in the kinematic sense*, would be to admit the existence of absolute space.

But if the one reveals true relations that the other hides from us, we can nevertheless regard it as physically more true than the other, since it has a richer content. Now in this regard no doubt is possible.

Behold the apparent diurnal motion of the stars, and the diurnal motion of the other heavenly bodies, and besides, the flattening of the earth, the rotation of Foucault's pendulum, the gyration of cyclones, the trade-winds, what not else? For the

24

Ptolemaist all these phenomena have no bond between them; for the Copernican they are produced by the one same cause. In saying, the earth turns round, I affirm that all these phenomena have an intimate relation, and *that is true,* and that remains true, although there is not and can not be absolute space.

So much for the rotation of the earth upon itself; what shall we say of its revolution around the sun? Here again, we have three phenomena which for the Ptolemaist are absolutely independent and which for the Copernican are referred back to the same origin; they are the apparent displacements of the planets on the celestial sphere, the aberration of the fixed stars, the parallax of these same stars. Is it by chance that all the planets admit an inequality whose period is a year, and that this period is precisely equal to that of aberration, precisely equal besides to that of parallax? To adopt Ptolemy's system is to answer, yes; to adopt that of Copernicus is to answer, no; this is to affirm that there is a bond between the three phenomena, and that also is true, although there is no absolute space.

In Ptolemy's system, the motions of the heavenly bodies can not be explained by the action of central forces, celestial mechanics is impossible. The intimate relations that celestial mechanics reveals to us between all the celestial phenomena are true relations; to affirm the immobility of the earth would be to deny these relations, that would be to fool ourselves.

The truth for which Galileo suffered remains, therefore, the truth, although it has not altogether the same meaning as for the vulgar, and its true meaning is much more subtle, more profound and more rich.

### 8. *Science for Its Own Sake*

Not against M. LeRoy do I wish to defend science for its own sake; maybe this is what he condemns, but this is what he cultivates, since he loves and seeks truth and could not live without it. But I have some thoughts to express.

We can not know all facts and it is necessary to choose those which are worthy of being known. According to Tolstoi, scientists make this choice at random, instead of making it, which would be reasonable, with a view to practical applications. On the contrary, scientists think that certain facts are more interesting than others, because they complete an unfinished harmony, or because they make one foresee a great number of other facts. If they are wrong, if this hierarchy of facts that they implicitly postulate is only an idle illusion, there could be no science for its own sake, and consequently there could be no science. As for me, I believe they are right, and, for example, I have shown above what is the high value of astronomical facts, not because they are capable of practical applications, but because they are the most instructive of all.

It is only through science and art that civilization is of value. Some have wondered at the formula: science for its own sake; and yet it is as good as life for its own sake, if life is only misery; and even as happiness for its own sake, if we do not believe that all pleasures are of the same quality, if we do not wish to admit that the goal of civilization is to furnish alcohol to people who love to drink.

Every act should have an aim. We must suffer, we must work, we must pay for our place at the game, but this is for seeing's sake; or at the very least that others may one day see.

All that is not thought is pure nothingness; since we can think only thoughts and all the words we use to speak of things can express only thoughts, to say there is something other than thought, is therefore an affirmation which can have no meaning.

And yet—strange contradiction for those who believe in time—geologic history shows us that life is only a short episode between two eternities of death, and that, even in this episode, conscious thought has lasted and will last only a moment. Thought is only a gleam in the midst of a long night.

But it is this gleam which is everything.

# SCIENCE AND METHOD

# INTRODUCTION

I BRING together here different studies relating more or less directly to questions of scientific methodology. The scientific method consists in observing and experimenting; if the scientist had at his disposal infinite time, it would only be necessary to say to him: ' Look and notice well '; but, as there is not time to see everything, and as it is better not to see than to see wrongly, it is necessary for him to make choice. The first question, therefore, is how he should make this choice. This question presents itself as well to the physicist as to the historian; it presents itself equally to the mathematician, and the principles which should guide each are not without analogy. The scientist conforms to them instinctively, and one can, reflecting on these principles, foretell the future of mathematics.

We shall understand them better yet if we observe the scientist at work, and first of all it is necessary to know the psychologic mechanism of invention and, in particular, that of mathematical creation. Observation of the processes of the work of the mathematician is particularly instructive for the psychologist.

In all the sciences of observation account must be taken of the errors due to the imperfections of our senses and our instruments. Luckily, we may assume that, under certain conditions, these errors are in part self-compensating, so as to disappear in the average; this compensation is due to chance. But what is chance? This idea is difficult to justify or even to define; and yet what I have just said about the errors of observation, shows that the scientist can not neglect it. It therefore is necessary to give a definition as precise as possible of this concept, so indispensable yet so illusive.

These are generalities applicable in sum to all the sciences; and for example the mechanism of mathematical invention does not differ sensibly from the mechanism of invention in general. Later I attack questions relating more particularly to certain special sciences and first to pure mathematics.

In the chapters devoted to these, I have to treat subjects a little more abstract. I have first to speak of the notion of space; every one knows space is relative, or rather every one says so, but many think still as if they believed it absolute; it suffices to reflect a little however to perceive to what contradictions they are exposed.

The questions of teaching have their importance, first in themselves, then because reflecting on the best way to make new ideas penetrate virgin minds is at the same time reflecting on how these notions were acquired by our ancestors, and consequently on their true origin, that is to say, in reality on their true nature. Why do children usually understand nothing of the definitions which satisfy scientists? Why is it necessary to give them others? This is the question I set myself in the succeeding chapter and whose solution should, I think, suggest useful reflections to the philosophers occupied with the logic of the sciences.

On the other hand, many geometers believe we can reduce mathematics to the rules of formal logic. Unheard-of efforts have been made to do this; to accomplish it, some have not hesitated, for example, to reverse the historic order of the genesis of our conceptions and to try to explain the finite by the infinite. I believe I have succeeded in showing, for all those who attack the problem unprejudiced, that here there is a fallacious illusion. I hope the reader will understand the importance of the question and pardon me the aridity of the pages devoted to it.

The concluding chapters relative to mechanics and astronomy will be easier to read.

Mechanics seems on the point of undergoing a complete revolution. Ideas which appeared best established are assailed by bold innovators. Certainly it would be premature to decide in their favor at once simply because they are innovators.

But it is of interest to make known their doctrines, and this is what I have tried to do. As far as possible I have followed the historic order; for the new ideas would seem too astonishing unless we saw how they arose.

Astronomy offers us majestic spectacles and raises gigantic problems. We can not dream of applying to them directly the experimental method; our laboratories are too small. But analogy with phenomena these laboratories permit us to attain may nevertheless guide the astronomer. The Milky Way, for example, is an assemblage of suns whose movements seem at first capricious. But may not this assemblage be compared to that of the molecules of a gas, whose properties the kinetic theory of gases has made known to us? It is thus by a roundabout way that the method of the physicist may come to the aid of the astronomer.

Finally I have endeavored to give in a few lines the history of the development of French geodesy; I have shown through what persevering efforts, and often what dangers, the geodesists have procured for us the knowledge we have of the figure of the earth. Is this then a question of method? Yes, without doubt, this history teaches us in fact by what precautions it is necessary to surround a serious scientific operation and how much time and pains it costs to conquer one new decimal.

# BOOK I

## SCIENCE AND THE SCIENTIST

### CHAPTER I

#### The Choice of Facts

TOLSTOI somewhere explains why 'science for its own sake' is
in his eyes an absurd conception. We can not know *all* facts,
since their number is practically infinite. It is necessary to
choose; then we may let this choice depend on the pure caprice
of our curiosity; would it not be better to let ourselves be guided
by utility, by our practical and above all by our moral needs;
have we nothing better to do than to count the number of lady-
bugs on our planet?

It is clear the word utility has not for him the sense men of
affairs give it, and following them most of our contemporaries.
Little cares he for industrial applications, for the marvels of
electricity or of automobilism, which he regards rather as obstacles to moral progress; utility for him is solely what can make
man better.

For my part, it need scarce be said, I could never be content
with either the one or the other ideal; I want neither that plutoc-
racy grasping and mean, nor that democracy goody and mediocre,
occupied solely in turning the other cheek, where would dwell
sages without curiosity, who, shunning excess, would not die of
disease, but would surely die of ennui. But that is a matter of
taste and is not what I wish to discuss.

The question nevertheless remains and should fix our attention;
if our choice can only be determined by caprice or by immediate
utility, there can be no science for its own sake, and consequently
no science. But is that true? That a choice must be made is
incontestable; whatever be our activity, facts go quicker than we,
and we can not catch them; while the scientist discovers one fact,

there happen milliards of milliards in a cubic millimeter of his body. To wish to comprise nature in science would be to want to put the whole into the part.

But scientists believe there is a hierarchy of facts and that among them may be made a judicious choice. They are right, since otherwise there would be no science, yet science exists. One need only open the eyes to see that the conquests of industry which have enriched so many practical men would never have seen the light, if these practical men alone had existed and if they had not been preceded by unselfish devotees who died poor, who never thought of utility, and yet had a guide far other than caprice.

As Mach says, these devotees have spared their successors the trouble of thinking. Those who might have worked solely in view of an immediate application would have left nothing behind them, and, in face of a new need, all must have been begun over again. Now most men do not love to think, and this is perhaps fortunate when instinct guides them, for most often, when they pursue an aim which is immediate and ever the same, instinct guides them better than reason would guide a pure intelligence. But instinct is routine, and if thought did not fecundate it, it would no more progress in man than in the bee or ant. It is needful then to think for those who love not thinking, and, as they are numerous, it is needful that each of our thoughts be as often useful as possible, and this is why a law will be the more precious the more general it is.

This shows us how we should choose: the most interesting facts are those which may serve many times; these are the facts which have a chance of coming up again. We have been so fortunate as to be born in a world where there are such. Suppose that instead of 60 chemical elements there were 60 milliards of them, that they were not some common, the others rare, but that they were uniformly distributed. Then, every time we picked up a new pebble there would be great probability of its being formed of some unknown substance; all that we knew of other pebbles would be worthless for it; before each new object we should be as the new-born babe; like it we could only obey our caprices or our needs. Biologists would be just as much at a loss if there were only individuals and no species and if heredity did not make sons like their fathers.

In such a world there would be no science; perhaps thought and even life would be impossible, since evolution could not there develop the preservational instincts. Happily it is not so; like all good fortune to which we are accustomed, this is not appreciated at its true worth.

Which then are the facts likely to reappear? They are first the simple facts. It is clear that in a complex fact a thousand circumstances are united by chance, and that only a chance still much less probable could reunite them anew. But are there any simple facts? And if there are, how recognize them? What assurance is there that a thing we think simple does not hide a dreadful complexity? All we can say is that we ought to prefer the facts which *seem* simple to those where our crude eye discerns unlike elements. And then one of two things: either this simplicity is real, or else the elements are so intimately mingled as not to be distinguishable. In the first case there is chance of our meeting anew this same simple fact, either in all its purity or entering itself as element in a complex manifold. In the second case this intimate mixture has likewise more chances of recurring than a heterogeneous assemblage; chance knows how to mix, it knows not how to disentangle, and to make with multiple elements a well-ordered edifice in which something is distinguishable, it must be made expressly. The facts which appear simple, even if they are not so, will therefore be more easily revived by chance. This it is which justifies the method instinctively adopted by the scientist, and what justifies it still better, perhaps, is that oft-recurring facts appear to us simple, precisely because we are used to them.

But where is the simple fact? Scientists have been seeking it in the two extremes, in the infinitely great and in the infinitely small. The astronomer has found it because the distances of the stars are immense, so great that each of them appears but as a point, so great that the qualitative differences are effaced, and because a point is simpler than a body which has form and qualities. The physicist on the other hand has sought the elementary phenomenon in fictively cutting up bodies into infinitesimal cubes, because the conditions of the problem, which undergo slow and continuous variation in passing from one point of the body to another, may be regarded as constant in the interior of each of these little cubes. In the same way the biologist has been instinctively led to regard the cell as more interesting than the whole animal, and the outcome has shown his wisdom, since cells belonging to organisms the most different are more alike, for the one who can recognize their resemblances, than are these organisms themselves. The sociologist is more embarrassed; the elements, which for him are men, are too unlike, too variable, too capricious, in a word, too complex; besides, history never begins over again. How then choose the interesting fact, which is that which begins again? Method is precisely the choice of facts; it is needful then to be occupied first with creating a method, and many have been imagined, since none imposes itself, so that sociology is the science which has the most methods and the fewest results.

Therefore it is by the regular facts that it is proper to begin; but after the rule is well established, after it is beyond all doubt, the facts in full conformity with it are erelong without interest since they no longer teach us anything new. It is then the exception which becomes important. We cease to seek resemblances; we devote ourselves above all to the differences, and among the differences are chosen first the most accentuated, not only because they are the most striking, but because they will be the most instructive. A simple example will make my thought plainer: Suppose one wishes to determine a curve by observing some of its points. The practician who concerns himself only with immediate utility would observe only the points he might need for some special object. These points would be badly distributed on the curve; they would be crowded in certain regions, rare in others, so that it would be impossible to join them by a continuous line, and they would be unavailable for other applications. The scientist will proceed differently; as he wishes to study the curve for itself, he will distribute regularly the points to be observed, and when enough are known he will join them by a regular line and then he will have the entire curve. But for that how does he proceed? If he has determined an extreme point of the curve, he does not stay near this extremity, but goes first to the other end; after the two extremities the most instructive point will be the mid-point, and so on.

So when a rule is established we should first seek the cases where this rule has the greatest chance of failing.  Thence, among other reasons, come the interest of astronomic facts, and the interest of the geologic past; by going very far away in space or very far away in time, we may find our usual rules entirely overturned, and these grand overturnings aid us the better to see or the better to understand the little changes which may happen nearer to us, in the little corner of the world where we are called to live and act.  We shall better know this corner for having traveled in distant countries with which we have nothing to do.

But what we ought to aim at is less the ascertainment of resemblances and differences than the recognition of likenesses hidden under apparent divergences.  Particular rules seem at first discordant, but looking more closely we see in general that they resemble each other; different as to matter, they are alike as to form, as to the order of their parts.  When we look at them with this bias, we shall see them enlarge and tend to embrace everything.  And this it is which makes the value of certain facts which come to complete an assemblage and to show that it is the faithful image of other known assemblages.

I will not further insist, but these few words suffice to show that the scientist does not choose at random the facts he observes.  He does not, as Tolstoi says, count the lady-bugs, because, however interesting lady-bugs may be, their number is subject to capricious variations.  He seeks to condense much experience and much thought into a slender volume; and that is why a little book on physics contains so many past experiences and a thousand times as many possible experiences whose result is known beforehand.

But we have as yet looked at only one side of the question.  The scientist does not study nature because it is useful; he studies it because he delights in it, and he delights in it because it is beautiful.  If nature were not beautiful, it would not be worth knowing, and if nature were not worth knowing, life would not be worth living.  Of course I do not here speak of that beauty which strikes the senses, the beauty of qualities and of appearances; not that I undervalue such beauty, far from it, but it has nothing to do with science; I mean that profounder beauty which comes from the harmonious order of the parts and which a pure intelligence can grasp. This it is which gives body, a structure so to speak, to the iridescent appearances which flatter our senses, and without this support the beauty of these fugitive dreams would be only imperfect, because it would be vague and always fleeting. On the contrary, intellectual beauty is sufficient unto itself, and it is for its sake, more perhaps than for the future good of humanity, that the scientist devotes himself to long and difficult labors.

It is, therefore, the quest of this especial beauty, the sense of the harmony of the cosmos, which makes us choose the facts most fitting to contribute to this harmony, just as the artist chooses from among the features of his model those which perfect the picture and give it character and life. And we need not fear that this instinctive and unavowed prepossession will turn the scientist aside from the search for the true. One may dream a harmonious world, but how far the real world will leave it behind! The greatest artists that ever lived, the Greeks, made their heavens; how shabby it is beside the true heavens, ours!

And it is because simplicity, because grandeur, is beautiful, that we preferably seek simple facts, sublime facts, that we delight now to follow the majestic course of the stars, now to examine with the microscope that prodigious littleness which is also a grandeur, now to seek in geologic time the traces of a past which attracts because it is far away.

We see too that the longing for the beautiful leads us to the same choice as the longing for the useful. And so it is that this economy of thought, this economy of effort, which is, according to Mach, the constant tendency of science, is at the same time a source of beauty and a practical advantage. The edifices that we admire are those where the architect has known how to proportion the means to the end, where the columns seem to carry gaily, without effort, the weight placed upon them, like the gracious caryatids of the Erechtheum.

Whence comes this concordance? Is it simply that the things which seem to us beautiful are those which best adapt themselves to our intelligence, and that consequently they are at the same time the implement this intelligence knows best how to use?

Or is there here a play of evolution and natural selection? Have the peoples whose ideal most conformed to their highest interest exterminated the others and taken their place? All pursued their ideals without reference to consequences, but while this quest led some to destruction, to others it gave empire. One is tempted to believe it. If the Greeks triumphed over the barbarians and if Europe, heir of Greek thought, dominates the world, it is because the savages loved loud colors and the clamorous tones of the drum which occupied only their senses, while the Greeks loved the intellectual beauty which hides beneath sensuous beauty, and this intellectual beauty it is which makes intelligence sure and strong.

Doubtless such a triumph would horrify Tolstoi, and he would not like to acknowledge that it might be truly useful. But this disinterested quest of the true for its own beauty is sane also and able to make man better. I well know that there are mistakes, that the thinker does not always draw thence the serenity he should find therein, and even that there are scientists of bad character. Must we, therefore, abandon science and study only morals? What! Do you think the moralists themselves are irreproachable when they come down from their pedestal?

# CHAPTER II

To foresee the future of mathematics, the true method is to study its history and its present state.

Is this not for us mathematicians in a way a professional procedure? We are accustomed to *extrapolate*, which is a means of deducing the future from the past and present, and as we well know what this amounts to, we run no risk of deceiving ourselves about the range of the results it gives us.

We have had hitherto prophets of evil. They blithely reiterate that all problems capable of solution have already been solved, and that nothing is left but gleaning. Happily the case of the past reassures us. Often it was thought all problems were solved or at least an inventory was made of all admitting solution. And then the sense of the word solution enlarged, the insoluble problems became the most interesting of all, and others unforeseen presented themselves. For the Greeks a good solution was one employing only ruler and compasses; then it became one obtained by the extraction of roots, then one using only algebraic or logarithmic functions. The pessimists thus found themselves always outflanked, always forced to retreat, so that at present I think there are no more.

My intention, therefore, is not to combat them, as they are dead; we well know that mathematics will continue to develop, but the question is how, in what direction? You will answer, 'in every direction,' and that is partly true; but if it were wholly true it would be a little appalling. Our riches would soon become encumbering and their accumulation would produce a medley as impenetrable as the unknown true was for the ignorant.

The historian, the physicist, even, must make a choice among facts; the head of the scientist, which is only a corner of the universe, could never contain the universe entire; so that among the innumerable facts nature offers, some will be passed by, others retained.

25                    369

Just so, *a fortiori*, in mathematics; no more can the geometer hold fast pell-mell all the facts presenting themselves to him; all the more because he it is, almost I had said his caprice, that creates these facts. He constructs a wholly new combination by putting together its elements; nature does not in general give it to him ready made.

Doubtless it sometimes happens that the mathematician undertakes a problem to satisfy a need in physics; that the physicist or engineer asks him to calculate a number for a certain application. Shall it be said that we geometers should limit ourselves to awaiting orders, and, in place of cultivating our science for our own delectation, try only to accommodate ourselves to the wants of our patrons? If mathematics has no other object besides aiding those who study nature, it is from these we should await orders. Is this way of looking at it legitimate? Certainly not; if we had not cultivated the exact sciences for themselves, we should not have created mathematics the instrument, and the day the call came from the physicist we should have been helpless.

Nor do the physicists wait to study a phenomenon until some urgent need of material life has made it a necessity for them; and they are right. If the scientists of the eighteenth century had neglected electricity as being in their eyes only a curiosity without practical interest, we should have had in the twentieth century neither telegraphy, nor electro-chemistry, nor electro-technics. The physicists, compelled to choose, are therefore not guided in their choice solely by utility. How then do they choose between the facts of nature? We have explained it in the preceding chapter: the facts which interest them are those capable of leading to the discovery of a law, and so they are analogous to many other facts which do not seem to us isolated, but closely grouped with others. The isolated fact attracts all eyes, those of the layman as well as of the scientist. But what the genuine physicist alone knows how to see, is the bond which unites many facts whose analogy is profound but hidden. The story of Newton's apple is probably not true, but it is symbolic; let us speak of it then as if it were true. Well then, we must believe that before Newton plenty of men had seen apples fall; not one knew how to conclude anything therefrom. Facts would be sterile were there not minds capable of choosing among them, discerning those behind which something was hidden, and of recognizing what is hiding, minds which under the crude fact perceive the soul of the fact.

We find just the same thing in mathematics. From the varied elements at our disposal we can get millions of different combinations; but one of these combinations, in so far as it is isolated, is absolutely void of value. Often we have taken great pains to construct it, but it serves no purpose, if not perhaps to furnish a task in secondary education. Quite otherwise will it be when this combination shall find place in a class of analogous combinations and we shall have noticed this analogy. We are no longer in the presence of a fact, but of a law. And upon that day the real discoverer will not be the workman who shall have patiently built up certain of these combinations; it will be he who brings to light their kinship. The first will have seen merely the crude fact, only the other will have perceived the soul of the fact. Often to fix this kinship it suffices him to make a new word, and this word is creative. The history of science furnishes us a crowd of examples familiar to all.

The celebrated Vienna philosopher Mach has said that the rôle of science is to produce economy of thought, just as machines produce economy of effort. And that is very true. The savage reckons on his fingers or by heaping pebbles. In teaching children the multiplication table we spare them later innumerable pebble bunchings. Some one has already found out, with pebbles or otherwise, that 6 times 7 is 42 and has had the idea of noting the result, and so we need not do it over again. He did not waste his time even if he reckoned for pleasure: his operation took him only two minutes; it would have taken in all two milliards if a milliard men had had to do it over after him.

The importance of a fact then is measured by its yield, that is to say, by the amount of thought it permits us to spare.

In physics the facts of great yield are those entering into a very general law, since from it they enable us to foresee a great number of others, and just so it is in mathematics. Suppose I have undertaken a complicated calculation and laboriously reached a result: I shall not be compensated for my trouble if thereby I have not become capable of foreseeing the results of other analogous calculations and guiding them with a certainty that avoids the gropings to which one must be resigned in a first attempt.  On the other hand, I shall not have wasted my time if these gropings themselves have ended by revealing to me the profound analogy of the problem just treated with a much more extended class of other problems; if they have shown me at once the resemblances and differences of these, if in a word they have made me perceive the possibility of a generalization. Then it is not a new result I have won, it is a new power.

The simple example that comes first to mind is that of an algebraic formula which gives us the solution of a type of numeric problems when finally we replace the letters by numbers.  Thanks to it, a single algebraic calculation saves us the pains of ceaselessly beginning over again new numeric calculations.  But this is only a crude example; we all know there are analogies inexpressible by a formula and all the more precious.

A new result is of value, if at all, when in unifying elements long known but hitherto separate and seeming strangers one to another it suddenly introduces order where apparently disorder reigned.  It then permits us to see at a glance each of these elements and its place in the assemblage.  This new fact is not merely precious by itself, but it alone gives value to all the old facts it combines.  Our mind is weak as are the senses; it would lose itself in the world's complexity were this complexity not harmonious; like a near-sighted person, it would see only the details and would be forced to forget each of these details before examining the following, since it would be incapable of embracing all. The only facts worthy our attention are those which introduce order into this complexity and so make it accessible.

Mathematicians attach great importance to the elegance of their methods and their results.  This is not pure dilettantism. What is it indeed that gives us the feeling of elegance in a solution, in a demonstration?  It is the harmony of the diverse parts, their symmetry, their happy balance; in a word it is all that introduces order, all that gives unity, that permits us to see clearly and to comprehend at once both the *ensemble* and the

details. But this is exactly what yields great results; in fact the
more we see this aggregate clearly and at a single glance, the
better we perceive its analogies with other neighboring objects,
consequently the more chances we have of divining the possible
generalizations. Elegance may produce the feeling of the unforeseen by the unexpected meeting of objects we are not accustomed
to bring together; there again it is fruitful, since it thus unveils
for us kinships before unrecognized. It is fruitful even when it
results only from the contrast between the simplicity of the
means and the complexity of the problem set; it makes us then
think of the reason for this contrast and very often makes us
see that chance is not the reason; that it is to be found in some
unexpected law. In a word, the feeling of mathematical elegance is only the satisfaction due to any adaptation of the solution to the needs of our mind, and it is because of this very
adaptation that this solution can be for us an instrument. Consequently this œsthetic satisfaction is bound up with the economy of thought. Again the comparison of the Erechtheum
comes to my mind, but I must not use it too often.

It is for the same reason that, when a rather long calculation
has led to some simple and striking result, we are not satisfied
until we have shown that we should have been *able to foresee,*
if not this entire result, at least its most characteristic traits.
Why? What prevents our being content with a calculation
which has told us, it seems, all we wished to know? It is because, in analogous cases, the long calculation might not again
avail, and that this is not so about the reasoning often half intuitive which would have enabled us to foresee. This reasoning
being short, we see at a single glance all its parts, so that we immediately perceive what must be changed to adapt it to all the
problems of the same nature which can occur. And then it
enables us to foresee if the solution of these problems will be
simple, it shows us at least if the calculation is worth undertaking.

What we have just said suffices to show how vain it would be
to seek to replace by any mechanical procedure the free initiative
of the mathematician. To obtain a result of real value, it is not
enough to grind out calculations, or to have a machine to put

things in order; it is not order alone, it is unexpected order, which is worth while. The machine may gnaw on the crude fact, the soul of the fact will always escape it.

Since the middle of the last century, mathematicians are more and more desirous of attaining absolute rigor; they are right, and this tendency will be more and more accentuated. In mathematics rigor is not everything, but without it there is nothing. A demonstration which is not rigorous is nothingness. I think no one will contest this truth. But if it were taken too literally, we should be led to conclude that before 1820, for example, there was no mathematics; this would be manifestly excessive; the geometers of that time understood voluntarily what we explain by prolix discourse. This does not mean that they did not see it at all; but they passed over it too rapidly, and to see it well would have necessitated taking the pains to say it.

But is it always needful to say it so many times; those who were the first to emphasize exactness before all else have given us arguments that we may try to imitate; but if the demonstrations of the future are to be built on this model, mathematical treatises will be very long; and if I fear the lengthenings, it is not solely because I deprecate encumbering libraries, but because I fear that in being lengthened out, our demonstrations may lose that appearance of harmony whose usefulness I have just explained.

The economy of thought is what we should aim at, so it is not enough to supply models for imitation. It is needful for those after us to be able to dispense with these models and, in place of repeating an argument already made, summarize it in a few words. And this has already been attained at times. For instance, there was a type of reasoning found everywhere, and everywhere alike. They were perfectly exact but long. Then all at once the phrase ' uniformity of convergence ' was hit upon and this phrase made those arguments needless; we were no longer called upon to repeat them, since they could be understood. Those who conquer difficulties then do us a double service: first they teach us to do as they at need, but above all they enable us as often as possible to avoid doing as they, yet without sacrifice of exactness.

We have just seen by one example the importance of words in mathematics, but many others could be cited. It is hard to believe how much a well-chosen word can economize thought, as Mach says. Perhaps I have already said somewhere that mathematics is the art of giving the same name to different things. It is proper that these things, differing in matter, be alike in form, that they may, so to speak, run in the same mold. When the language has been well chosen, we are astonished to see that all the proofs made for a certain object apply immediately to many new objects; there is nothing to change, not even the words, since the names have become the same.

A well-chosen word usually suffices to do away with the exceptions from which the rules stated in the old way suffer; this is why we have created negative quantities, imaginaries, points at infinity, and what not. And exceptions, we must not forget, are pernicious because they hide the laws.

Well, this is one of the characteristics by which we recognize the facts which yield great results. They are those which allow of these happy innovations of language. The crude fact then is often of no great interest; we may point it out many times without having rendered great service to science. It takes value only when a wiser thinker perceives the relation for which it stands, and symbolizes it by a word.

Moreover the physicists do just the same. They have invented the word 'energy,' and this word has been prodigiously fruitful, because it also made the law by eliminating the exceptions, since it gave the same name to things differing in matter and like in form.

Among words that have had the most fortunate influence I would select 'group' and 'invariant.' They have made us see the essence of many mathematical reasonings; they have shown us in how many cases the old mathematicians considered groups without knowing it, and how, believing themselves far from one another, they suddenly found themselves near without knowing why.

To-day we should say that they had dealt with isomorphic groups. We now know that in a group the matter is of little interest, the form alone counts, and that when we know a group we thus know all the isomorphic groups; and thanks to these words 'group' and 'isomorphism,' which condense in a few syllables this subtile rule and quickly make it familiar to all minds, the transition is immediate and can be done with every economy of thought effort. The idea of group besides attaches to that of transformation. Why do we put such a value on the invention of a new transformation? Because from a single theorem it enables us to get ten or twenty; it has the same value as a zero adjoined to the right of a whole number.

This then it is which has hitherto determined the direction of mathematical advance, and just as certainly will determine it in the future. But to this end the nature of the problems which come up contributes equally. We can not forget what must be our aim. In my opinion this aim is double. Our science borders upon both philosophy and physics, and we work for our two neighbors; so we have always seen and shall still see mathematicians advancing in two opposite directions.

On the one hand, mathematical science must reflect upon itself, and that is useful since reflecting on itself is reflecting on the human mind which has created it, all the more because it is the very one of its creations for which it has borrowed least from without. This is why certain mathematical speculations are useful, such as those devoted to the study of the postulates, of unusual geometries, of peculiar functions. The more these speculations diverge from ordinary conceptions, and consequently from nature and applications, the better they show us what the human mind can create when it frees itself more and more from the tyranny of the external world, the better therefore they let us know it in itself.

But it is toward the other side, the side of nature, that we must direct the bulk of our army. There we meet the physicist or the engineer, who says to us: "Please integrate this differential equation for me; I might need it in a week in view of a construction which should be finished by that time." "This equation," we answer, "does not come under one of the integrable types; you know there are not many." "Yes, I know; but then what good are you?" Usually to understand each other is enough; the engineer in reality does not need the integral in finite terms;

he needs to know the general look of the integral function, or he simply wants a certain number which could readily be deduced from this integral if it were known. Usually it is not known, but the number can be calculated without it if we know exactly what number the engineer needs and with what approximation.

Formerly an equation was considered solved only when its solution had been expressed by aid of a finite number of known functions; but that is possible scarcely once in a hundred times. What we always can do, or rather what we should always seek to do, is to solve the problem *qualitatively* so to speak; that is to say, seek to know the general form of the curve which represents the unknown function.

It remains to find the *quantitative* solution of the problem; but if the unknown can not be determined by a finite calculation, it may always be represented by a convergent infinite series vhich enables us to calculate it. Can that be regarded as a true solution? We are told that Newton sent Leibnitz an anagram almost like this: aaaaabbbeeeeii, etc. Leibnitz naturally understood nothing at all of it; but we, who have the key, know that this anagram meant, translated into modern terms: ''I can integrate all differential equations''; and we are tempted to say that Newton had either great luck or strange delusions. He merely wished to say he could form (by the method of indeterminate coefficients) a series of powers formally satisfying the proposed equation.

Such a solution would not satisfy us to-day, and for two reasons: because the convergence is too slow and because the terms follow each other without obeying any law. On the contrary, the series ⊕ seems to us to leave nothing to be desired, first because it converges very quickly (this is for the practical man who wishes to get at a number as quickly as possible) and next because we see at a glance the law of the terms (this is to satisfy the esthetic need of the theorist).

But then there are no longer solved problems and others which are not; there are only problems *more or less* solved, according as they are solved by a series converging more or less rapidly, or ruled by a law more or less harmonious. It often happens however that an imperfect solution guides us toward a better one. Sometimes the series converges so slowly that the computation is impracticable and we have only succeeded in proving the possibility of the problem.

And then the engineer finds this a mockery, and justly, since it will not aid him to complete his construction by the date fixed. He little cares to know if it will benefit engineers of the twenty-second century. But as for us, we think differently and we are sometimes happier to have spared our grandchildren a day's work than to have saved our contemporaries an hour.

Sometimes by groping, empirically, so to speak, we reach a formula sufficiently convergent. "What more do you want?" says the engineer. And yet, in spite of all, we are not satisfied; we should have liked *to foresee* that convergence. Why? Because if we had known how to foresee it once, we would know how to foresee it another time. We have succeeded; that is a small matter in our eyes if we can not validly expect to do so again.

In proportion as science develops, its total comprehension becomes more difficult; then we seek to cut it in pieces and to be satisfied with one of these pieces: in a word, to specialize. If we went on in this way, it would be a grievous obstacle to the progress of science. As we have said, it is by unexpected union between its diverse parts that it progresses. To specialize too much would be to forbid these drawings together. It is to be hoped that congresses like those of Heidelberg and Rome, by putting us in touch with one another, will open for us vistas over neighboring domains and oblige us to compare them with our own, to range somewhat abroad from our own little village; thus they will be the best remedy for the danger just mentioned.

But I have lingered too long over generalities; it is time to enter into detail.

Let us pass in review the various special sciences which combined make mathematics; let us see what each has accomplished, whither it tends and what we may hope from it. If the preceding views are correct, we should see that the greatest advances in the past have happened when two of these sciences have united, when we have become conscious of the similarity of their form, despite the difference of their matter, when they have so modeled themselves upon each other that each could profit by the other's conquests. We should at the same time foresee in combinations of the same sort the progress of the future.

## Arithmetic

Progress in arithmetic has been much slower than in algebra and analysis, and it is easy to see why. The feeling of continuity is a precious guide which the arithmetician lacks; each whole number is separated from the others,—it has, so to speak, its own individuality. Each of them is a sort of exception and this is why general theorems are rarer in the theory of numbers; this is also why those which exist are more hidden and longer elude the searchers.

If arithmetic is behind algebra and analysis, the best thing for it to do is to seek to model itself upon these sciences so as to profit by their advance. The arithmetician ought therefore to take as guide the analogies with algebra. These analogies are numerous and if, in many cases, they have not yet been studied sufficiently closely to become utilizable, they at least have long been foreseen, and even the language of the two sciences shows they have been recognized. Thus we speak of transcendent numbers and thus we account for the future classification of these numbers already having as model the classification of transcendent functions, and still we do not as yet very well see how to pass from one classification to the other; but had it been seen, it would already have been accomplished and would no longer be the work of the future.

The first example that comes to my mind is the theory of congruences, where is found a perfect parallelism to the theory of algebraic equations. Surely we shall succeed in completing this parallelism, which must hold for instance between the theory of algebraic curves and that of congruences with two variables. And when the problems relative to congruences with several variables shall be solved, this will be a first step toward the solution of many questions of indeterminate analysis.

## Algebra

The theory of algebraic equations will still long hold the attention of geometers; numerous and very different are the sides whence it may be attacked.

We need not think algebra is ended because it gives us rules to form all possible combinations; it remains to find the interesting combinations, those which satisfy such and such a condition. Thus will be formed a sort of indeterminate analysis where the unknowns will no longer be whole numbers, but polynomials. This time it is algebra which will model itself upon arithmetic, following the analogy of the whole number to the integral polynomial with any coefficients or to the integral polynomial with integral coefficients.

### GEOMETRY

It looks as if geometry could contain nothing which is not already included in algebra or analysis; that geometric facts are only algebraic or analytic facts expressed in another language. It might then be thought that after our review there would remain nothing more for us to say relating specially to geometry. This would be to fail to recognize the importance of well-constructed language, not to comprehend what is added to the things themselves by the method of expressing these things and consequently of grouping them.

First the geometric considerations lead us to set ourselves new problems; these may be, if you choose, analytic problems, but such as we never would have set ourselves in connection with analysis. Analysis profits by them however, as it profits by those it has to solve to satisfy the needs of physics.

A great advantage of geometry lies in the fact that in it the senses can come to the aid of thought, and help find the path to follow, and many minds prefer to put the problems of analysis into geometric form. Unhappily our senses can not carry us very far, and they desert us when we wish to soar beyond the classical three dimensions. Does this mean that, beyond the restricted domain wherein they seem to wish to imprison us, we should rely only on pure analysis and that all geometry of more than three dimensions is vain and objectless? The greatest masters of a preceding generation would have answered 'yes'; to-day we are so familiarized with this notion that we can speak of it, even in a university course, without arousing too much astonishment.

But what good is it? That is easy to see: First it gives us a very convenient terminology, which expresses concisely what the ordinary analytic language would say in prolix phrases. Moreover, this language makes us call like things by the same name and emphasize analogies it will never again let us forget. It enables us therefore still to find our way in this space which is too big for us and which we can not see, always recalling visible space, which is only an imperfect image of it doubtless, but which is nevertheless an image. Here again, as in all the preceding examples, it is analogy with the simple which enables us to comprehend the complex.

This geometry of more than three dimensions is not a simple analytic geometry; it is not purely quantitative, but qualitative also, and it is in this respect above all that it becomes interesting. There is a science called *analysis situs* and which has for its object the study of the positional relations of the different elements of a figure, apart from their sizes. This geometry is purely qualitative; its theorems would remain true if the figures, instead of being exact, were roughly imitated by a child. We may also make an *analysis situs* of more than three dimensions. The importance of *analysis situs* is enormous and can not be too much emphasized; the advantage obtained from it by Riemann, one of its chief creators, would suffice to prove this. We must achieve its complete construction in the higher spaces; then we shall have an instrument which will enable us really to see in hyperspace and supplement our senses.

The problems of *analysis situs* would perhaps not have suggested themselves if the analytic language alone had been spoken; or rather, I am mistaken, they would have occurred surely, since their solution is essential to a crowd of questions in analysis, but they would have come singly, one after another, and without our being able to perceive their common bond.

### CANTORISM

I have spoken above of our need to go back continually to the first principles of our science, and of the advantage of this for the study of the human mind. This need has inspired two endeavors which have taken a very prominent place in the most recent annals of mathematics. The first is Cantorism, which has

rendered our science such conspicuous service.  Cantor introduced into science a new way of considering mathematical infinity.  One of the characteristic traits of Cantorism is that in
place of going up to the general by building up constructions
more and more complicated and defining by construction, it starts
from the *genus supremum* and defines only, as the scholastics
would have said, *per genus proximum et differentiam specificam.*
Thence comes the horror it has sometimes inspired in certain
minds, for instance in Hermite, whose favorite idea was to compare the mathematical to the natural sciences.  With most of
us these prejudices have been dissipated, but it has come to
pass that we have encountered certain paradoxes, certain apparent contradictions that would have delighted Zeno the Eleatic
and the school of Megara.  And then each must seek the remedy.
For my part, I think, and I am not the only one, that the important thing is never to introduce entities not completely definable
in a finite number of words.  Whatever be the cure adopted, we
may promise ourselves the joy of the doctor called in to follow
a beautiful pathologic case.

### THE INVESTIGATION OF THE POSTULATES

On the other hand, efforts have been made to enumerate the
axioms and postulates, more or less hidden, which serve as foundation to the different theories of mathematics.  Professor Hilbert
has obtained the most brilliant results.  It seems at first that this
domain would be very restricted and there would be nothing
more to do when the inventory should be ended, which could not
take long.  But when we shall have enumerated all, there will be
many ways of classifying all; a good librarian always finds something to do, and each new classification will be instructive for
the philosopher.

Here I end this review which I could not dream of making
complete.  I think these examples will suffice to show by what
mechanism the mathematical sciences have made their progress
in the past and in what direction they must advance in the future.

# CHAPTER III

## MATHEMATICAL CREATION

THE genesis of mathematical creation is a problem which should intensely interest the psychologist. It is the activity in which the human mind seems to take least from the outside world, in which it acts or seems to act only of itself and on itself, so that in studying the procedure of geometric thought we may hope to reach what is most essential in man's mind.

This has long been appreciated, and some time back the journal called *L'enseignement mathématique,* edited by Laisant and Fehr, began an investigation of the mental habits and methods of work of different mathematicians. I had finished the main outlines of this article when the results of that inquiry were published, so I have hardly been able to utilize them and shall confine myself to saying that the majority of witnesses confirm my conclusions; I do not say all, for when the appeal is to universal suffrage unanimity is not to be hoped.

A first fact should surprise us, or rather would surprise us if we were not so used to it. How does it happen there are people who do not understand mathematics? If mathematics invokes only the rules of logic, such as are accepted by all normal minds; if its evidence is based on principles common to all men, and that none could deny without being mad, how does it come about that so many persons are here refractory?

That not every one can invent is nowise mysterious. That not every one can retain a demonstration once learned may also pass. But that not every one can understand mathematical reasoning when explained appears very surprising when we think of it. And yet those who can follow this reasoning only with difficulty are in the majority: that is undeniable, and will surely not be gainsaid by the experience of secondary-school teachers.

And further: how is error possible in mathematics? A sane mind should not be guilty of a logical fallacy, and yet there are

383

very fine minds who do not trip in brief reasoning such as occurs in the ordinary doings of life, and who are incapable of following or repeating without error the mathematical demonstrations which are longer, but which after all are only an accumulation of brief reasonings wholly analogous to those they make so easily. Need we add that mathematicians themselves are not infallible?

The answer seems to me evident. Imagine a long series of syllogisms, and that the conclusions of the first serve as premises of the following: we shall be able to catch each of these syllogisms, and it is not in passing from premises to conclusion that we are in danger of deceiving ourselves. But between the moment in which we first meet a proposition as conclusion of one syllogism, and that in which we reencounter it as premise of another syllogism occasionally some time will elapse, several links of the chain will have unrolled; so it may happen that we have forgotten it, or worse, that we have forgotten its meaning. So it may happen that we replace it by a slightly different proposition, or that, while retaining the same enunciation, we attribute to it a slightly different meaning, and thus it is that we are exposed to error.

Often the mathematician uses a rule. Naturally he begins by demonstrating this rule; and at the time when this proof is fresh in his memory he understands perfectly its meaning and its bearing, and he is in no danger of changing it. But subsequently he trusts his memory and afterward only applies it in a mechanical way; and then if his memory fails him, he may apply it all wrong. Thus it is, to take a simple example, that we sometimes make slips in calculation because we have forgotten our multiplication table.

According to this, the special aptitude for mathematics would be due only to a very sure memory or to a prodigious force of attention. It would be a power like that of the whist-player who remembers the cards played; or, to go up a step, like that of the chess-player who can visualize a great number of combinations and hold them in his memory. Every good mathematician ought to be a good chess-player, and inversely; likewise he should be a good computer. Of course that sometimes happens; thus Gauss was at the same time a geometer of genius and a very precocious and accurate computer.

But there are exceptions; or rather I err; I can not call them exceptions without the exceptions being more than the rule. Gauss it is, on the contrary, who was an exception. As for myself, I must confess, I am absolutely incapable even of adding without mistakes. In the same way I should be but a poor chess-player; I would perceive that by a certain play I should expose myself to a certain danger; I would pass in review several other plays, rejecting them for other reasons, and then finally I should make the move first examined, having meantime forgotten the danger I had foreseen.

In a word, my memory is not bad, but it would be insufficient to make me a good chess-player. Why then does it not fail me in a difficult piece of mathematical reasoning where most chess-players would lose themselves? Evidently because it is guided by the general march of the reasoning. A mathematical demonstration is not a simple juxtaposition of syllogisms, it is syllogisms *placed in a certain order,* and the order in which these elements are placed is much more important than the elements themselves. If I have the feeling, the intuition, so to speak, of this order, so as to perceive at a glance the reasoning as a whole, I need no longer fear lest I forget one of the elements, for each of them will take its allotted place in the array, and that without any effort of memory on my part.

It seems to me then, in repeating a reasoning learned, that I could have invented it. This is often only an illusion; but even then, even if I am not so gifted as to create it by myself, I myself re-invent it in so far as I repeat it.

We know that this feeling, this intuition of mathematical order, that makes us divine hidden harmonies and relations, can not be possessed by every one. Some will not have either this delicate feeling so difficult to define, or a strength of memory and attention beyond the ordinary, and then they will be absolutely incapable of understanding higher mathematics. Such are the majority. Others will have this feeling only in a slight degree, but they will be gifted with an uncommon memory and a great power of attention. They will learn by heart the details

26

one after another; they can understand mathematics and sometimes make applications, but they cannot create. Others, finally,
will possess in a less or greater degree the special intuition
referred to, and then not only can they understand mathematics
even if their memory is nothing extraordinary, but they may
become creators and try to invent with more or less success
according as this intuition is more or less developed in them.

In fact, what is mathematical creation? It does not consist
in making new combinations with mathematical entities already
known. Any one could do that, but the combinations so made
would be infinite in number and most of them absolutely without interest. To create consists precisely in not making useless
combinations and in making those which are useful and which
are only a small minority. Invention is discernment, choice.

How to make this choice I have before explained; the mathematical facts worthy of being studied are those which, by their
analogy with other facts, are capable of leading us to the knowledge of a mathematical law just as experimental facts lead us to
the knowledge of a physical law. They are those which reveal
to us unsuspected kinship between other facts, long known, but
wrongly believed to be strangers to one another.

Among chosen combinations the most fertile will often be those
formed of elements drawn from domains which are far apart.
Not that I mean as sufficing for invention the bringing together
of objects as disparate as possible; most combinations so formed
would be entirely sterile. But certain among them, very rare,
are the most fruitful of all.

To invent, I have said, is to choose; but the word is perhaps
not wholly exact. It makes one think of a purchaser before whom
are displayed a large number of samples, and who examines
them, one after the other, to make a choice. Here the samples
would be so numerous that a whole lifetime would not suffice to
examine them. This is not the actual state of things. The sterile
combinations do not even present themselves to the mind of the
inventor. Never in the field of his consciousness do combinations appear that are not really useful, except some that he rejects
but which have to some extent the characteristics of useful combinations. All goes on as if the inventor were an examiner for

the second degree who would only have to question the candidates who had passed a previous examination.

But what I have hitherto said is what may be observed or inferred in reading the writings of the geometers, reading reflectively.

It is time to penetrate deeper and to see what goes on in the very soul of the mathematician. For this, I believe, I can do best by recalling memories of my own. But I shall limit myself to telling how I wrote my first memoir on Fuchsian functions. I beg the reader's pardon; I am about to use some technical expressions, but they need not frighten him, for he is not obliged to understand them. I shall say, for example, that I have found the demonstration of such a theorem under such circumstances. This theorem will have a barbarous name, unfamiliar to many, but that is unimportant; what is of interest for the psychologist is not the theorem but the circumstances.

For fifteen days I strove to prove that there could not be any functions like those I have since called Fuchsian functions. I was then very ignorant; every day I seated myself at my work table, stayed an hour or two, tried a great number of combinations and reached no results. One evening, contrary to my custom, I drank black coffee and could not sleep. Ideas rose in crowds; I felt them collide until pairs interlocked, so to speak, making a stable combination. By the next morning I had established the existence of a class of Fuchsian functions, those which come from the hypergeometric series; I had only to write out the results, which took but a few hours.

Then I wanted to represent these functions by the quotient of two series; this idea was perfectly conscious and deliberate, the analogy with elliptic functions guided me. I asked myself what properties these series must have if they existed, and I succeeded without difficulty in forming the series I have called theta-Fuchsian.

Just at this time I left Caen, where I was then living, to go on a geologic excursion under the auspices of the school of mines. The changes of travel made me forget my mathematical work. 'Having reached Coutances, we entered an omnibus to go some place or other. At the moment when I put my foot on the step the idea came to me, without anything in my former thoughts seeming to have paved the way for it, that the transformations I had used to define the Fuchsian functions were identical with those of non-Euclidean geometry. I did not verify the idea; I should not have had time, as, upon taking my seat in the omnibus, I went on with a conversation already commenced, but I felt a perfect certainty. On my return to Caen, for conscience' sake I verified the result at my leisure.

Then I turned my attention to the study of some arithmetical questions apparently without much success and without a suspicion of any connection with my preceding researches. Disgusted with my failure, I went to spend a few days at the seaside, and thought of something else. One morning, walking on the bluff, the idea came to me, with just the same characteristics of brevity, suddenness and immediate certainty, that the arithmetic transformations of indeterminate ternary quadratic forms were identical with those of non-Euclidean geometry.

Returned to Caen, I meditated on this result and deduced the consequences. The example of quadratic forms showed me that there were Fuchsian groups other than those corresponding to the hypergeometric series; I saw that I could apply to them the theory of theta-Fuchsian series and that consequently there existed Fuchsian functions other than those from the hypergeometric series, the ones I then knew. Naturally I set myself to form all these functions. I made a systematic attack upon them and carried all the outworks, one after another. There was one however that still held out, whose fall would involve that of the whole place. But all my efforts only served at first the better to show me the difficulty, which indeed was something. All this work was perfectly conscious.

Thereupon I left for Mont-Valérien, where I was to go through my military service; so I was very differently occupied. One day, going along the street, the solution of the difficulty which had stopped me suddenly appeared to me. I did not try to go deep into it immediately, and only after my service did I again take up the question. I had all the elements and had only to arrange them and put them together. So I wrote out my final memoir at a single stroke and without difficulty.

I shall limit myself to this single example; it is useless to multiply them. In regard to my other researches I would have to say analogous things, and the observations of other mathematicians given in *L'enseignement mathématique* would only confirm them.

Most striking at first is this appearance of sudden illumination, a manifest sign of long, unconscious prior work. The rôle of this unconscious work in mathematical invention appears to me incontestable, and traces of it would be found in other cases where it is less evident. Often when one works at a hard question, nothing good is accomplished at the first attack. Then one takes a rest, longer or shorter, and sits down anew to the work. During the first half-hour, as before, nothing is found, and then all of a sudden the decisive idea presents itself to the mind. It might be said that the conscious work has been more fruitful because it has been interrupted and the rest has given back to the mind its force and freshness. But it is more probable that this rest has been filled out with unconscious work and that the result of this work has afterward revealed itself to the geometer just as in the cases I have cited; only the revelation, instead of coming during a walk or a journey, has happened during a period of conscious work, but independently of this work which plays at most a rôle of excitant, as if it were the goad stimulating the results already reached during rest, but remaining unconscious, to assume the conscious form.

There is another remark to be made about the conditions of this unconscious work: it is possible, and of a certainty it is only fruitful, if it is on the one hand preceded and on the other hand followed by a period of conscious work. These sudden inspirations (and the examples already cited sufficiently prove this) never happen except after some days of voluntary effort which has appeared absolutely fruitless and whence nothing good seems to have come, where the way taken seems totally astray. These efforts then have not been as sterile as one thinks; they have set agoing the unconscious machine and without them it would not have moved and would have produced nothing.

The need for the second period of conscious work, after the inspiration, is still easier to understand. It is necessary to put in shape the results of this inspiration, to deduce from them the immediate consequences, to arrange them, to word the demonstrations, but above all is verification necessary. I have spoken of the feeling of absolute certitude accompanying the inspiration; in the cases cited this feeling was no deceiver, nor is it usually. But do not think this a rule without exception; often this feeling deceives us without being any the less vivid, and we only find it out when we seek to put on foot the demonstration. I have especially noticed this fact in regard to ideas coming to me in the morning or evening in bed while in a semi-hypnagogic state.

Such are the realities; now for the thoughts they force upon us. The unconscious, or, as we say, the subliminal self plays an important rôle in mathematical creation; this follows from what we have said. But usually the subliminal self is considered as purely automatic. Now we have seen that mathematical work is not simply mechanical, that it could not be done by a machine, however perfect. It is not merely a question of applying rules, of making the most combinations possible according to certain fixed laws. The combinations so obtained would be exceedingly numerous, useless and cumbersome. The true work of the inventor consists in choosing among these combinations so as to eliminate the useless ones or rather to avoid the trouble of making them, and the rules which must guide this choice are extremely fine and delicate. It is almost impossible to state them precisely; they are felt rather than formulated. Under these conditions, how imagine a sieve capable of applying them mechanically?

A first hypothesis now presents itself: the subliminal self is in no way inferior to the conscious self; it is not purely automatic; it is capable of discernment; it has tact, delicacy; it knows how to choose, to divine. What do I say? It knows better how to divine than the conscious self, since it succeeds where that has failed. In a word, is not the subliminal self superior to the conscious self? You recognize the full importance of this question. Boutroux in a recent lecture has shown how it came up on a very different occasion, and what consequences would follow an affirmative answer. (See also, by the same author, *Science et Religion*, pp. 313 ff.)

Is this affirmative answer forced upon us by the facts I have just given? I confess that, for my part, I should hate to accept it. Reexamine the facts then and see if they are not compatible with another explanation.

It is certain that the combinations which present themselves to the mind in a sort of sudden illumination, after an unconscious working somewhat prolonged, are generally useful and fertile combinations, which seem the result of a first impression. Does it follow that the subliminal self, having divined by a delicate intuition that these combinations would be useful, has formed only these, or has it rather formed many others which were lacking in interest and have remained unconscious?

In this second way of looking at it, all the combinations would be formed in consequence of the automatism of the subliminal self, but only the interesting ones would break into the domain of consciousness. And this is still very mysterious. What is the cause that, among the thousand products of our unconscious activity, some are called to pass the threshold, while others remain below? Is it a simple chance which confers this privilege? Evidently not; among all the stimuli of our senses, for example, only the most intense fix our attention, unless it has been drawn to them by other causes. More generally the privileged unconscious phenomena, those susceptible of becoming conscious, are those which, directly or indirectly, affect most profoundly our emotional sensibility.

It may be surprising to see emotional sensibility invoked *à propos* of mathematical demonstrations which, it would seem, can interest only the intellect. This would be to forget the feeling of mathematical beauty, of the harmony of numbers and forms, of geometric elegance. This is a true esthetic feeling that all real mathematicians know, and surely it belongs to emotional sensibility.

Now, what are the mathematic entities to which we attribute this character of beauty and elegance, and which are capable of developing in us a sort of esthetic emotion? They are those whose elements are harmoniously disposed so that the mind without effort can embrace their totality while realizing the details. This harmony is at once a satisfaction of our esthetic needs and an aid to the mind, sustaining and guiding. And at the same time, in putting under our eyes a well-ordered whole, it makes us foresee a mathematical law. Now, as we have said above, the only mathematical facts worthy of fixing our attention and capable of being useful are those which can teach us a mathematical law. So that we reach the following conclusion: The useful combinations are precisely the most beautiful, I mean those best able to charm this special sensibility that all mathematicians know, but of which the profane are so ignorant as often to be tempted to smile at it.

What happens then? Among the great numbers of combinations blindly formed by the subliminal self, almost all are without interest and without utility; but just for that reason they are also without effect upon the esthetic sensibility. Consciousness will never know them; only certain ones are harmonious, and, consequently, at once useful and beautiful. They will be capable of touching this special sensibility of the geometer of which I have just spoken, and which, once aroused, will call our attention to them, and thus give them occasion to become conscious.

This is only a hypothesis, and yet here is an observation which may confirm it: when a sudden illumination seizes upon the mind of the mathematician, it usually happens that it does not deceive him, but it also sometimes happens, as I have said, that it does not stand the test of verification; well, we almost always notice that this false idea, had it been true, would have gratified our natural feeling for mathematical elegance.

Thus it is this special esthetic sensibility which plays the rôle of the delicate sieve of which I spoke, and that sufficiently explains why the one lacking it will never be a real creator.

Yet all the difficulties have not disappeared. The conscious self is narrowly limited, and as for the subliminal self we know not its limitations, and this is why we are not too reluctant in supposing that it has been able in a short time to make more different combinations than the whole life of a conscious being could encompass. Yet these limitations exist. Is it likely that it is able to form all the possible combinations, whose number would frighten the imagination? Nevertheless that would seem necessary, because if it produces only a small part of these combinations, and if it makes them at random, there would be small chance that the *good,* the one we should choose, would be found among them.

Perhaps we ought to seek the explanation in that preliminary period of conscious work which always precedes all fruitful unconscious labor. Permit me a rough comparison. Figure the future elements of our combinations as something like the hooked atoms of Epicurus. During the complete repose of the mind, these atoms are motionless, they are, so to speak, hooked to the wall; so this complete rest may be indefinitely prolonged without the atoms meeting, and consequently without any combination between them.

On the other hand, during a period of apparent rest and unconscious work, certain of them are detached from the wall and put in motion. They flash in every direction through the space (I was about to say the room) where they are enclosed, as would, for example, a swarm of gnats or, if you prefer a more learned comparison, like the molecules of gas in the kinematic theory of gases. Then their mutual impacts may produce new combinations.

What is the rôle of the preliminary conscious work? It is evidently to mobilize certain of these atoms, to unhook them from the wall and put them in swing. We think we have done no good, because we have moved these elements a thousand different ways in seeking to assemble them, and have found no satisfactory aggregate. But, after this shaking up imposed upon them by our will, these atoms do not return to their primitive rest. They freely continue their dance.

Now, our will did not choose them at random; it pursued a perfectly determined aim. The mobilized atoms are therefore not any atoms whatsoever; they are those from which we might reasonably expect the desired solution. Then the mobilized atoms undergo impacts which make them enter into combinations among themselves or with other atoms at rest which they struck against in their course. Again I beg pardon, my comparison is very rough, but I scarcely know how otherwise to make my thought understood.

However it may be, the only combinations that have a chance of forming are those where at least one of the elements is one of those atoms freely chosen by our will. Now, it is evidently among these that is found what I called the *good combination*. Perhaps this is a way of lessening the paradoxical in the original hypothesis.

Another observation. It never happens that the unconscious work gives us the result of a somewhat long calculation *all made*, where we have only to apply fixed rules. We might think the wholly automatic subliminal self particularly apt for this sort of work, which is in a way exclusively mechanical. It seems that thinking in the evening upon the factors of a multiplication we might hope to find the product ready made upon our awakening, or again that an algebraic calculation, for example a verification, would be made unconsciously. Nothing of the sort, as observation proves. All one may hope from these inspirations, fruits of unconscious work, is a point of departure for such calculations. As for the calculations themselves, they must be made in the second period of conscious work, that which follows the inspiration, that in which one verifies the results of this inspiration and deduces their consequences. The rules of these calculations are strict and complicated. They require discipline, attention, will, and therefore consciousness. In the subliminal self, on the contrary, reigns what I should call liberty, if we might give this name to the simple absence of discipline and to the disorder born of chance. Only, this disorder itself permits unexpected combinations.

I shall make a last remark: when above I made certain personal observations, I spoke of a night of excitement when I worked in spite of myself. Such cases are frequent, and it is not necessary that the abnormal cerebral activity be caused by a physical excitant as in that I mentioned. It seems, in such cases, that one is present at his own unconscious work, made partially perceptible to the over-excited consciousness, yet without having changed its nature. Then we vaguely comprehend what distinguishes the two mechanisms or, if you wish, the working methods of the two egos. And the psychologic observations I have been able thus to make seem to me to confirm in their general outlines the views I have given.

Surely they have need of it, for they are and remain in spite of all very hypothetical: the interest of the questions is so great that I do not repent of having submitted them to the reader.

# CHAPTER IV

## CHANCE

### I

"How dare we speak of the laws of chance? Is not chance the antithesis of all law?" So says Bertrand at the beginning of his *Calcul des probabilités*. Probability is opposed to certitude; so it is what we do not know and consequently it seems what we could not calculate. Here is at least apparently a contradiction, and about it much has already been written.

And first, what is chance? The ancients distinguished between phenomena seemingly obeying harmonious laws, established once for all, and those which they attributed to chance; these were the ones unpredictable because rebellious to all law. In each domain the precise laws did not decide everything, they only drew limits between which chance might act. In this conception the word chance had a precise and objective meaning: what was chance for one was also chance for another and even for the gods.

But this conception is not ours to-day. We have become absolute determinists, and even those who want to reserve the rights of human free will let determinism reign undividedly in the inorganic world at least. Every phenomenon, however minute, has a cause; and a mind infinitely powerful, infinitely well-informed about the laws of nature, could have foreseen it from the beginning of the centuries. If such a mind existed, we could not play with it at any game of chance; we should always lose.

In fact for it the word chance would not have any meaning, or rather there would be no chance. It is because of our weakness and our ignorance that the word has a meaning for us. And, even without going beyond our feeble humanity, what is chance for the ignorant is not chance for the scientist. Chance is only the measure of our ignorance. Fortuitous phenomena are, by definition, those whose laws we do not know.

But is this definition altogether satisfactory? When the first

Chaldean shepherds followed with their eyes the movements of the stars, they knew not as yet the laws of astronomy; would they have dreamed of saying that the stars move at random? If a modern physicist studies a new phenomenon, and if he discovers its law Tuesday, would he have said Monday that this phenomenon was fortuitous? Moreover, do we not often invoke what Bertrand calls the laws of chance, to predict a phenomenon? For example, in the kinetic theory of gases we obtain the known laws of Mariotte and of Gay-Lussac by means of the hypothesis that the velocities of the molecules of gas vary irregularly, that is to say at random. All physicists will agree that the observable laws would be much less simple if the velocities were ruled by any simple elementary law whatsoever, if the molecules were, as we say, *organized*, if they were subject to some discipline. It is due to chance, that is to say, to our ignorance, that we can draw our conclusions; and then if the word chance is simply synonymous with ignorance what does that mean? Must we therefore translate as follows?

"You ask me to predict for you the phenomena about to happen. If, unluckily, I knew the laws of these phenomena I could make the prediction only by inextricable calculations and would have to renounce attempting to answer you; but as I have the good fortune not to know them, I will answer you at once. And what is most surprising, my answer will be right."

So it must well be that chance is something other than the name we give our ignorance, that among phenomena whose causes are unknown to us we must distinguish fortuitous phenomena about which the calculus of probabilities will provisionally give information, from those which are not fortuitous and of which we can say nothing so long as we shall not have determined the laws governing them. For the fortuitous phenomena themselves, it is clear that the information given us by the calculus of probabilities will not cease to be true upon the day when these phenomena shall be better known.

The director of a life insurance company does not know when each of the insured will die, but he relies upon the calculus of probabilities and on the law of great numbers, and he is not deceived, since he distributes dividends to his stockholders. These dividends would not vanish if a very penetrating and very indiscrete physician should, after the policies were signed, reveal to the director the life chances of the insured. This doctor would dissipate the ignorance of the director, but he would have no influence on the dividends, which evidently are not an outcome of this ignorance.

## II

To find a better definition of chance we must examine some of the facts which we agree to regard as fortuitous, and to which the calculus of probabilities seems to apply; we then shall investigate what are their common characteristics.

The first example we select is that of unstable equilibrium; if a cone rests upon its apex, we know well that it will fall, but we do not know toward what side; it seems to us chance alone will decide. If the cone were perfectly symmetric, if its axis were perfectly vertical, if it were acted upon by no force other than gravity, it would not fall at all. But the least defect in symmetry will make it lean slightly toward one side or the other, and if it leans, however little, it will fall altogether toward that side. Even if the symmetry were perfect, a very slight tremor, a breath of air could make it incline some seconds of arc; this will be enough to determine its fall and even the sense of its fall which will be that of the initial inclination.

A very slight cause, which escapes us, determines a considerable effect which we can not help seeing, and then we say this effect is due to chance. If we could know exactly the laws of nature and the situation of the universe at the initial instant, we should be able to predict exactly the situation of this same universe at a subsequent instant. But even when the natural laws should have no further secret for us, we could know the initial situation only *approximately*. If that permits us to foresee the subsequent situation *with the same degree of approximation*, this is all we require, we say the phenomenon has been predicted, that it is ruled by laws. But this is not always the case; it may happen that slight differences in the initial conditions produce very great differences in the final phenomena; a slight error in the former would make an enormous error in the latter.  Prediction becomes impossible and we have the fortuitous phenomenon.

Our second example will be very analogous to the first and we shall take it from meteorology.  Why have the meteorologists such difficulty in predicting the weather with any certainty?  Why do the rains, the tempests themselves seem to us to come by chance, so that many persons find it quite natural to pray for rain or shine, when they would think it ridiculous to pray for an eclipse?  We see that great perturbations generally happen in regions where the atmosphere is in unstable equilibrium.  The meteorologists are aware that this equilibrium is unstable, that a cyclone is arising somewhere; but where they can not tell; one-tenth of a degree more or less at any point, and the cyclone bursts here and not there, and spreads its ravages over countries it would have spared.  This we could have foreseen if we had known that tenth of a degree, but the observations were neither sufficiently close nor sufficiently precise, and for this reason all seems due to the agency of chance.  Here again we find the same contrast between a very slight cause, unappreciable to the ob-server, and important effects, which are sometimes tremendous disasters.

Let us pass to another example, the distribution of the minor planets on the zodiac.  Their initial longitudes may have been any longitudes whatever; but their mean motions were different and they have revolved for so long a time that we may say they are now distributed *at random* along the zodiac.  Very slight initial differences between their distances from the sun, or, what comes to the same thing, between their mean motions, have ended by giving enormous differences between their present longitudes.  An excess of the thousandth of a second in the daily mean motion will give in fact a second in three years, a degree in ten thousand years, an entire circumference in three or four million years, and what is that to the time which has passed since the minor planets detached themselves from the nebula of Laplace?  Again therefore we see a slight cause and a great effect; or better, slight differences in the cause and great differ-ences in the effect.

The game of roulette does not take us as far as might seem

from the preceding example. Assume a needle to be turned on a pivot over a dial divided into a hundred sectors alternately red and black. If it stops on a red sector I win; if not, I lose. Evidently all depends upon the initial impulse I give the needle. The needle will make, suppose, ten or twenty turns, but it will stop sooner or not so soon, according as I shall have pushed it more or less strongly. It suffices that the impulse vary only by a thousandth or a two thousandth to make the needle stop over a black sector or over the following red one. These are differences the muscular sense can not distinguish and which elude even the most delicate instruments. So it is impossible for me to foresee what the needle I have started will do, and this is why my heart throbs and I hope everything from luck. The difference in the cause is imperceptible, and the difference in the effect is for me of the highest importance, since it means my whole stake.

### III

Permit me, in this connection, a thought somewhat foreign to my subject. Some years ago a philosopher said that the future is determined by the past, but not the past by the future; or, in other words, from knowledge of the present we could deduce the future, but not the past; because, said he, a cause can have only one effect, while the same effect might be produced by several different causes. It is clear no scientist can subscribe to this conclusion. The laws of nature bind the antecedent to the consequent in such a way that the antecedent is as well determined by the consequent as the consequent by the antecedent. But whence came the error of this philosopher? We know that in virtue of Carnot's principle physical phenomena are irreversible and the world tends toward uniformity. When two bodies of different temperature come in contact, the warmer gives up heat to the colder; so we may foresee that the temperature will equalize. But once equal, if asked about the anterior state, what can we answer? We might say that one was warm and the other cold, but not be able to divine which formerly was the warmer.

And yet in reality the temperatures will never reach perfect equality. The difference of the temperatures only tends asymptotically toward zero. There comes a moment when our ther-

mometers are powerless to make it known.   But if we had ther-
mometers a thousand times, a hundred thousand times as sensitive, we should recognize that there still is a slight difference, and
that one of the bodies remains a little warmer than the other, and
so we could say this it is which formerly was much the warmer.

So then there are, contrary to what we found in the former
examples, great differences in cause and slight differences in
effect.   Flammarion once imagined an observer going away from
the earth with a velocity greater than that of light; for him time
would have changed sign.   History would be turned about, and
Waterloo would precede Austerlitz.   Well, for this observer,
effects and causes would be inverted; unstable equilibrium would
no longer be the exception.   Because of the universal irreversibility, all would seem to him to come out of a sort of chaos in
unstable equilibrium.   All nature would appear to him delivered
over to chance.

IV

Now for other examples where we shall see somewhat different
characteristics.   Take first the kinetic theory of gases.   How
should we picture a receptacle filled with gas?   Innumerable
molecules, moving at high speeds, flash through this receptacle
in every direction.   At every instant they strike against its walls
or each other, and these collisions happen under the most diverse
conditions.   What above all impresses us here is not the littleness of the causes, but their complexity, and yet the former element is still found here and plays an important rôle.   If a molecule deviated right or left from its trajectory, by a very small
quantity, comparable to the radius of action of the gaseous molecules, it would avoid a collision or sustain it under different conditions, and that would vary the direction of its velocity after
the impact, perhaps by ninety degrees or by a hundred and
eighty degrees.

And this is not all; we have just seen that it is necessary to
deflect the molecule before the clash by only an infinitesimal, to
produce its deviation after the collision by a finite quantity.   If
then the molecule undergoes two successive shocks, it will suffice
to deflect it before the first by an infinitesimal of the second
order, for it to deviate after the first encounter by an infinites-

imal of the first order, and after the second hit, by a finite quantity. And the molecule will not undergo merely two shocks; it will undergo a very great number per second. So that if the first shock has multiplied the deviation by a very large number $A$, after $n$ shocks it will be multiplied by $A^n$. It will therefore become very great not merely because $A$ is large, that is to say because little causes produce big effects, but because the exponent $n$ is large, that is to say because the shocks are very numerous and the causes very complex.

Take a second example. Why do the drops of rain in a shower seem to be distributed at random? This is again because of the complexity of the causes which determine their formation. Ions are distributed in the atmosphere. For a long while they have been subjected to air-currents constantly changing, they have been caught in very small whirlwinds, so that their final distribution has no longer any relation to their initial distribution. Suddenly the temperature falls, vapor condenses, and each of these ions becomes the center of a drop of rain. To know what will be the distribution of these drops and how many will fall on each paving-stone, it would not be sufficient to know the initial situation of the ions, it would be necessary to compute the effect of a thousand little capricious air-currents.

And again it is the same if we put grains of powder in suspension in water. The vase is ploughed by currents whose law we know not, we only know it is very complicated. At the end of a certain time the grains will be distributed at random, that is to say uniformly, in the vase; and this is due precisely to the complexity of these currents. If they obeyed some simple law, if, for example the vase revolved and the currents circulated around the axis of the vase, describing circles, it would no longer be the same, since each grain would retain its initial altitude and its initial distance from the axis.

We should reach the same result in considering the mixing of two liquids or of two fine-grained powders. And to take a grosser example, this is also what happens when we shuffle playing-cards. At each stroke the cards undergo a permutation (analogous to that studied in the theory of substitutions). What will happen? The probability of a particular permutation (for

27

example, that bringing to the $n$th place the card occupying the $\phi(n)$th place before the permutation) depends upon the player's habits. But if this player shuffles the cards long enough, there will be a great number of successive permutations, and the resulting final order will no longer be governed by aught but chance; I mean to say that all possible orders will be equally probable. It is to the great number of successive permutations, that is to say to the complexity of the phenomenon, that this result is due.

A final word about the theory of errors. Here it is that the causes are complex and multiple. To how many snares is not the observer exposed, even with the best instrument! He should apply himself to finding out the largest and avoiding them. These are the ones giving birth to systematic errors. But when he has eliminated those, admitting that he succeeds, there remain many small ones which, their effects accumulating, may become dangerous. Thence come the accidental errors; and we attribute them to chance because their causes are too complicated and too numerous. Here again we have only little causes, but each of them would produce only a slight effect; it is by their union and their number that their effects become formidable.

## V

We may take still a third point of view, less important than the first two and upon which I shall lay less stress. When we seek to foresee an event and examine its antecedents, we strive to search into the anterior situation. This could not be done for all parts of the universe and we are content to know what is passing in the neighborhood of the point where the event should occur, or what would appear to have some relation to it. An examination can not be complete and we must know how to choose. But it may happen that we have passed by circumstances which at first sight seemed completely foreign to the foreseen happening, to which one would never have dreamed of attributing any influence and which nevertheless, contrary to all anticipation, come to play an important rôle.

A man passes in the street going to his business; some one knowing the business could have told why he started at such a time and went by such a street. On the roof works a tiler. The contractor employing him could in a certain measure foresee what he would do. But the passer-by scarcely thinks of the tiler, nor the tiler of him; they seem to belong to two worlds completely foreign to one another. And yet the tiler drops a tile which kills the man, and we do not hesitate to say this is chance.

Our weakness forbids our considering the entire universe and makes us cut it up into slices. We try to do this as little artificially as possible. And yet it happens from time to time that two of these slices react upon each other. The effects of this mutual action then seem to us to be due to chance.

Is this a third way of conceiving chance? Not always; in fact most often we are carried back to the first or the second. Whenever two worlds usually foreign to one another come thus to react upon each other, the laws of this reaction must be very complex. On the other hand, a very slight change in the initial conditions of these two worlds would have been sufficient for the reaction not to have happened. How little was needed for the man to pass a second later or the tiler to drop his tile a second sooner.

## VI

All we have said still does not explain why chance obeys laws. Does the fact that the causes are slight or complex suffice for our foreseeing, if not their effects *in each case,* at least what their effects will be, *on the average?* To answer this question we had better take up again some of the examples already cited.

I shall begin with that of the roulette. I have said that the point where the needle will stop depends upon the initial push given it. What is the probability of this push having this or that value? I know nothing about it, but it is difficult for me not to suppose that this probability is represented by a continuous analytic function. The probability that the push is comprised between $\alpha$ and $\alpha + \epsilon$ will then be sensibly equal to the probability of its being comprised between $\alpha + \epsilon$ and $\alpha + 2\epsilon$, *provided $\epsilon$ be very small.* This is a property common to all analytic functions. Minute variations of the function are proportional to minute variations of the variable.

But we have assumed that an exceedingly slight variation of the push suffices to change the color of the sector over which the needle finally stops. From $\alpha$ to $\alpha + \epsilon$ it is red, from $\alpha + \epsilon$ to $\alpha + 2\epsilon$ it is black; the probability of each red sector is therefore the same as of the following black, and consequently the total probability of red equals the total probability of black.

The datum of the question is the analytic function representing the probability of a particular initial push. But the theorem remains true whatever be this datum, since it depends upon a property common to all analytic functions. From this it follows finally that we no longer need the datum.

What we have just said for the case of the roulette applies also to the example of the minor planets. The zodiac may be regarded as an immense roulette on which have been tossed many little balls with different initial impulses varying according to some law. Their present distribution is uniform and independent of this law, for the same reason as in the preceding case. Thus we see why phenomena obey the laws of chance when slight differences in the causes suffice to bring on great differences in the effects. The probabilities of these slight differences may then be regarded as proportional to these differences themselves, just because these differences are minute, and the infinitesimal increments of a continuous function are proportional to those of the variable.

Take an entirely different example, where intervenes especially the complexity of the causes. Suppose a player shuffles a pack of cards. At each shuffle he changes the order of the cards, and he may change them in many ways. To simplify the exposition, consider only three cards. The cards which before the shuffle occupied respectively the places 123, may after the shuffle occupy the places

$$123, \ 231, \ 312, \ 321, \ 132, \ 213.$$

Each of these six hypotheses is possible and they have respectively for probabilities:

$$p_1, \ p_2, \ p_3, \ p_4, \ p_5, \ p_6.$$

The sum of these six numbers equals 1; but this is all we know of them; these six probabilities depend naturally upon the habits of the player which we do not know.

At the second shuffle and the following, this will recommence, and under the same conditions; I mean that $p_4$ for example represents always the probability that the three cards which occupied after the $n$th shuffle and before the $n + 1$th the places 123, occupy the places 321 after the $n + 1$th shuffle. And this remains true whatever be the number $n$, since the habits of the player and his way of shuffling remain the same.

But if the number of shuffles is very great, the cards which before the first shuffle occupied the places 123 may, after the last shuffle, occupy the places

$$123, \ 231, \ 312, \ 321, \ 132, \ 213$$

and the probability of these six hypotheses will be sensibly the same and equal to 1/6; and this will be true whatever be the numbers $p_1 \ldots p_6$ which we do not know. The great number of shuffles, that is to say the complexity of the causes, has produced uniformity.

This would apply without change if there were more than three cards, but even with three cards the demonstration would be complicated; let it suffice to give it for only two cards. Then we have only two possibilities 12, 21 with the probabilities $p_1$ and $p_2 = 1 - p_1$.

Suppose $n$ shuffles and suppose I win one franc if the cards are finally in the initial order and lose one if they are finally inverted. Then, my mathematical expectation will be $(p_1 - p_2)^n$.

The difference $p_1 - p_2$ is certainly less than 1; so that if $n$ is very great my expectation will be zero; we need not learn $p_1$ and $p_2$ to be aware that the game is equitable.

There would always be an exception if one of the numbers $p_1$ and $p_2$ was equal to 1 and the other naught. *Then it would not apply because our initial hypotheses would be too simple.*

What we have just seen applies not only to the mixing of cards, but to all mixings, to those of powders and of liquids; and even to those of the molecules of gases in the kinetic theory of gases.

To return to this theory, suppose for a moment a gas whose molecules can not mutually clash, but may be deviated by hitting the insides of the vase wherein the gas is confined. If the form of the vase is sufficiently complex the distribution of the molecules and that of the velocities will not be long in becoming uniform. But this will not be so if the vase is spherical or if it has the shape of a cuboid. Why? Because in the first case the distance from the center to any trajectory will remain constant; in the second case this will be the absolute value of the angle of each trajectory with the faces of the cuboid.

So we see what should be understood by conditions *too simple;* they are those which conserve something, which leave an invariant remaining. Are the differential equations of the problem too simple for us to apply the laws of chance? This question would seem at first view to lack precise meaning; now we know what it means. They are too simple if they conserve something, if they admit a uniform integral. If something in the initial conditions remains unchanged, it is clear the final situation can no longer be independent of the initial situation.

We come finally to the theory of errors. We know not to what are due the accidental errors, and precisely because we do not know, we are aware they obey the law of Gauss. Such is the paradox. The explanation is nearly the same as in the preceding cases. We need know only one thing: that the errors are very numerous, that they are very slight, that each may be as well negative as positive. What is the curve of probability of each of them? We do not know; we only suppose it is symmetric. We prove then that the resultant error will follow Gauss's law, and this resulting law is independent of the particular laws which we do not know. Here again the simplicity of the result is born of the very complexity of the data.

## VII

But we are not through with paradoxes. I have just recalled the figment of Flammarion, that of the man going quicker than light, for whom time changes sign. I said that for him all phenomena would seem due to chance. That is true from a certain point of view, and yet all these phenomena at a given moment would not be distributed in conformity with the laws of chance, since the distribution would be the same as for us, who, seeing them unfold harmoniously and without coming out of a primal chaos, do not regard them as ruled by chance.

What does that mean? For Lumen, Flammarion's man, slight causes seem to produce great effects; why do not things go on as for us when we think we see grand effects due to little causes? Would not the same reasoning be applicable in his case?

Let us return to the argument. When slight differences in the causes produce vast differences in the effects, why are these effects distributed according to the laws of chance? Suppose a difference of a millimeter in the cause produces a difference of a kilometer in the effect. If I win in case the effect corresponds to a kilometer bearing an even number, my probability of winning will be 1/2. Why? Because to make that, the cause must correspond to a millimeter with an even number. Now, according to all appearance, the probability of the cause varying between certain limits will be proportional to the distance apart of these limits, provided this distance be very small. If this hypothesis were not admitted there would no longer be any way of representing the probability by a continuous function.

What now will happen when great causes produce small effects? This is the case where we should not attribute the phenomenon to chance and where on the contrary Lumen would attribute it to chance. To a difference of a kilometer in the cause would correspond a difference of a millimeter in the effect. Would the probability of the cause being comprised between two limits $n$ kilometers apart still be proportional to $n$? We have no reason to suppose so, since this distance, $n$ kilometers, is great. But the probability that the effect lies between two limits $n$ millimeters apart will be precisely the same, so it will not be proportional to $n$, even though this distance, $n$ millimeters, be small. There is no way therefore of representing the law of probability of effects by a continuous curve. This curve, understand, may remain continuous in the *analytic* sense of the word; to *infinitesimal* variations of the abscissa will correspond infinitesimal variations of the ordinate. But *practically* it will not be continuous, since *very small* variations of the ordinate would not correspond to very small variations of the abscissa. It would become impossible to trace the curve with an ordinary pencil; that is what I mean.

So what must we conclude? Lumen has no right to say that

the probability of the cause (*his* cause, our effect) should be represented necessarily by a continuous function. But then why have we this right? It is because this state of unstable equilibrium which we have been calling initial is itself only the final outcome of a long previous history. In the course of this history complex causes have worked a great while: they have contributed to produce the mixture of elements and they have tended to make everything uniform at least within a small region; they have rounded off the corners, smoothed down the hills and filled up the valleys. However capricious and irregular may have been the primitive curve given over to them, they have worked so much toward making it regular that finally they deliver over to us a continuous curve. And this is why we may in all confidence assume its continuity.

Lumen would not have the same reasons for such a conclusion. For him complex causes would not seem agents of equalization and regularity, but on the contrary would create only inequality and differentiation. He would see a world more and more varied come forth from a sort of primitive chaos. The changes he could observe would be for him unforeseen and impossible to foresee. They would seem to him due to some caprice or another; but this caprice would be quite different from our chance, since it would be opposed to all law, while our chance still has its laws. All these points call for lengthy explications, which perhaps would aid in the better comprehension of the irreversibility of the universe.

## VIII

We have sought to define chance, and now it is proper to put a question. Has chance thus defined, in so far as this is possible, objectivity?

It may be questioned. I have spoken of very slight or very complex causes. But what is very little for one may be very big for another, and what seems very complex to one may seem simple to another. In part I have already answered by saying precisely in what cases differential equations become too simple for the laws of chance to remain applicable. But it is fitting to examine the matter a little more closely, because we may take still other points of view.

What means the phrase 'very slight'? To understand it we need only go back to what has already been said. A difference is very slight, an interval is very small, when within the limits of this interval the probability remains sensibly constant. And why may this probability be regarded as constant within a small interval? It is because we assume that the law of probability is represented by a continuous curve, continuous not only in the analytic sense, but *practically* continuous, as already explained. This means that it not only presents no absolute hiatus, but that it has neither salients nor reentrants too acute or too accentuated.

And what gives us the right to make this hypothesis? We have already said it is because, since the beginning of the ages, there have always been complex causes ceaselessly acting in the same way and making the world tend toward uniformity without ever being able to turn back. These are the causes which little by little have flattened the salients and filled up the reentrants, and this is why our probability curves now show only gentle undulations. In milliards of milliards of ages another step will have been made toward uniformity, and these undulations will be ten times as gentle; the radius of mean curvature of our curve will have become ten times as great. And then such a length as seems to us to-day not very small, since on our curve an arc of this length can not be regarded as rectilineal, should on the contrary at that epoch be called very little, since the curvature will have become ten times less and an arc of this length may be sensibly identified with a sect.

Thus the phrase 'very slight' remains relative; but it is not relative to such or such a man, it is relative to the actual state of the world. It will change its meaning when the world shall have become more uniform, when all things shall have blended still more. But then doubtless men can no longer live and must give place to other beings—should I say far smaller or far larger? So that our criterion, remaining true for all men, retains an objective sense.

And on the other hand what means the phrase 'very complex'? I have already given one solution, but there are others. Complex causes we have said produce a blend more and more intimate, but after how long a time will this blend satisfy us? When will it have accumulated sufficient complexity? When shall we have sufficiently shuffled the cards? If we mix two powders, one blue, the other white, there comes a moment when the tint of the mixture seems to us uniform because of the feebleness of our senses; it will be uniform for the presbyte, forced to gaze from afar, before it will be so for the myope. And when it has become uniform for all eyes, we still could push back the limit by the use of instruments. There is no chance for any man ever to discern the infinite variety which, if the kinetic theory is true, hides under the uniform appearance of a gas. And yet if we accept Gouy's ideas on the Brownian movement, does not the microscope seem on the point of showing us something analogous?

This new criterion is therefore relative like the first; and if it retains an objective character, it is because all men have approximately the same senses, the power of their instruments is limited, and besides they use them only exceptionally.

## IX

It is just the same in the moral sciences and particularly in history. The historian is obliged to make a choice among the events of the epoch he studies; he recounts only those which seem to him the most important. He therefore contents himself with relating the most momentous events of the sixteenth century, for example, as likewise the most remarkable facts of the seventeenth century. If the first suffice to explain the second, we say these conform to the laws of history. But if a great event of the seventeenth century should have for cause a small fact of the sixteenth century which no history reports, which all the world has neglected, then we say this event is due to chance. This word has therefore the same sense as in the physical sciences; it means that slight causes have produced great effects.

The greatest bit of chance is the birth of a great man. It is only by chance that meeting of two germinal cells, of different sex, containing precisely, each on its side, the mysterious elements whose mutual reaction must produce the genius. One will agree that these elements must be rare and that their meeting is still more rare. How slight a thing it would have required to deflect from its route the carrying spermatozoon. It would have sufficed to deflect it a tenth of a millimeter and Napoleon would not have been born and the destinies of a continent would have been changed. No example can better make us understand the veritable characteristics of chance.

One more word about the paradoxes brought out by the application of the calculus of probabilities to the moral sciences. It has been proved that no Chamber of Deputies will ever fail to contain a member of the opposition, or at least such an event would be so improbable that we might without fear wager the contrary, and bet a million against a sou.

Condorcet has striven to calculate how many jurors it would require to make a judicial error practically impossible. If we had used the results of this calculation, we should certainly have been exposed to the same disappointments as in betting, on the faith of the calculus, that the opposition would never be without a representative.

The laws of chance do not apply to these questions. If justice be not always meted out to accord with the best reasons, it uses less than we think the method of Bridoye. This is perhaps to be regretted, for then the system of Condorcet would shield us from judicial errors.

What is the meaning of this? We are tempted to attribute facts of this nature to chance because their causes are obscure; but this is not true chance. The causes are unknown to us, it is true, and they are even complex; but they are not sufficiently so, since they conserve something. We have seen that this it is which distinguishes causes 'too simple.' When men are brought together they no longer decide at random and independently one of another; they influence one another. Multiplex causes come into action. They worry men, dragging them to right or left, but one thing there is they can not destroy, this is their Panurge flock-of-sheep habits. And this is an invariant.

## X

Difficulties are indeed involved in the application of the calculus of probabilities to the exact sciences. Why are the decimals of a table of logarithms, why are those of the number $\pi$ distributed in accordance with the laws of chance? Elsewhere I have already studied the question in so far as it concerns logarithms, and there it is easy.   It is clear that a slight difference of argument will give a slight difference of logarithm, but a great difference in the sixth decimal of the logarithm.  Always we find again the same criterion.

But as for the number $\pi$, that presents more difficulties, and I have at the moment nothing worth while to say.

There would be many other questions to resolve, had I wished to attack them before solving that which I more specially set myself.  When we reach a simple result, when we find for example a round number, we say that such a result can not be due to chance, and we seek, for its explanation, a non-fortuitous cause.   And in fact there is only a very slight probability that among 10,000 numbers chance will give a round number; for example, the number 10,000.   This has only one chance in 10,000.  But there is only one chance in 10,000 for the occurrence of any other one number; and yet this result will not astonish us, nor will it be hard for us to attribute it to chance; and that simply because it will be less striking.

Is this a simple illusion of ours, or are there cases where this way of thinking is legitimate?  We must hope so, else were all science impossible.  When we wish to check a hypothesis, what do we do?  We can not verify all its consequences, since they would be infinite in number; we content ourselves with verifying certain ones and if we succeed we declare the hypothesis confirmed, because so much success could not be due to chance. And this is always at bottom the same reasoning.

I can not completely justify it here, since it would take too much time; but I may at least say that we find ourselves confronted by two hypotheses, either a simple cause or that aggregate of complex causes we call chance.  We find it natural to suppose that the first should produce a simple result, and then, if we find that simple result, the round number for example, it seems more likely to us to be attributable to the simple cause which must give it almost certainly, than to chance which could only give it once in 10,000 times.   It will not be the same if we find a result which is not simple; chance, it is true, will not give this more than once in 10,000 times; but neither has the simple cause any more chance of producing it.

# BOOK II

------

## MATHEMATICAL REASONING

### CHAPTER I

#### THE RELATIVITY OF SPACE

#### I

IT is impossible to represent to oneself empty space; all our efforts to imagine a pure space, whence should be excluded the changing images of material objects, can result only in a representation where vividly colored surfaces, for example, are replaced by lines of faint coloration, and we can not go to the very end in his way without all vanishing and terminating in nothingness. Thence comes the irreducible relativity of space.

Whoever speaks of absolute space uses a meaningless phrase. This is a truth long proclaimed by all who have reflected upon the matter, but which we are too often led to forget.

I am at a determinate point in Paris, place du Panthéon for instance, and I say: I shall come back *here* to-morrow. If I be asked: Do you mean you will return to the same point of space, I shall be tempted to answer: yes; and yet I shall be wrong, since by to-morrow the earth will have journeyed hence, carrying with it the place du Panthéon, which will have traveled over more than two million kilometers. And if I tried to speak more precisely, I should gain nothing, since our globe has run over these two million kilometers in its motion with relation to the sun, while the sun in its turn is displaced with reference to the Milky Way, while the Milky Way itself is doubtless in motion without our being able to perceive its velocity. So that we are completely ignorant, and always shall be, of how much the place du Panthéon is displaced in a day.

In sum, I meant to say: To-morrow I shall see again the dome and the pediment of the Panthéon, and if there were no Panthéon my phrase would be meaningless and space would vanish.

This is one of the most commonplace forms of the principle of the relativity of space; but there is another, upon which Delbeuf has particularly insisted. Suppose that in the night all the dimensions of the universe become a thousand times greater: the world will have remained *similar* to itself, giving to the word *similitude* the same meaning as in Euclid, Book VI. Only what was a meter long will measure thenceforth a kilometer, what was a millimeter long will become a meter. The bed whereon I lie and my body itself will be enlarged in the same proportion.

When I awake to-morrow morning, what sensation shall I feel in presence of such an astounding transformation? Well, I shall perceive nothing at all. The most precise measurements will be incapable of revealing to me anything of this immense convulsion, since the measures I use will have varied precisely in the same proportion as the objects I seek to measure. In reality, this convulsion exists only for those who reason as if space were absolute. If I for a moment have reasoned as they do, it is the better to bring out that their way of seeing implies contradiction. In fact it would be better to say that, space being relative, nothing at all has happened, which is why we have perceived nothing.

Has one the right, therefore, to say he knows the distance between two points? No, since this distance could undergo enormous variations without our being able to perceive them, provided the other distances have varied in the same proportion. We have just seen that when I say: I shall be here to-morrow, this does not mean: To-morrow I shall be at the same point of space where I am to-day, but rather: To-morrow I shall be at the same distance from the Panthéon as to-day. And we see that this statement is no longer sufficient and that I should say: To-morrow and to-day my distance from the Panthéon will be equal to the same number of times the height of my body.

But this is not all; I have supposed the dimensions of the world to vary, but that at least the world remained always similar to itself. We might go much further, and one of the most astonishing theories of modern physics furnishes us the occasion.

According to Lorentz and Fitzgerald, all the bodies borne along in the motion of the earth undergo a deformation.

This deformation is, in reality, very slight, since all dimensions parallel to the movement of the earth diminish by a hundred millionth, while the dimensions perpendicular to this movement are unchanged. But it matters little that it is slight, that it exists suffices for the conclusion I am about to draw. And besides, I have said it was slight, but in reality I know nothing about it; I have myself been victim of the tenacious illusion which makes us believe we conceive an absolute space; I have thought of the motion of the earth in its elliptic orbit around the sun, and I have allowed thirty kilometers as its velocity. But its real velocity (I mean, this time, not its absolute velocity, which is meaningless, but its velocity with relation to the ether), I do not know that, and have no means of knowing it: it is perhaps 10, 100 times greater, and then the deformation will be 100, 10,000 times more.

Can we show this deformation? Evidently not; here is a cube with edge one meter; in consequence of the earth's displacement it is deformed, one of its edges, that parallel to the motion, becomes smaller, the others do not change. If I wish to assure myself of it by aid of a meter measure, I shall measure first one of the edges perpendicular to the motion and shall find that my standard meter fits this edge exactly; and in fact neither of these two lengths is changed, since both are perpendicular to the motion. Then I wish to measure the other edge, that parallel to the motion; to do this I displace my meter and turn it so as to apply it to the edge. But the meter, having changed orientation and become parallel to the motion, has undergone, in its turn, the deformation, so that though the edge be not a meter long, it will fit exactly, I shall find out nothing.

You ask then of what use is the hypothesis of Lorentz and of Fitzgerald if no experiment can permit of its verification? It is my exposition that has been incomplete; I have spoken only of measurements that can be made with a meter; but we can also measure a length by the time it takes light to traverse it, on condition we suppose the velocity of light constant and independent of direction. Lorentz could have accounted for the facts by supposing the velocity of light greater in the direction of the earth's motion than in the perpendicular direction. He preferred to suppose that the velocity is the same in these different directions, but that the bodies are smaller in the one than in the other. If the wave surfaces of light had undergone the same deformations as the material bodies we should never have perceived the Lorentz-Fitzgerald deformation.

In either case, it is not a question of absolute magnitude, but of the measure of this magnitude by means of some instrument; this instrument may be a meter, or the path traversed by light; it is only the relation of the magnitude to the instrument that we measure; and if this relation is altered, we have no way of knowing whether it is the magnitude or the instrument which has changed.

But what I wish to bring out is, that in this deformation the world has not remained similar to itself; squares have become rectangles, circles ellipses, spheres ellipsoids. And yet we have no way of knowing whether this deformation be real.

Evidently one could go much further: in place of the Lorentz-Fitzgerald deformation, whose laws are particularly simple, we could imagine any deformation whatsoever. Bodies could be deformed according to any laws, as complicated as we might wish, we never should notice it provided all bodies without exception were deformed according to the same laws. In saying, all bodies without exception, I include of course our own body and the light rays emanating from different objects.

If we look at the world in one of those mirrors of complicated shape which deform objects in a bizarre way, the mutual relations of the different parts of this world would not be altered; if, in fact two real objects touch, their images likewise seem to touch. Of course when we look in such a mirror we see indeed the deformation, but this is because the real world subsists alongside of its deformed image; and then even were this real world hidden from us, something there is could not be hidden, ourself; we could not cease to see, or at least to feel, our body and our limbs which have not been deformed and which continue to serve us as instruments of measure.

But if we imagine our body itself deformed in the same way

as if seen in the mirror, these instruments of measure in their turn will fail us and the deformation will no longer be ascertainable.

Consider in the same way two worlds images of one another; to each object $P$ of the world $A$ corresponds in the world $B$ an object $P'$, its image; the coordinates of this image $P'$ are determinate functions of those of the object $P$; moreover these functions may be any whatsoever; I only suppose them chosen once for all. Between the position of $P$ and that of $P'$ there is a constant relation; what this relation is, matters not; enough that it be constant.

Well, these two worlds will be indistinguishable one from the other. I mean the first will be for its inhabitants what the second is for its. And so it will be as long as the two worlds remain strangers to each other. Suppose we live in world $A$, we shall have constructed our science and in particular our geometry; during this time the inhabitants of world $B$ will have constructed a science, and as their world is the image of ours, their geometry will also be the image of ours or, better, it will be the same. But if for us some day a window is opened upon world $B$, how we shall pity them: "Poor things," we shall say, "they think they have made a geometry, but what they call so is only a grotesque image of ours; their straights are all twisted, their circles are humped, their spheres have capricious inequalities." And we shall never suspect they say the same of us, and one never will know who is right.

We see in how broad a sense should be understood the relativity of space; space is in reality amorphous and the things which are therein alone give it a form. What then should be thought of that direct intuition we should have of the straight or of distance? So little have we intuition of distance in itself that in the night, as we have said, a distance might become a thousand times greater without our being able to perceive it, if all other distances had undergone the same alteration. And even in a night the world $B$ might be substituted for the world $A$ without our having any way of knowing it, and then the straight lines of yesterday would have ceased to be straight and we should never notice.

28

One part of space is not by itself and in the absolute sense of the word equal to another part of space; because if so it is for us, it would not be for the dwellers in world $B$; and these have just as much right to reject our opinion as we to condemn theirs.

I have elsewhere shown what are the consequences of these facts from the viewpoint of the idea we should form of non-Euclidean geometry and other analogous geometries; to that I do not care to return; and to-day I shall take a somewhat different point of view.

## II

If this intuition of distance, of direction, of the straight line, if this direct intuition of space in a word does not exist, whence comes our belief that we have it? If this is only an illusion, why is this illusion so tenacious? It is proper to examine into this. We have said there is no direct intuition of size and we can only arrive at the relation of this magnitude to our instruments of measure. We should therefore not have been able to construct space if we had not had an instrument to measure it; well, this instrument to which we relate everything, which we use instinctively, it is our own body. It is in relation to our body that we place exterior objects, and the only spatial relations of these objects that we can represent are their relations to our body. It is our body which serves us, so to speak, as system of axes of coordinates.

For example, at an instant $\alpha$, the presence of the object $A$ is revealed to me by the sense of sight; at another instant, $\beta$, the presence of another object, $B$, is revealed to me by another sense, that of hearing or of touch, for instance. I judge that this object $B$ occupies the same place as the object $A$. What does that mean? First that does not signify that these two objects occupy, at two different moments, the same point of an absolute space, which even if it existed would escape our cognition, since, between the instants $\alpha$ and $\beta$, the solar system has moved and we can not know its displacement. That means these two objects occupy the same relative position with reference to our body.

But even this, what does it mean? The impressions that have come to us from these objects have followed paths absolutely different, the optic nerve for the object $A$, the acoustic nerve for the object $B$. They have nothing in common from the qualitative point of view. The representations we are able to make of these two objects are absolutely heterogeneous, irreducible one to the other. Only I know that to reach the object $A$ I have just to extend the right arm in a certain way; even when I abstain from doing it, I represent to myself the muscular sensations and other analogous sensations which would accompany this extension, and this representation is associated with that of the object $A$.

Now, I likewise know I can reach the object $B$ by extending my right arm in the same manner. an extension accompanied by the same train of muscular sensations. And when I say these two objects occupy the same place, I mean nothing more.

I also know I could have reached the object $A$ by another appropriate motion of the left arm and I represent to myself the muscular sensations which would have accompanied this movement; and by this same motion of the left arm, accompanied by the same sensations, I likewise could have reached the object $B$.

And that is very important, since thus I can defend myself against dangers menacing me from the object $A$ or the object $B$. With each of the blows we can be hit, nature has associated one or more parries which permit of our guarding ourselves. The same parry may respond to several strokes; and so it is, for instance, that the same motion of the right arm would have allowed us to guard at the instant $\alpha$ against the object $A$ and at the instant $\beta$ against the object $B$. Just so, the same stroke can be parried in several ways, and we have said, for instance, the object $A$ could be reached indifferently either by a certain movement of the right arm or by a certain movement of the left arm.

All these parries have nothing in common except warding off the same blow, and this it is, and nothing else, which is meant when we say they are movements terminating at the same point of space. Just so, these objects, of which we say they occupy the same point of space, have nothing in common, except that the same parry guards against them.

Or, if you choose, imagine innumerable telegraph wires, some centripetal, others centrifugal. The centripetal wires warn us of accidents happening without; the centrifugal wires carry the reparation.   Connections are so established that when a cen- tripetal wire is traversed by a current this acts on a relay and so starts a current in one of the centrifugal wires, and things are so arranged that several centripetal wires may act on the same centrifugal wire if the same remedy suits several ills, and that a centripetal wire may agitate different centrifugal wires, either simultaneously or in lieu one of the other when the same ill may be cured by several remedies.

It is this complex system of associations, it is this table of distri- bution, so to speak, which is all our geometry or, if you wish, all in our geometry that is instinctive.   What we call our intui- tion of the straight line or of distance is the consciousness we have of these associations and of their imperious character.

And it is easy to understand whence comes this imperious character itself.   An association will seem to us by so much the more indestructible as it is more ancient.   But these associations are not, for the most part, conquests of the individual, since their trace is seen in the new-born babe : they are conquests of the race. Natural selection had to bring about these conquests by so much the more quickly as they were the more necessary.

On this account, those of which we speak must have been of the earliest in date, since without them the defense of the organ- ism would have been impossible.   From the time when the cell- ules were no longer merely juxtaposed, but were called upon to give mutual aid, it was needful that a mechanism organize anal- ogous to what we have described, so that this aid miss not its way, but forestall the peril.

When a frog is decapitated, and a drop of acid is placed on a point of its skin, it seeks to wipe off the acid with the nearest foot, and, if this foot be amputated, it sweeps it off with the foot of the opposite side.   There we have the double parry of which I have just spoken, allowing the combating of an ill by a second remedy, if the first fails.   And it is this multiplicity of parries, and the resulting coordination, which is space.

We see to what depths of the unconscious we must descend to find the first traces of these spatial associations, since only the inferior parts of the nervous system are involved.   Why be astonished then at the resistance we oppose to every attempt made to dissociate what so long has been associated? Now, it is just this resistance that we call the evidence for the geometric truths; this evidence is nothing but the repugnance we feel toward breaking with very old habits which have always proved good.

## III

The space so created is only a little space extending no farther than my arm can reach; the intervention of the memory is necessary to push back its limits. There are points which will remain out of my reach, whatever effort I make to stretch forth my hand; if I were fastened to the ground like a hydra polyp, for instance, which can only extend its tentacles, all these points would be outside of space, since the sensations we could experience from the action of bodies there situated, would be associated with the idea of no movement allowing us to reach them, of no appropriate parry. These sensations would not seem to us to have any spatial character and we should not seek to localize them.

But we are not fixed to the ground like the lower animals; we can, if the enemy be too far away, advance toward him first and extend the hand when we are sufficiently near. This is still a parry, but a parry at long range. On the other hand, it is a complex parry, and into the representation we make of it enter the representation of the muscular sensations caused by the movements of the legs, that of the muscular sensations caused by the final movement of the arm, that of the sensations of the semicircular canals, etc. We must, besides, represent to ourselves, not a complex of simultaneous sensations, but a complex of successive sensations, following each other in a determinate order, and this is why I have just said the intervention of memory was necessary. Notice moreover that, to reach the same point, I may approach nearer the mark to be attained, so as to have to stretch my arm less. What more? It is not one, it is a thousand parries I can oppose to the same danger. All these parries are made of sensations which may have nothing in common and yet we regard them as defining the same point of space, since they may respond to the same danger and are all associated with the notion of this danger. It is the potentiality of warding off the same stroke which makes the unity of these different parries, as it is the possibility of being parried in the same way which makes the unity of the strokes so different in kind, which may menace us from the same point of space. It is this double unity which makes the individuality of each point of space, and, in the notion of point, there is nothing else.

The space before considered, which might be called *restricted space,* was referred to coordinate axes bound to my body; these axes were fixed, since my body did not move and only my members were displaced. What are the axes to which we naturally refer the *extended space?* that is to say the new space just defined. We·define a point by the sequence of movements to be made to reach it, starting from a certain initial position of the body. The axes are therefore fixed to this initial position of the body.

But the position I call initial may be arbitrarily chosen among all the positions my body has successively occupied; if the memory more or less unconscious of these successive positions is necessary for the genesis of the notion of space, this memory may go back more or less far into the past. Thence results in the definition itself of space a certain indetermination, and it is precisely this indetermination which constitutes its relativity.

There is no absolute space, there is only space relative to a certain initial position of the body. For a conscious being fixed to the ground like the lower animals, and consequently knowing only restricted space, space would still be relative (since it would have reference to his body), but this being would not be conscious of this relativity, because the axes of reference for this restricted space would be unchanging! Doubtless the rock to which this being would be fettered would not be motionless, since it would be carried along in the movement of our planet; for us consequently these axes would change at each instant; but for him they would be changeless. We have the faculty of referring our extended space now to the position *A* of our body, considered as initial, again to the position *B*, which it had some moments afterward, and which we are free to regard in its turn as initial; we make therefore at each instant unconscious transformations of coordinates. This faculty would be lacking in our imaginary being, and from not having traveled, he would think space absolute. At every instant, his system of axes would be imposed upon him; this system would have to change greatly in reality, but for him it would be always the same, since it would be always the *only* system. Quite otherwise is it with us, who at each instant have many systems between which we may choose at will, on condition of going back by memory more or less far into the past.

This is not all; restricted space would not be homogeneous; the different points of this space could not be regarded as equivalent, since some could be reached only at the cost of the greatest efforts, while others could be easily attained. On the contrary, our extended space seems to us homogeneous, and we say all its points are equivalent. What does that mean?

If we start from a certain place $A$, we can, from this position, make certain movements, $M$, characterized by a certain complex of muscular sensations. But, starting from another position, $B$, we make movements $M'$ characterized by the same muscular sensations. Let $a$, then, be the situation of a certain point of the body, the end of the index finger of the right hand for example, in the initial position $A$, and $b$ the situation of this same index when, starting from this position $A$, we have made the motions $M$. Afterwards, let $a'$ be the situation of this index in the position $B$, and $b'$ its situation when, starting from the position $B$, we have made the motions $M'$.

Well, I am accustomed to say that the points of space $a$ and $b$ are related to each other just as the points $a'$ and $b'$, and this simply means that the two series of movements $M$ and $M'$ are accompanied by the same muscular sensations. And as I am conscious that, in passing from the position $A$ to the position $B$, my body has remained capable of the same movements, I know there is a point of space related to the point $a'$ just as any point $b$ is to the point $a$, so that the two points $a$ and $a'$ are equivalent. This is what is called the homogeneity of space. And, at the same time, this is why space is relative, since its properties remain the same whether it be referred to the axes $A$ or to the axes $B$. So that the relativity of space and its homogeneity are one sole and same thing.

Now, if I wish to pass to the great space, which no longer serves only for me, but where I may lodge the universe, I get there by an act of imagination. I imagine how a giant would feel who could reach the planets in a few steps; or, if you choose, what I myself should feel in presence of a miniature world where these planets were replaced by little balls, while on one of these little balls moved a liliputian I should call myself. But this act of imagination would be impossible for me had I not previously constructed my restricted space and my extended space for my own use.

IV

Why now have all these spaces three dimensions? Go back to the "table of distribution" of which we have spoken. We have on the one side the list of the different possible dangers; designate them by $A1$, $A2$, etc.; and, on the other side, the list of the different remedies which I shall call in the same way $B1$, $B2$, etc. We have then connections between the contact studs or push buttons of the first list and those of the second, so that when, for instance, the announcer of danger $A3$ functions, it will put or may put in action the relay corresponding to the parry $B4$.

As I have spoken above of centripetal or centrifugal wires, I fear lest one see in all this, not a simple comparison, but a description of the nervous system. Such is not my thought, and that for several reasons: first I should not permit myself to put forth an opinion on the structure of the nervous system which I do not know, while those who have studied it speak only circumspectly; again because, despite my incompetence, I well know this scheme would be too simplistic; and finally because on my list of parries, some would figure very complex, which might even, in the case of extended space, as we have seen above, consist of many steps followed by a movement of the arm. It is not a question then of physical connection between two real conductors, but of psychologic association between two series of sensations.

If $A1$ and $A2$ for instance are both associated with the parry $B1$, and if $A1$ is likewise associated with the parry $B2$, it will generally happen that $A2$ and $B2$ will also themselves be associated. If this fundamental law were not generally true, there would exist only an immense confusion and there would be nothing resembling a conception of space or a geometry. How in fact have we defined a point of space. We have done it in two ways: it is on the one hand the aggregate of announcers $A$ in connection with the same parry $B$; it is on the other hand the aggregate of parries $B$ in connection with the same announcer $A$. If our law was not true, we should say $A1$ and $A2$ correspond to the same point since they are both in connection with $B1$; but we should likewise say they do not correspond to the same point, since $A1$ would be in connection with $B2$ and the same would not be true of $A2$. This would be a contradiction.

But, from another side, if the law were rigorously and always true, space would be very different from what it is. We should have categories strongly contrasted between which would be portioned out on the one hand the announcers $A$, on the other hand the parries $B$; these categories would be excessively numerous, but they would be entirely separated one from another. Space would be composed of points very numerous, but discrete; it would be *discontinuous*. There would be no reason for ranging these points in one order rather than another, nor consequently for attributing to space three dimensions.

But it is not so; permit me to resume for a moment the language of those who already know geometry; this is quite proper since this is the language best understood by those I wish to make understand me.

When I desire to parry the stroke, I seek to attain the point whence comes this blow, but it suffices that I approach quite near. Then the parry $B1$ may answer for $A1$ and for $A2$, if the point which corresponds to $B1$ is sufficiently near both to that corresponding to $A1$ and to that corresponding to $A2$. But it may happen that the point corresponding to another parry $B2$ may be sufficiently near the point corresponding to $A1$ and not sufficiently near the point corresponding to $A2$; so that the parry $B2$ may answer for $A1$ without answering for $A2$. For one who does not yet know geometry, this translates itself simply by a derogation of the law stated above. And then things will happen thus:

Two parries $B1$ and $B2$ will be associated with the same warning $A1$ and with a large number of warnings which we shall range in the same category as $A1$ and which we shall make correspond to the same point of space.    But we may find warnings $A2$ which will be associated with $B2$ without being associated with $B1$, and which in compensation will be associated with $B3$, which $B3$ was not associated with $A1$, and so forth, so that we may write the series

$$B1, \quad A1, \quad B2, \quad A2, \quad B3, \quad A3, \quad B4, \quad A4,$$

where each term is associated with the following and the preceding, but not with the terms several places away.

Needless to add that each of the terms of these series is not isolated, but forms part of a very numerous category of other warnings or of other parries which have the same connections as it, and which may be regarded as belonging to the same point of space.

The fundamental law, though admitting of exceptions, remains therefore almost always true.    Only, in consequence of these exceptions, these categories, in place of being entirely separated, encroach partially one upon another and mutually penetrate in a certain measure, so that space becomes continuous.

On the other hand, the order in which these categories are to be ranged is no longer arbitrary, and if we refer to the preceding series, we see it is necessary to put $B2$ between $A1$ and $A2$ and consequently between $B1$ and $B3$ and that we could not for instance put it between $B3$ and $B4$.

There is therefore an order in which are naturally arranged our categories which correspond to the points of space, and experience teaches us that this order presents itself under the form of a table of triple entry, and this is why space has three dimensions.

## V

So the characteristic property of space, that of having three dimensions, is only a property of our table of distribution, an internal property of the human intelligence, so to speak.    It would suffice to destroy certain of these connections, that is to say of the associations of ideas to give a different table of distribution, and that might be enough for space to acquire a fourth dimension.

Some persons will be astonished at such a result. The external world, they will think, should count for something. If the number of dimensions comes from the way we are made, there might be thinking beings living in our world, but who might be made differently from us and who would believe space has more or less than three dimensions. Has not M. de Cyon said that the Japanese mice, having only two pair of semicircular canals, believe that space is two-dimensional? And then this thinking being, if he is capable of constructing a physics, would he not make a physics of two or of four dimensions, and which in a sense would still be the same as ours, since it would be the description of the same world in another language?

It seems in fact that it would be possible to translate our physics into the language of geometry of four dimensions; to attempt this translation would be to take great pains for little profit, and I shall confine myself to citing the mechanics of Hertz where we have something analogous. However, it seems that the translation would always be less simple than the text, and that it would always have the air of a translation, that the language of three dimensions seems the better fitted to the description of our world, although this description can be rigorously made in another idiom. Besides, our table of distribution was not made at random. There is connection between the warning $A1$ and the parry $B1$, this is an internal property of our intelligence; but why this connection? It is because the parry $B1$ affords means effectively to guard against the danger $A1$; and this is a fact exterior to us, this is a property of the exterior world. Our table of distribution is therefore only the translation of an aggregate of exterior facts; if it has three dimensions, this is because it has adapted itself to a world having certain properties; and the chief of these properties is that there exist natural solids whose displacements follow sensibly the laws we call laws of motion of rigid solids. If therefore the language of three dimensions is that which permits us most easily to describe our world, we should not be astonished; this language is copied from our table of distribution; and it is in order to be able to live in this world that this table has been established.

I have said we could conceive, living in our world, thinking beings whose table of distribution would be four-dimensional and who consequently would think in hyperspace. It is not certain however that such beings, admitting they were born there, could live there and defend themselves against the thousand dangers by which they would there be assailed.

## VI

A few remarks to end with. There is a striking contrast between the roughness of this primitive geometry, reducible to what I call a table of distribution, and the infinite precision of the geometers' geometry. And yet this is born of that; but not of that alone; it must be made fecund by the faculty we have of constructing mathematical concepts, such as that of group, for instance; it was needful to seek among the pure concepts that which best adapts itself to this rough space whose genesis I have sought to explain and which is common to us and the higher animals.

The evidence for certain geometric postulates, we have said, is only our repugnance to renouncing very old habits. But these postulates are infinitely precise, while these habits have something about them essentially pliant. When we wish to think, we need postulates infinitely precise, since this is the only way to avoid contradiction; but among all the possible systems of postulates, there are some we dislike to choose because they are not sufficiently in accord with our habits; however pliant, however elastic they may be, these have a limit of elasticity.

We see that if geometry is not an experimental science, it is a science born apropos of experience; that we have created the space it studies, but adapting it to the world wherein we live. We have selected the most convenient space, but experience has guided our choice; as this choice has been unconscious, we think it has been imposed upon us; some say experience imposes it, others that we are born with our space ready made; we see from the preceding considerations, what in these two opinions is the part of truth, what of error.

In this progressive education whose outcome has been the construction of space, it is very difficult to determine what is the part of the individual, what the part of the race. How far could one of us, transported from birth to an entirely different world, where were dominant, for instance, bodies moving in conformity to the laws of motion of non-Euclidean solids, renounce the ancestral space to build a space completely new?

The part of the race seems indeed preponderant; yet if to it we owe rough space, the soft space I have spoken of, the space of the higher animals, is it not to the unconscious experience of the individual we owe the infinitely precise space of the geometer? This is a question not easy to solve. Yet we cite a fact showing that the space our ancestors have bequeathed us still retains a certain plasticity. Some hunters learn to shoot fish under water, though the image of these fish be turned up by refraction. Besides they do it instinctively: they therefore have learned to modify their old instinct of direction; or, if you choose, to substitute for the association $A1$, $B1$, another association $A1$, $B2$, because experience showed them the first would not work.

# CHAPTER II

## MATHEMATICAL DEFINITIONS AND TEACHING

1. I SHOULD speak here of general definitions in mathematics; at least that is the title, but it will be impossible to confine myself to the subject as strictly as the rule of unity of action would require; I shall not be able to treat it without touching upon a few other related questions, and if thus I am forced from time to time to walk on the bordering flower-beds on the right or left, I pray you bear with me.

What is a good definition? For the philosopher or the scientist it is a definition which applies to all the objects defined, and only those; it is the one satisfying the rules of logic. But in teaching it is not that; a good definition is one understood by the scholars.

How does it happen that so many refuse to understand mathematics? Is that not something of a paradox? Lo and behold! a science appealing only to the fundamental principles of logic, to the principle of contradiction, for instance, to that which is the skeleton, so to speak, of our intelligence, to that of which we can not divest ourselves without ceasing to think, and there are people who find it obscure! and they are even in the majority! That they are incapable of inventing may pass, but that they do not understand the demonstrations shown them, that they remain blind when we show them a light which seems to us flashing pure flame, this it is which is altogether prodigious.

And yet there is no need of a wide experience with examinations to know that these blind men are in no wise exceptional beings. This is a problem not easy to solve, but which should engage the attention of all those wishing to devote themselves to teaching.

What is it, to understand? Has this word the same meaning for all the world? To understand the demonstration of a theorem, is that to examine successively each of the syllogisms composing it and to ascertain its correctness, its conformity to the rules of the game? Likewise, to understand a definition, is this merely to recognize that one already knows the meaning of all the terms employed and to ascertain that it implies no contradiction?

For some, yes; when they have done this, they will say: I understand.

For the majority, no. Almost all are much more exacting; they wish to know not merely whether all the syllogisms of a demonstration are correct, but why they link together in this order rather than another. In so far as to them they seem engendered by caprice and not by an intelligence always conscious of the end to be attained, they do not believe they understand.

Doubtless they are not themselves just conscious of what they crave and they could not formulate their desire, but if they do not get satisfaction, they vaguely feel that something is lacking. Then what happens? In the beginning they still perceive the proofs one puts under their eyes; but as these are connected only by too slender a thread to those which precede and those which follow, they pass without leaving any trace in their head; they are soon forgotten; a moment bright, they quickly vanish in night eternal. When they are farther on, they will no longer see even this ephemeral light, since the theorems lean one upon another and those they would need are forgotten; thus it is they become incapable of understanding mathematics.

This is not always the fault of their teacher; often their mind, which needs to perceive the guiding thread, is too lazy to seek and find it. But to come to their aid, we first must know just what hinders them.

Others will always ask of what use is it; they will not have understood if they do not find about them, in practise or in nature, the justification of such and such a mathematical concept. Under each word they wish to put a sensible image; the definition must evoke this image, so that at each stage of the demonstration they may see it transform and evolve. Only upon this condition do they comprehend and retain. Often these deceive themselves; they do not listen to the reasoning, they look at the figures; they think they have understood and they have only seen.

2. How many different tendencies! Must we combat them? Must we use them? And if we wish to combat them, which should

be favored? Must we show those content with the pure logic that they have seen only one side of the matter? Or need we say to those not so cheaply satisfied that what they demand is not necessary?

In other words, should we constrain the young people to change the nature of their minds? Such an attempt would be vain; we do not possess the philosopher's stone which would enable us to transmute one into another the metals confided to us; all we can do is to work with them, adapting ourselves to their properties.

Many children are incapable of becoming mathematicians, to whom however it is necessary to teach mathematics; and the mathematicians themselves are not all cast in the same mold. To read their works suffices to distinguish among them two sorts of minds, the logicians like Weierstrass for example, the intuitives like Riemann. There is the same difference among our students. The one sort prefer to treat their problems 'by analysis' as they say, the others 'by geometry.'

It is useless to seek to change anything of that, and besides would it be desirable? It is well that there are logicians and that there are intuitives; who would dare say whether he preferred that Weierstrass had never written or that there never had been a Riemann. We must therefore resign ourselves to the diversity of minds, or better we must rejoice in it.

3. Since the word understand has many meanings, the definitions which will be best understood by some will not be best suited to others. We have those which seek to produce an image, and those where we confine ourselves to combining empty forms, perfectly intelligible, but purely intelligible, which abstraction has deprived of all matter.

I know not whether it be necessary to cite examples. Let us cite them, anyhow, and first the definition of fractions will furnish us an extreme case. In the primary schools, to define a fraction, one cuts up an apple or a pie; it is cut up mentally of course and not in reality, because I do not suppose the budget of the primary instruction allows of such prodigality. At the Normal School, on the other hand, or at the college, it is said: a fraction is the combination of two whole numbers separated by a horizontal bar; we define by conventions the operations to which these symbols may be submitted; it is proved that the rules of these operations are the same as in calculating with whole numbers, and we ascertain finally that multiplying the fraction, according to these rules, by the denominator gives the numerator. This is all very well because we are addressing young people long familiarized with the notion of fractions through having cut up apples or other objects, and whose mind, matured by a hard mathematical education, has come little by little to desire a purely logical definition. But the débutant to whom one should try to give it, how dumfounded!

Such also are the definitions found in a book justly admired and greatly honored, the *Foundations of Geometry* by Hilbert. See in fact how he begins: *We think three systems of* THINGS *which we shall call points, straights and planes.* What are these 'things'?

We know not, nor need we know; it would even be a pity to seek to know; all we have the right to know of them is what the assumptions tell us; this for example: *Two distinct points always determine a straight,* which is followed by this remark: *in place of determine, we may say the two points are on the straight, or the straight goes through these two points or joins the two points.*

Thus 'to be on a straight' is simply defined as synonymous with 'determine a straight.' Behold a book of which I think much good, but which I should not recommend to a school boy. Yet I could do so without fear, he would not read much of it. I have taken extreme examples and no teacher would dream of going that far. But even stopping short of such models, does he not already expose himself to the same danger?

Suppose we are in a class; the professor dictates: the circle is the locus of points of the plane equidistant from an interior point called the center. The good scholar writes this phrase in his note-book; the bad scholar draws faces; but neither understands; then the professor takes the chalk and draws a circle on the board. "Ah!" think the scholars, "why did he not say at once: a circle is a ring, we should have understood." Doubtless the professor is right. The scholars' definition would have been of no avail, since it could serve for no demonstration, since besides it would

29

not give them the salutary habit of analyzing their conceptions. But one should show them that they do not comprehend what they think they know, lead them to be conscious of the roughness of their primitive conception, and of themselves to wish it purified and made precise.

4. I shall return to these examples; I only wished to show you the two opposed conceptions; they are in violent contrast. This contrast the history of science explains. If we read a book written fifty years ago, most of the reasoning we find there seems lacking in rigor. Then it was assumed a continuous function can change sign only by vanishing; to-day we prove it. It was assumed the ordinary rules of calculation are applicable to incommensurable numbers; to-day we prove it. Many other things were assumed which sometimes were false.

We trusted to intuition; but intuition can not give rigor, nor even certainty; we see this more and more. It tells us for instance that every curve has a tangent, that is to say that every continuous function has a derivative, and that is false. And as we sought certainty, we had to make less and less the part of intuition.

What has made necessary this evolution? We have not been slow to perceive that rigor could not be established in the reasonings, if it were not first put into the definitions.

The objects occupying mathematicians were long ill defined; we thought we knew them because we represented them with the senses or the imagination; but we had of them only a rough image and not a precise concept upon which reasoning could take hold. It is there that the logicians would have done well to direct their efforts.

So for the incommensurable number, the vague idea of continuity, which we owe to intuition, has resolved itself into a complicated system of inequalities bearing on whole numbers. Thus have finally vanished all those difficulties which frightened our fathers when they reflected upon the foundations of the infinitesimal calculus. To-day only whole numbers are left in analysis, or systems finite or infinite of whole numbers, bound by a plexus of equalities and inequalities. Mathematics we say is arithmetized.

5. But do you think mathematics has attained absolute rigor without making any sacrifice? Not at all; what it has gained in rigor it has lost in objectivity. It is by separating itself from reality that it has acquired this perfect purity. We may freely run over its whole domain, formerly bristling with obstacles, but these obstacles have not disappeared. They have only been moved to the frontier, and it would be necessary to vanquish them anew if we wished to break over this frontier to enter the realm of the practical.

We had a vague notion, formed of incongruous elements, some *a priori*, others coming from experiences more or less digested; we thought we knew, by intuition, its principal properties. To-day we reject the empiric elements, retaining only the *a priori*; one of the properties serves as definition and all the others are deduced from it by rigorous reasoning. This is all very well, but it remains to be proved that this property, which has become a definition, pertains to the real objects which experience had made known to us and whence we drew our vague intuitive notion. To prove that, it would be necessary to appeal to experience, or to make an effort of intuition, and if we could not prove it, our theorems would be perfectly rigorous, but perfectly useless.

Logic sometimes makes monsters. Since half a century we have seen arise a crowd of bizarre functions which seem to try to resemble as little as possible the honest functions which serve some purpose. No longer continuity, or perhaps continuity, but no derivatives, etc. Nay more, from the logical point of view, it is these strange functions which are the most general, those one meets without seeking no longer appear except as particular case. There remains for them only a very small corner.

Heretofore when a new function was invented, it was for some practical end; to-day they are invented expressly to put at fault the reasonings of our fathers, and one never will get from them anything more than that.

If logic were the sole guide of the teacher, it would be necessary to begin with the most general functions, that is to say with the most bizarre. It is the beginner that would have to be set grappling with this teratologic museum.  If you do not do it, the logicians might say, you will achieve rigor only by stages.

6. Yes, perhaps, but we can not make so cheap of reality, and I mean not only the reality of the sensible world, which however has its worth, since it is to combat against it that nine tenths of your students ask of you weapons.  There is a reality more subtile, which makes the very life of the mathematical beings, and which is quite other than logic.

Our body is formed of cells, and the cells of atoms; are these cells and these atoms then all the reality of the human body? The way these cells are arranged, whence results the unity of the individual, is it not also a reality and much more interesting?

A naturalist who never had studied the elephant except in the microscope, would he think he knew the animal adequately? It is the same in mathematics.  When the logician shall have broken up each demonstration into a multitude of elementary operations, all correct, he still will not possess the whole reality; this I know not what which makes the unity of the demonstration will completely escape him.

In the edifices built up by our masters, of what use to admire the work of the mason if we can not comprehend the plan of the architect?  Now pure logic can not give us this appreciation of the total effect; this we must ask of intuition.

Take for instance the idea of continuous function.  This is at first only a sensible image, a mark traced by the chalk on the blackboard.  Little by little it is refined; we use it to construct a complicated system of inequalities, which reproduces all the features of the primitive image; when all is done, we have *removed the centering,* as after the construction of an arch; this rough representation, support thenceforth useless, has disappeared and there remains only the edifice itself, irreproachable in the eyes of the logician.  And yet, if the professor did not recall the primitive image, if he did not restore momentarily the *centering,* how could the student divine by what caprice all these inequalities have been scaffolded in this fashion one upon another? The definition would be logically correct, but it would not show him the veritable reality.

7. So back we must return; doubtless it is hard for a master

to teach what does not entirely satisfy him; but the satisfaction of the master is not the unique object of teaching; we should first give attention to what the mind of the pupil is and to what we wish it to become.

Zoologists maintain that the embryonic development of an animal recapitulates in brief the whole history of its ancestors throughout geologic time. It seems it is the same in the development of minds. The teacher should make the child go over the path his fathers trod; more rapidly, but without skipping stations. For this reason, the history of science should be our first guide.

Our fathers thought they knew what a fraction was, or continuity, or the area of a curved surface; we have found they did not know it. Just so our scholars think they know it when they begin the serious study of mathematics. If without warning I tell them: "No, you do not know it; what you think you understand, you do not understand; I must prove to you what seems to you evident," and if in the demonstration I support myself upon premises which to them seem less evident than the conclusion, what shall the unfortunates think? They will think that the science of mathematics is only an arbitrary mass of useless subtilities; either they will be disgusted with it, or they will play it as a game and will reach a state of mind like that of the Greek sophists.

Later, on the contrary, when the mind of the scholar, familiarized with mathematical reasoning, has been matured by this long frequentation, the doubts will arise of themselves and then your demonstration will be welcome. It will awaken new doubts, and the questions will arise successively to the child, as they arose successively to our fathers, until perfect rigor alone can satisfy him. To doubt everything does not suffice, one must know why he doubts.

8. The principal aim of mathematical teaching is to develop certain faculties of the mind, and among them intuition is not the least precious. It is through it that the mathematical world remains in contact with the real world, and if pure mathematics could do without it, it would always be necessary to have recourse to it to fill up the chasm which separates the symbol from reality.

The practician will always have need of it, and for one pure geometer there should be a hundred practicians.

The engineer should receive a complete mathematical education, but for what should it serve him?

To see the different aspects of things and see them quickly; he has no time to hunt mice. It is necessary that, in the complex physical objects presented to him, he should promptly recognize the point where the mathematical tools we have put in his hands can take hold. How could he do it if we should leave between instruments and objects the deep chasm hollowed out by the logicians?

9. Besides the engineers, other scholars, less numerous, are in their turn to become teachers; they therefore must go to the very bottom; a knowledge deep and rigorous of the first principles is for them before all indispensable. But this is no reason not to cultivate in them intuition; for they would get a false idea of the science if they never looked at it except from a single side, and besides they could not develop in their students a quality they did not themselves possess.

For the pure geometer himself, this faculty is necessary; it is by logic one demonstrates, by intuition one invents. To know how to criticize is good, to know how to create is better. You know how to recognize if a combination is correct; what a predicament if you have not the art of choosing among all the possible combinations. Logic tells us that on such and such a way we are sure not to meet any obstacle; it does not say which way leads to the end. For that it is necessary to see the end from afar, and the faculty which teaches us to see is intuition. Without it the geometer would be like a writer who should be versed in grammar but had no ideas. Now how could this faculty develop if, as soon as it showed itself, we chase it away and proscribe it, if we learn to set it at naught before knowing the good of it.

And here permit a parenthesis to insist upon the importance of written exercises. Written compositions are perhaps not sufficiently emphasized in certain examinations, at the polytechnic school, for instance. I am told they would close the door against very good scholars who have mastered the course, thoroughly understanding it, and who nevertheless are incapable of making the slightest application. I have just said the word understand has several meanings: such students only understand in the first way, and we have seen that suffices neither to make an engineer nor a geometer. Well, since choice must be made, I prefer those who understand completely.

10. But is the art of sound reasoning not also a precious thing, which the professor of mathematics ought before all to cultivate? I take good care not to forget that. It should occupy our attention and from the very beginning. I should be distressed to see geometry degenerate into I know not what tachymetry of low grade and I by no means subscribe to the extreme doctrines of certain German Oberlehrer. But there are occasions enough to exercise the scholars in correct reasoning in the parts of mathematics where the inconveniences I have pointed out do not present themselves. There are long chains of theorems where absolute logic has reigned from the very first and, so to speak, quite naturally, where the first geometers have given us models we should constantly imitate and admire.

It is in the exposition of first principles that it is necessary to avoid too much subtility; there it would be most discouraging and moreover useless. We can not prove everything and we can not define everything; and it will always be necessary to borrow from intuition; what does it matter whether it be done a little sooner or a little later, provided that in using correctly premises it has furnished us, we learn to reason soundly.

11. Is it possible to fulfill so many opposing conditions? Is this possible in particular when it is a question of giving a definition? How find a concise statement satisfying at once the uncompromising rules of logic, our desire to grasp the place of the new notion in the totality of the science, our need of thinking with images? Usually it will not be found, and this is why it is not enough to state a definition; it must be prepared for and justified.

What does that mean? You know it has often been said: every definition implies an assumption, since it affirms the existence of the object defined. The definition then will not be justified, from the purely logical point of view, until one shall have *proved* that it involves no contradiction, neither in the terms, nor with the verities previously admitted.

But this is not enough; the definition is stated to us as a convention; but most minds will revolt if we wish to impose it upon them as an *arbitrary* convention. They will be satisfied only when you have answered numerous questions.

Usually mathematical definitions, as M. Liard has shown, are veritable constructions built up wholly of more simple notions. But why assemble these elements in this way when a thousand other combinations were possible?

Is it by caprice? If not, why had this combination more right to exist than all the others? To what need does it respond? How was it foreseen that it would play an important rôle in the development of the science, that it would abridge our reasonings and our calculations? Is there in nature some familiar object which is so to speak the rough and vague image of it?

This is not all; if you answer all these questions in a satisfactory manner, we shall see indeed that the new-born had the right to be baptized; but neither is the choice of a name arbitrary; it is needful to explain by what analogies one has been guided and that if analogous names have been given to different things, these things at least differ only in material and are allied in form; that their properties are analogous and so to say parallel.

At this cost we may satisfy all inclinations. If the statement is correct enough to please the logician, the justification will satisfy the intuitive. But there is still a better procedure; wherever possible, the justification should precede the statement and prepare for it; one should be led on to the general statement by the study of some particular examples.

Still another thing: each of the parts of the statement of a definition has as aim to distinguish the thing to be defined from a class of other neighboring objects. The definition will be understood only when you have shown, not merely the object defined, but the neighboring objects from which it is proper to distinguish it, when you have given a grasp of the difference and when you have added explicitly: this is why in stating the definition I have said this or that.

But it is time to leave generalities and examine how the somewhat abstract principles I have expounded may be applied in arithmetic, geometry, analysis and mechanics.

### ARITHMETIC

12. The whole number is not to be defined; in return, one ordinarily defines the operations upon whole numbers; I believe the scholars learn these definitions by heart and attach no meaning to them. For that there are two reasons: first they are made to learn them too soon, when their mind as yet feels no need of them; then these definitions are not satisfactory from the logical point of view. A good definition for addition is not to be found just simply because we must stop and can not define everything. It is not defining addition to say it consists in adding. All that can be done is to start from a certain number of concrete examples and say: the operation we have performed is called addition.

For subtraction it is quite otherwise; it may be logically defined as the operation inverse to addition; but should we begin in that way? Here also start with examples, show on these examples the reciprocity of the two operations; thus the definition will be prepared for and justified.

Just so again for multiplication; take a particular problem; show that it may be solved by adding several equal numbers; then show that we reach the result more quickly by a multiplication, an operation the scholars already know how to do by routine and out of that the logical definition will issue naturally.

Division is defined as the operation inverse to multiplication; but begin by an example taken from the familiar notion of partition and show on this example that multiplication reproduces the dividend.

There still remain the operations on fractions. The only difficulty is for multiplication. It is best to expound first the theory of proportion; from it alone can come a logical definition; but to make acceptable the definitions met at the beginning of this theory, it is necessary to prepare for them by numerous examples taken from classic problems of the rule of three, taking pains to introduce fractional data.

Neither should we fear to familiarize the scholars with the

notion of proportion by geometric images, either by appealing to what they remember if they have already studied geometry, or in having recourse to direct intuition, if they have not studied it, which besides will prepare them to study it. Finally I shall add that after defining multiplication of fractions, it is needful to justify this definition by showing that it is commutative, associative and distributive, and calling to the attention of the auditors that this is established to justify the definition.

One sees what a rôle geometric images play in all this; and this rôle is justified by the philosophy and the history of the science. If arithmetic had remained free from all admixture of geometry, it would have known only the whole number; it is to adapt itself to the needs of geometry that it invented anything else.

GEOMETRY

In geometry we meet forthwith the notion of the straight line. Can the straight line be defined? The well-known definition, the shortest path from one point to another, scarcely satisfies me. I should start simply with the *ruler* and show at first to the scholar how one may verify a ruler by turning; this verification is the true definition of the straight line; the straight line is an axis of rotation. Next he should be shown how to verify the ruler by sliding and he would have one of the most important properties of the straight line.

As to this other property of being the shortest path from one point to another, it is a theorem which can be demonstrated apodictically, but the demonstration is too delicate to find a place in secondary teaching. It will be worth more to show that a ruler previously verified fits on a stretched thread. In presence of difficulties like these one need not dread to multiply assumptions, justifying them by rough experiments.

It is needful to grant these assumptions, and if one admits a few more of them than is strictly necessary, the evil is not very great; the essential thing is to learn to reason soundly on the assumptions admitted. Uncle Sarcey, who loved to repeat, often said that at the theater the spectator accepts willingly all the postulates imposed upon him at the beginning, but the curtain once raised, he becomes uncompromising on the logic. Well, it is just the same in mathematics.

For the circle, we may start with the compasses; the scholars will recognize at the first glance the curve traced; then make them observe that the distance of the two points of the instrument remains constant, that one of these points is fixed and the other movable, and so we shall be led naturally to the logical definition.

The definition of the plane implies an axiom and this need not be hidden. Take a drawing board and show that a moving ruler may be kept constantly in complete contact with this plane and yet retain three degrees of freedom. Compare with the cylinder and the cone, surfaces on which an applied straight retains only two degrees of freedom; next take three drawing boards; show first that they will glide while remaining applied to one another and this with three degrees of freedom; and finally to distinguish the plane from the sphere, show that two of these boards which fit a third will fit each other.

Perhaps you are surprised at this incessant employment of moving things; this is not a rough artifice; it is much more philosophic than one would at first think. What is geometry for the philosopher? It is the study of a group. And what group? That of the motions of solid bodies. How define this group then without moving some solids?

Should we retain the classic definition of parallels and say parallels are two coplanar straights which do not meet, however far they be prolonged? No, since this definition is negative, since it is unverifiable by experiment, and consequently can not be regarded as an immediate datum of intuition. No, above all because it is wholly strange to the notion of group, to the consideration of the motion of solid bodies which is, as I have said, the true source of geometry. Would it not be better to define first the rectilinear translation of an invariable figure, as a motion wherein all the points of this figure have rectilinear trajectories; to show that such a translation is possible by making a square glide on a ruler?

From this experimental ascertainment, set up as an assumption, it would be easy to derive the notion of parallel and Euclid's postulate itself.

## Mechanics

I need not return to the definition of velocity, or acceleration, or other kinematic notions; they may be advantageously connected with that of the derivative.

I shall insist, on the other hand, upon the dynamic notions of force and mass.

I am struck by one thing: how very far the young people who have received a high-school education are from applying to the real world the mechanical laws they have been taught. It is not only that they are incapable of it; they do not even think of it. For them the world of science and the world of reality are separated by an impervious partition wall.

If we try to analyze the state of mind of our scholars, this will astonish us less. What is for them the real definition of force? Not that which they recite, but that which, crouching in a nook of their mind, from there directs it wholly. Here is the definition: forces are arrows with which one makes parallelograms. These arrows are imaginary things which have nothing to do with anything existing in nature. This would not happen if they had been shown forces in reality before representing them by arrows.

How shall we define force?

I think I have elsewhere sufficiently shown there is no good logical definition. There is the anthropomorphic definition, the sensation of muscular effort; this is really too rough and nothing useful can be drawn from it.

Here is how we should go: first, to make known the genus force, we must show one after the other all the species of this genus; they are very numerous and very different; there is the pressure of fluids on the insides of the vases wherein they are contained; the tension of threads; the elasticity of a spring; the gravity working on all the molecules of a body; friction; the normal mutual action and reaction of two solids in contact.

This is only a qualitative definition; it is necessary to learn to measure force. For that begin by showing that one force may be replaced by another without destroying equilibrium; we may find the first example of this substitution in the balance and Borda's double weighing.

Then show that a weight may be replaced, not only by another weight, but by force of a different nature: for instance, Prony's brake permits replacing weight by friction.

From all this arises the notion of the equivalence of two forces.

The direction of a force must be defined. If a force $F$ is equivalent to another force $F'$ applied to the body considered by means of a stretched string, so that $F$ may be replaced by $F'$ without affecting the equilibrium, then the point of attachment of the string will be by definition the point of application of the force $F'$, and that of the equivalent force $F$; the direction of the string will be the direction of the force $F'$ and that of the equivalent force $F$.

From that, pass to the comparison of the magnitude of forces. If a force can replace two others with the same direction, it equals their sum; show for example that a weight of 20 grams may replace two 10-gram weights.

Is this enough? Not yet. We now know how to compare the intensity of two forces which have the same direction and same point of application; we must learn to do it when the directions are different. For that, imagine a string stretched by a weight and passing over a pulley; we shall say that the tensor of the two legs of the string is the same and equal to the tension weight.

This definition of ours enables us to compare the tensions of the two pieces of our string, and, using the preceding definitions, to compare any two forces having the same direction as these two pieces. It should be justified by showing that the tension of the last piece of the string remains the same for the same tensor weight, whatever be the number and the disposition of the reflecting pulleys. It has still to be completed by showing this is only true if the pulleys are frictionless.

Once master of these definitions, it is to be shown that the point of application, the direction and the intensity suffice to determine a force; that two forces for which these three elements are the same are *always* equivalent and may *always* be replaced by one another, whether in equilibrium or in movement, and this whatever be the other forces acting.

It must be shown that two concurrent forces may always be replaced by a unique resultant; and that *this resultant remains the same,* whether the body be at rest or in motion and whatever be the other forces applied to it.

Finally it must be shown that forces thus defined satisfy the principle of the equality of action and reaction.

Experiment it is, and experiment alone, which can teach us all that. It will suffice to cite certain common experiments, which the scholars make daily without suspecting it, and to perform before them a few experiments, simple and well chosen.

It is after having passed through all these meanders that one may represent forces by arrows, and I should even wish that in the development of the reasonings return were made from time to time from the symbol to the reality. For instance it would not be difficult to illustrate the parallelogram of forces by aid of an apparatus formed of three strings, passing over pulleys, stretched by weights and in equilibrium while pulling on the same point.

Knowing force, it is easy to define mass; this time the definition should be borrowed from dynamics; there is no way of doing otherwise, since the end to be attained is to give understanding of the distinction between mass and weight. Here again, the definition should be led up to by experiments; there is in fact a machine which seems made expressly to show what mass is, Atwood's machine; recall also the laws of the fall of bodies, that the acceleration of gravity is the same for heavy as for light bodies, and that it varies with the latitude, etc.

Now, if you tell me that all the methods I extol have long been applied in the schools, I shall rejoice over it more than be surprised at it. I know that on the whole our mathematical teaching is good. I do not wish it overturned; that would even distress me. I only desire betterments slowly progressive. This teaching should not be subjected to brusque oscillations under the capricious blast of ephemeral fads. In such tempests its high educative value would soon founder. A good and sound logic should continue to be its basis. The definition by example is always necessary, but it should prepare the way for the logical definition, it should not replace it; it should at least make this wished for, in the cases where the true logical definition can be advantageously given only in advanced teaching.

Understand that what I have here said does not imply giving up what I have written elsewhere. I have often had occasion to criticize certain definitions I extol to-day. These criticisms hold good completely. These definitions can only be provisory. But it is by way of them that we must pass.

# CHAPTER III

## MATHEMATICS AND LOGIC

### INTRODUCTION

CAN mathematics be reduced to logic without having to appeal to principles peculiar to mathematics? There is a whole school, abounding in ardor and full of faith, striving to prove it. They have their own special language, which is without words, using only signs. This language is understood only by the initiates, so that commoners are disposed to bow to the trenchant affirmations of the adepts. It is perhaps not unprofitable to examine these affirmations somewhat closely, to see if they justify the peremptory tone with which they are presented.

But to make clear the nature of the question it is necessary to enter upon certain historical details and in particular to recall the character of the works of Cantor.

Since long ago the notion of infinity had been introduced into mathematics; but this infinite was what philosophers call a *becoming*. The mathematical infinite was only a quantity capable of increasing beyond all limit: it was a variable quantity of which it could not be said that it *had passed* all limits, but only that it *could pass* them.

Cantor has undertaken to introduce into mathematics an *actual infinite*, that is to say a quantity which not only is capable of passing all limits, but which is regarded as having already passed them. He has set himself questions like these: Are there more points in space than whole numbers? Are there more points in space than points in a plane? etc.

And then the number of whole numbers, that of the points of space, etc., constitutes what he calls a *transfinite cardinal number*, that is to say a cardinal number greater than all the ordinary cardinal numbers. And he has occupied himself in comparing these transfinite cardinal numbers. In arranging in a proper order the elements of an aggregate containing an infinity of

448

them, he has also imagined what he calls transfinite ordinal numbers upon which I shall not dwell.

Many mathematicians followed his lead and set a series of questions of the sort. They so familiarized themselves with transfinite numbers that they have come to make the theory of finite numbers depend upon that of Cantor's cardinal numbers. In their eyes, to teach arithmetic in a way truly logical, one should begin by establishing the general properties of transfinite cardinal numbers, then distinguish among them a very small class, that of the ordinary whole numbers. Thanks to this détour, one might succeed in proving all the propositions relative to this little class (that is to say all our arithmetic and our algebra) without using any principle foreign to logic. This method is evidently contrary to all sane psychology; it is certainly not in this way that the human mind proceeded in constructing mathematics; so its authors do not dream, I think, of introducing it into secondary teaching. But is it at least logic, or, better, is it correct? It may be doubted.

The geometers who have employed it are however very numerous. They have accumulated formulas and they have thought to free themselves from what was not pure logic by writing memoirs where the formulas no longer alternate with explanatory discourse as in the books of ordinary mathematics, but where this discourse has completely disappeared.

Unfortunately they have reached contradictory results, what are called the *cantorian antinomies,* to which we shall have occasion to return. These contradictions have not discouraged them and they have tried to modify their rules so as to make those disappear which had already shown themselves, without being sure, for all that, that new ones would not manifest themselves.

It is time to administer justice on these exaggerations. I do not hope to convince them; for they have lived too long in this atmosphere. Besides, when one of their demonstrations has been refuted, we are sure to see it resurrected with insignificant alterations, and some of them have already risen several times from their ashes. Such long ago was the Lernæan hydra with its famous heads which always grew again. Hercules got through, since his hydra had only nine heads, or eleven; but here there are too many, some in England, some in Germany, in Italy, in France, and he would have to give up the struggle. So I appeal only to men of good judgment unprejudiced.

## I

In these latter years numerous works have been published on pure mathematics and the philosophy of mathematics, trying to separate and isolate the logical elements of mathematical reasoning. These works have been analyzed and expounded very clearly by M. Couturat in a book entitled: *The Principles of Mathematics.*

For M. Couturat, the new works, and in particular those of Russell and Peano, have finally settled the controversy, so long pending between Leibnitz and Kant. They have shown that there are no synthetic judgments *a priori* (Kant's phrase to designate judgments which can neither be demonstrated analytically, nor reduced to identities, nor established experimentally), they have shown that mathematics is entirely reducible to logic and that intuition here plays no rôle.

This is what M. Couturat has set forth in the work just cited; this he says still more explicitly in his Kant jubilee discourse, so that I heard my neighbor whisper: "I well see this is the centenary of Kant's *death*."

Can we subscribe to this conclusive condemnation? I think not, and I shall try to show why.

## II

What strikes us first in the new mathematics is its purely formal character: "We think," says Hilbert, "three sorts of *things*, which we shall call points, straights and planes. We convene that a straight shall be determined by two points, and that in place of saying this straight is determined by these two points, we may say it passes through these two points, or that these two points are situated on this straight." What these *things* are, not only we do not know, but we should not seek to know. We have no need to, and one who never had seen either point or straight or plane could geometrize as well as we. That the phrase *to pass through,* or the phrase *to be situated upon* may arouse in us no image, the first is simply a synonym of *to be determined* and the second of *to determine.*

Thus, be it understood, to demonstrate a theorem, it is neither necessary nor even advantageous to know what it means. The geometer might be replaced by the *logic piano* imagined by Stanley Jevons; or, if you choose, a machine might be imagined where the assumptions were put in at one end, while the theorems came out at the other, like the legendary Chicago machine where the pigs go in alive and come out transformed into hams and sausages. No more than these machines need the mathematician know what he does.

I do not make this formal character of his geometry a reproach to Hilbert. This is the way he should go, given the problem he set himself. He wished to reduce to a minimum the number of the fundamental assumptions of geometry and completely enumerate them; now, in reasonings where our mind remains active, in those where intuition still plays a part, in living reasonings, so to speak, it is difficult not to introduce an assumption or a postulate which passes unperceived. It is therefore only after having carried back all the geometric reasonings to a form purely mechanical that he could be sure of having accomplished his design and finished his work.

What Hilbert did for geometry, others have tried to do for arithmetic and analysis. Even if they had entirely succeeded, would the Kantians be finally condemned to silence? Perhaps not, for in reducing mathematical thought to an empty form, it is certainly mutilated.

Even admitting it were established that all the theorems could be deduced by procedures purely analytic, by simple logical combinations of a finite number of assumptions, and that these assumptions are only conventions; the philosopher would still have the right to investigate the origins of these conventions, to see why they have been judged preferable to the contrary conventions.

And then the logical correctness of the reasonings leading from the assumptions to the theorems is not the only thing which should occupy us. The rules of perfect logic, are they the whole of mathematics? As well say the whole art of playing chess reduces to the rules of the moves of the pieces. Among all the constructs which can be built up of the materials furnished by logic, choice must be made; the true geometer makes this choice judiciously because he is guided by a sure instinct, or by some vague consciousness of I know not what more profound and more hidden geometry, which alone gives value to the edifice constructed.

To seek the origin of this instinct, to study the laws of this deep geometry, felt, not stated, would also be a fine employment for the philosophers who do not want logic to be all. But it is not at this point of view I wish to put myself, it is not thus I wish to consider the question. The instinct mentioned is necessary for the inventor, but it would seem at first we might do without it in studying the science once created. Well, what I wish to investigate is if it be true that, the principles of logic once admitted, one can, I do not say discover, but demonstrate, all the mathematical verities without making a new appeal to intuition.

### III

I once said no to this question:[1] should our reply be modified by the recent works? My saying no was because "the principle of complete induction" seemed to me at once necessary to the mathematician and irreducible to logic. The statement of this principle is: "If a property be true of the number 1, and if we establish that it is true of $n + 1$ provided it be of $n$, it will be true of all the whole numbers." Therein I see the mathematical reasoning par excellence. I did not mean to say, as has been supposed, that all mathematical reasonings can be reduced to an application of this principle. Examining these reasonings closely, we there should see applied many other analogous principles, presenting the same essential characteristics. In this category of principles, that of complete induction is only the simplest of all and this is why I have chosen it as type.

The current name, principle of complete induction, is not justified. This mode of reasoning is none the less a true mathe-

---

[1] See *Science and Hypothesis,* chapter I.

matical induction which differs from ordinary induction only by its certitude.

## IV

### DEFINITIONS AND ASSUMPTIONS

The existence of such principles is a difficulty for the uncompromising logicians; how do they pretend to get out of it? The principle of complete induction, they say, is not an assumption properly so called or a synthetic judgment *a priori;* it is just simply the definition of whole number. It is therefore a simple convention. To discuss this way of looking at it, we must examine a little closely the relations between definitions and assumptions.

Let us go back first to an article by M. Couturat on mathematical definitions which appeared in *l'Enseignement mathématique,* a magazine published by Gauthier-Villars and by Georg at Geneva. We shall see there a distinction between the *direct definition* and the *definition by postulates.*

"The definition by postulates," says M. Couturat, "applies, not to a single notion, but to a system of notions; it consists in enumerating the fundamental relations which unite them and which enable us to demonstrate all their other properties; these relations are postulates."

If previously have been defined all these notions but one, then this last will be by definition the thing which verifies these postulates. Thus certain indemonstrable assumptions of mathematics would be only disguised definitions. This point of view is often legitimate; and I have myself admitted it in regard for instance to Euclid's postulate.

The other assumptions of geometry do not suffice to completely define distance; the distance then will be, by definition, among all the magnitudes which satisfy these other assumptions, that which is such as to make Euclid's postulate true.

Well the logicians suppose true for the principle of complete induction what I admit for Euclid's postulate; they want to see in it only a disguised definition.

But to give them this right, two conditions must be fulfilled. Stuart Mill says every definition implies an assumption, that by which the existence of the defined object is affirmed. According to that, it would no longer be the assumption which might be a disguised definition, it would on the contrary be the definition which would be a disguised assumption. Stuart Mill meant the word existence in a material and empirical sense; he meant to say that in defining the circle we affirm there are round things in nature.

Under this form, his opinion is inadmissible. Mathematics is independent of the existence of material objects; in mathematics the word exist can have only one meaning, it means free from contradiction. Thus rectified, Stuart Mill's thought becomes exact; in defining a thing, we affirm that the definition implies no contradiction.

If therefore we have a system of postulates, and if we can demonstrate that these postulates imply no contradiction, we shall have the right to consider them as representing the definition of one of the notions entering therein. If we can not demonstrate that, it must be admitted without proof, and that then will be an assumption; so that, seeking the definition under the postulate, we should find the assumption under the definition.

Usually, to show that a definition implies no contradiction, we proceed *by example*, we try to make an *example* of a thing satisfying the definition. Take the case of a definition by postulates; we wish to define a notion $A$, and we say that, by definition, an $A$ is anything for which certain postulates are true. If we can prove directly that all these postulates are true of a certain object $B$, the definition will be justified; the object $B$ will be an *example* of an $A$. We shall be certain that the postulates are not contradictory, since there are cases where they are all true at the same time.

But such a direct demonstration by example is not always possible.

To establish that the postulates imply no contradiction, it is then necessary to consider all the propositions deducible from these postulates considered as premises, and to show that, among these propositions, no two are contradictory. If these propositions are finite in number, a direct verification is possible. This case is infrequent and uninteresting. If these propositions are infinite in number, this direct verification can no longer be made;

recourse must be had to procedures where in general it is necessary to invoke just this principle of complete induction which is precisely the thing to be proved.

This is an explanation of one of the conditions the logicians should satisfy, *and further on we shall see they have not done it.*

## V

There is a second. When we give a definition, it is to use it.

We therefore shall find in the sequel of the exposition the word defined; have we the right to affirm, of the thing represented by this word, the postulate which has served for definition? Yes, evidently, if the word has retained its meaning, if we do not attribute to it implicitly a different meaning. Now this is what sometimes happens and it is usually difficult to perceive it; it is needful to see how this word comes into our discourse, and if the gate by which it has entered does not imply in reality a definition other than that stated.

This difficulty presents itself in all the applications of mathematics. The mathematical notion has been given a definition very refined and very rigorous; and for the pure mathematician all doubt has disappeared; but if one wishes to apply it to the physical sciences for instance, it is no longer a question of this pure notion, but of a concrete object which is often only a rough image of it. To say that this object satisfies, at least approximately, the definition, is to state a new truth, which experience alone can put beyond doubt, and which no longer has the character of a conventional postulate.

But without going beyond pure mathematics, we also meet the same difficulty.

You give a subtile definition of numbers; then, once this definition given, you think no more of it; because, in reality, it is not it which has taught you what number is; you long ago knew that, and when the word number further on is found under your pen, you give it the same sense as the first comer. To know what is this meaning and whether it is the same in this phrase or that, it is needful to see how you have been led to speak of number and to introduce this word into these two phrases. I shall not for the moment dilate upon this point, because we shall have occasion to return to it.

Thus consider a word of which we have given explicitly a definition $A$; afterwards in the discourse we make a use of it which implicitly supposes another definition $B$. It is possible that these two definitions designate the same thing. But that this is so is a new truth which must either be demonstrated or admitted as an independent assumption.

*We shall see farther on that the logicians have not fulfilled the second condition any better than the first.*

## VI

The definitions of number are very numerous and very different; I forego the enumeration even of the names of their authors. We should not be astonished that there are so many. If one among them was satisfactory, no new one would be given. If each new philosopher occupying himself with this question has thought he must invent another one, this was because he was not satisfied with those of his predecessors, and he was not satisfied with them because he thought he saw a petitio principii.

I have always felt, in reading the writings devoted to this problem, a profound feeling of discomfort; I was always expecting to run against a petitio principii, and when I did not immediately perceive it, I feared I had overlooked it.

This is because it is impossible to give a definition without using a sentence, and difficult to make a sentence without using a number word, or at least the word several, or at least a word in the plural. And then the declivity is slippery and at each instant there is risk of a fall into petitio principii.

I shall devote my attention in what follows only to those of these definitions where the petitio principii is most ably concealed.

## VII

### PASIGRAPHY

The symbolic language created by Peano plays a very grand rôle in these new researches. It is capable of rendering some service, but I think M. Couturat attaches to it an exaggerated importance which must astonish Peano himself.

The essential element of this language is certain algebraic

signs which represent the different conjunctions: if, and, or, therefore. That these signs may be convenient is possible; but that they are destined to revolutionize all philosophy is a different matter. It is difficult to admit that the word *if* acquires, when written $\bigcirc$, a virtue it had not when written if. This invention of Peano was first called *pasigraphy*, that is to say the art of writing a treatise on mathematics without using a single word of ordinary language. This name defined its range very exactly. Later, it was raised to a more eminent dignity by conferring on it the title of *logistic*. This word is, it appears, employed at the Military Academy, to designate the art of the quartermaster of cavalry, the art of marching and cantoning troops; but here no confusion need be feared, and it is at once seen that this new name implies the design of revolutionizing logic.

We may see the new method at work in a mathematical memoir by Burali-Forti, intitled: *Una Questione sui numeri transfiniti*, inserted in Volume XI of the *Rendiconti del circolo matematico di Palermo*.

I begin by saying this memoir is very interesting, and my taking it here as example is precisely because it is the most important of all those written in the new language. Besides, the uninitiated may read it, thanks to an Italian interlinear translation.

What constitutes the importance of this memoir is that it has given the first example of those antinomies met in the study of transfinite numbers and making since some years the despair of mathematicians. The aim, says Burali-Forti, of this note is to show there may be two transfinite numbers (ordinals), *a* and *b*, such that *a* is neither equal to, greater than, nor less than *b*.

To reassure the reader, to comprehend the considerations which follow, he has no need of knowing what a transfinite ordinal number is.

Now, Cantor had precisely proved that between two transfinite numbers as between two finite, there can be no other relation than equality, or inequality in one sense or the other. But it is not of the substance of this memoir that I wish to speak here; that would carry me much too far from my subject; I only wish to consider the form, and just to ask if this form makes it gain much in rigor and whether it thus compensates for the efforts it imposes upon the writer and the reader.

First we see Burali-Forti define the number 1 as follows:

$$1 = \iota T' \{ K o \widehat{n}(u, \ h) \epsilon(u \epsilon Un) \},$$

a definition eminently fitted to give an idea of the number 1 to persons who had never heard speak of it.

I understand Peanian too ill to dare risk a critique, but still I fear this definition contains a petitio principii, considering that I see the figure 1 in the first member and Un in letters in the second.

However that may be, Burali-Forti starts from this definition and, after a short calculation, reaches the equation:

(27)                              $1 \epsilon No,$

which tells us that One is a number.

And since we are on these definitions of the first numbers, we recall that M. Couturat has also defined 0 and 1.

What is zero? It is the number of elements of the null class. And what is the null class? It is that containing no element.

To define zero by null, and null by no, is really to abuse the wealth of language; so M. Couturat has introduced an improvement in his definition, by writing:

$$0 = \iota\Lambda : \phi x = \Lambda \cdot \mathfrak{d} \cdot \Lambda = (x \epsilon \phi x),$$

which means: zero is the number of things satisfying a condition never satisfied.

But as never means *in no case* I do not see that the progress is great.

I hasten to add that the definition M. Couturat gives of the number 1 is more satisfactory.

One, says he in substance, is the number of elements in a class in which any two elements are identical.

It is more satisfactory, I have said, in this sense that to define 1, he does not use the word one; in compensation, he uses the word two. But I fear, if asked what is two, M. Couturat would have to use the word one.

## VIII

But to return to the memoir of Burali-Forti; I have said his conclusions are in direct opposition to those of Cantor. Now, one day M. Hadamard came to see me and the talk fell upon this antinomy.

"Burali-Forti's reasoning," I said, "does it not seem to you irreproachable?" "No, and on the contrary I find nothing to object to in that of Cantor. Besides, Burali-Forti had no right to speak of the aggregate of *all* the ordinal numbers."

"Pardon, he had the right, since he could always put

$$\Omega = T'(\text{No}, \bar{\epsilon}>).$$

I should like to know who was to prevent him, and can it be said a thing does not exist, when we have called it $\Omega$?"

It was in vain, I could not convince him (which besides would have been sad, since he was right). Was it merely because I do not speak the Peanian with enough eloquence? Perhaps; but between ourselves I do not think so.

Thus, despite all this pasigraphic apparatus, the question was not solved. What does that prove? In so far as it is a question only of proving one a number, pasigraphy suffices, but if a difficulty presents itself, if there is an antinomy to solve, pasigraphy becomes impotent.

# CHAPTER IV

## The New Logics

### I

### *The Russell Logic*

To justify its pretensions, logic had to change. We have seen new logics arise of which the most interesting is that of Russell. It seems he has nothing new to write about formal logic, as if Aristotle there had touched bottom. But the domain Russell attributes to logic is infinitely more extended than that of the classic logic, and he has put forth on the subject views which are original and at times well warranted.

First, Russell subordinates the logic of classes to that of propositions, while the logic of Aristotle was above all the logic of classes and took as its point of departure the relation of subject to predicate. The classic syllogism, "Socrates is a man," etc., gives place to the hypothetical syllogism: "If $A$ is true, $B$ is true; now if $B$ is true, $C$ is true," etc. And this is, I think, a most happy idea, because the classic syllogism is easy to carry back to the hypothetical syllogism, while the inverse transformation is not without difficulty.

And then this is not all. Russell's logic of propositions is the study of the laws of combination of the conjunctions *if, and, or,* and the negation *not.*

In adding here two other conjunctions *and* and *or,* Russell opens to logic a new field. The symbols *and, or* follow the same laws as the two signs $\times$ and $+$, that is to say the commutative associative and distributive laws. Thus *and* represents logical multiplication, while *or* represents logical addition. This also is very interesting.

Russell reaches the conclusion that any false proposition implies all other propositions true or false. M. Couturat says this conclusion will at first seem paradoxical. It is sufficient however to have corrected a bad thesis in mathematics to recognize how right Russell is. The candidate often is at great pains to get the first false equation; but that once obtained, it is only sport then for him to accumulate the most surprising results, some of which even may be true.

## II

We see how much richer the new logic is than the classic logic; the symbols are multiplied and allow of varied combinations *which are no longer limited in number*. Has one the right to give this extension to the meaning of the word *logic?* It would be useless to examine this question and to seek with Russell a mere quarrel about words. Grant him what he demands; but be not astonished if certain verities declared irreducible to logic in the old sense of the word find themselves now reducible to logic in the new sense—something very different.

A great number of new notions have been introduced, and these are not simply combinations of the old. Russell knows this, and not only at the beginning of the first chapter, 'The Logic of Propositions,' but at the beginning of the second and third, 'The Logic of Classes' and 'The Logic of Relations,' he introduces new words that he declares indefinable.

And this is not all; he likewise introduces principles he declares indemonstrable. But these indemonstrable principles are appeals to intuition, synthetic judgments *a priori*. We regard them as intuitive when we meet them more or less explicitly enunciated in mathematical treatises; have they changed character because the meaning of the word logic has been enlarged and we now find them in a book entitled *Treatise on Logic? They have not changed nature; they have only changed place.*

## III

Could these principles be considered as disguised definitions? It would then be necessary to have some way of proving that they imply no contradiction. It would be necessary to establish that, however far one followed the series of deductions, he would never be exposed to contradicting himself.

We might attempt to reason as follows: We can verify that

the operations of the new logic applied to premises exempt from contradiction can only give consequences equally exempt from contradiction. If therefore after $n$ operations we have not met contradiction, we shall not encounter it after $n + 1$. Thus it is impossible that there should be a moment when contradiction *begins,* which shows we shall never meet it. Have we the right to reason in this way? No, for this would be to make use of complete induction; and *remember, we do not yet know the principle of complete induction.*

We therefore have not the right to regard these assumptions as disguised definitions and only one resource remains for us, to admit a new act of intuition for each of them. Moreover I believe this is indeed the thought of Russell and M. Couturat.

Thus each of the nine indefinable notions and of the twenty indemonstrable propositions (I believe if it were I that did the counting, I should have found some more) which are the foundation of the new logic, logic in the broad sense, presupposes a new and independent act of our intuition and (why not say it?) a veritable synthetic judgment *a priori.* On this point all seem agreed, but what Russell claims, and *what seems to me doubtful, is that after these appeals to intuition, that will be the end of it; we need make no others and can build all mathematics without the intervention of any new element.*

## IV

M. Couturat often repeats that this new logic is altogether independent of the idea of number. I shall not amuse myself by counting how many numeral adjectives his exposition contains, both cardinal and ordinal, or indefinite adjectives such as several. We may cite, however, some examples:

"The logical product of *two* or *more* propositions is . . .";

"All propositions are capable only of *two* values, true and false";

"The relative product of *two* relations is a relation";

"A relation exists between *two* terms," etc., etc.

Sometimes this inconvenience would not be unavoidable, but sometimes also it is essential. A relation is incomprehensible without two terms; it is impossible. to have the intuition of the relation, without having at the same time that of its two terms, and without noticing they are two, because, if the relation is to be conceivable, it is necessary that there be two and only two.

## V

### *Arithmetic*

I reach what M. Couturat calls the *ordinal theory* which is the foundation of arithmetic properly so called. M. Couturat begins by stating Peano's five assumptions, which are independent, as has been proved by Peano and Padoa.

1. Zero is an integer.

2. Zero is not the successor of any integer.

3. The successor of an integer is an integer.

To this it would be proper to add,

Every integer has a successor.

4. Two integers are equal if their successors are.

The fifth assumption is the principle of complete induction.

M. Couturat considers these assumptions as disguised definitions; they constitute the definition by postulates of zero, of successor, and of integer.

But we have seen that for a definition by postulates to be acceptable we must be able to prove that it implies no contradiction.

Is this the case here? Not at all.

The demonstration can not be made *by example*. We can not take a part of the integers, for instance the first three, and prove they satisfy the definition.

If I take the series 0, 1, 2, I see it fulfils the assumptions 1, 2, 4 and 5; but to satisfy assumption 3 it still is necessary that 3 be an integer, and consequently that the series 0, 1, 2, 3, fulfil the assumptions; we might prove that it satisfies assumptions 1, 2, 4, 5, but assumption 3 requires besides that 4 be an integer and that the series 0, 1, 2, 3, 4 fulfil the assumptions, and so on.

It is therefore impossible to demonstrate the assumptions for certain integers without proving them for all; we must give up proof by example.

It is necessary then to take all the consequences of our assumptions and see if they contain no contradiction.

If these consequences were finite in number, this would be easy; but they are infinite in number; they are the whole of mathematics, or at least all arithmetic.

What then is to be done? Perhaps strictly we could repeat the reasoning of number III.

But as we have said, this reasoning is complete induction, and it is precisely the principle of complete induction whose justification would be the point in question.

## VI

### *The Logic of Hilbert*

I come now to the capital work of Hilbert which he communicated to the Congress of Mathematicians at Heidelberg, and of which a French translation by M. Pierre Boutroux appeared in *l'Enseignement mathématique,* while an English translation due to Halsted appeared in *The Monist.*[1] In this work, which contains profound thoughts, the author's aim is analogous to that of Russell, but on many points he diverges from his predecessor.

"But," he says (*Monist,* p. 340), "on attentive consideration we become aware that in the usual exposition of the laws of logic certain fundamental concepts of arithmetic are already employed; for example, the concept of the aggregate, in part also the concept of number.

"We fall thus into a vicious circle and therefore to avoid paradoxes a partly simultaneous development of the laws of logic and arithmetic is requisite."

We have seen above that what Hilbert says of the principles of logic *in the usual exposition* applies likewise to the logic of Russell. So for Russell logic is prior to arithmetic; for Hilbert they are 'simultaneous.' We shall find further on other differences still greater, but we shall point them out as we come to them. I prefer to follow step by step the development of Hilbert's thought, quoting textually the most important passages.

[1] 'The Foundations of Logic and Arithmetic,' *Monist,* XV., 338–352.

"Let us take as the basis of our consideration first of all a thought-thing 1 (one)" (p. 341). Notice that in so doing we in no wise imply the notion of number, because it is understood that 1 is here only a symbol and that we do not at all seek to know its meaning. "The taking of this thing together with itself respectively two, three or more times. . . ." Ah! this time it is no longer the same; if we introduce the words 'two,' 'three,' and above all 'more,' 'several,' we introduce the notion of number; and then the definition of finite whole number which we shall presently find, will come too late. Our author was too circumspect not to perceive this begging of the question. So at the end of his work he tries to proceed to a truly patching-up process.

Hilbert then introduces two simple objects 1 and =, and considers all the combinations of these two objects, all the combinations of their combinations, etc. It goes without saying that we must forget the ordinary meaning of these two signs and not attribute any to them.

Afterwards he separates these combinations into two classes, the class of the existent and the class of the non-existent, and till further orders this separation is entirely arbitrary. Every affirmative statement tells us that a certain combination belongs to the class of the existent; every negative statement tells us that a certain combination belongs to the class of the non-existent.

## VII

Note now a difference of the highest importance. For Russell any object whatsoever, which he designates by $x$, is an object absolutely undetermined and about which he supposes nothing; for Hilbert it is one of the combinations formed with the symbols 1 and =; he could not conceive of the introduction of anything other than combinations of objects already defined. Moreover Hilbert formulates his thought in the neatest way, and I think I must reproduce *in extenso* his statement (p. 348):

"In the assumptions the arbitraries (as equivalent for the concept 'every' and 'all' in the customary logic) represent only those thought-things and their combinations with one another, which at this stage are laid down as fundamental or are to be

31

newly defined. Therefore in the deduction of inferences from the assumptions, the arbitraries, which occur in the assumptions, can be replaced only by such thought-things and their combinations.

"Also we must duly remember, that through the super-addition and making fundamental of a new thought-thing the preceding assumptions undergo an enlargement of their validity, and where necessary, are to be subjected to a change in conformity with the sense."

The contrast with Russell's view-point is complete. For this philosopher we may substitute for $x$ not only objects already known, but anything.

Russell is faithful to his point of view, which is that of comprehension. He starts from the general idea of being, and enriches it more and more while restricting it, by adding new qualities. Hilbert on the contrary recognizes as possible beings only combinations of objects already known; so that (looking at only one side of his thought) we might say he takes the view-point of extension.

## VIII

Let us continue with the exposition of Hilbert's ideas. He introduces two assumptions which he states in his symbolic language but which signify, in the language of the uninitiated, that every quality is equal to itself and that every operation performed upon two identical quantities gives identical results.

So stated, they are evident, but thus to present them would be to misrepresent Hilbert's thought. For him mathematics has to combine only pure symbols, and a true mathematician should reason upon them without preconceptions as to their meaning. So his assumptions are not for him what they are for the common people.

He considers them as representing the definition by postulates of the symbol (=) heretofore void of all signification. But to justify this definition we must show that these two assumptions lead to no contradiction. For this Hilbert used the reasoning of our number III, without appearing to perceive that he is using complete induction.

## IX

The end of Hilbert's memoir is altogether enigmatic and I shall not lay stress upon it. Contradictions accumulate; we feel that the author is dimly conscious of the *petitio principii* he has committed, and that he seeks vainly to patch up the holes in his argument.

What does this mean? At the point of proving that the definition of the whole number by the assumption of complete induction implies no contradiction, Hilbert withdraws as Russell and Couturat withdrew, because the difficulty is too great.

## X

### *Geometry*

Geometry, says M. Couturat, is a vast body of doctrine wherein the principle of complete induction does not enter. That is true in a certain measure; we can not say it is entirely absent, but it enters very slightly. If we refer to the *Rational Geometry* of Dr. Halsted (New York, John Wiley and Sons, 1904) built up in accordance with the principles of Hilbert, we see the principle of induction enter for the first time on page 114 (unless I have made an oversight, which is quite possible).[2]

So geometry, which only a few years ago seemed the domain where the reign of intuition was uncontested, is to-day the realm where the logicians seem to triumph. Nothing could better measure the importance of the geometric works of Hilbert and the profound impress they have left on our conceptions.

But be not deceived. What is after all the fundamental theorem of geometry? It is that the assumptions of geometry imply no contradiction, and this we can not prove without the principle of induction.

How does Hilbert demonstrate this essential point? By leaning upon analysis and through it upon arithmetic and through it upon the principle of induction.

And if ever one invents another demonstration, it will still be necessary to lean upon this principle, since the possible consequences of the assumptions, of which it is necessary to show that they are not contradictory, are infinite in number.

[2] Second ed., 1907, p. 86; French ed., 1911, p. 97.  G. B. H.

## XI

### *Conclusion*

Our conclusion straightway is that the principle of induction can not be regarded as the disguised definition of the entire world.

Here are three truths: (1) The principle of complete induction; (2) Euclid's postulate; (3) the physical law according to which phosphorus melts at 44° (cited by M. Le Roy).

These are said to be three disguised definitions: the first, that of the whole number; the second, that of the straight line; the third, that of phosphorus.

I grant it for the second; I do not admit it for the other two. I must explain the reason for this apparent inconsistency.

First, we have seen that a definition is acceptable only on condition that it implies no contradiction. We have shown likewise that for the first definition this demonstration is impossible; on the other hand, we have just recalled that for the second Hilbert has given a complete proof.

As to the third, evidently it implies no contradiction. Does this mean that the definition guarantees, as it should, the existence of the object defined? We are here no longer in the mathematical sciences, but in the physical, and the word existence has no longer same meaning. It no longer signifies absence of contradiction; it means objective existence.

You already see a first reason for the distinction I made between the three cases; there is a second. In the applications we have to make of these three concepts, do they present themselves to us as defined by these three postulates?

The possible applications of the principle of induction are innumerable; take, for example, one of those we have expounded above, and where it is sought to prove that an aggregate of assumptions can lead to no contradiction. For this we consider one of the series of syllogisms we may go on with in starting from these assumptions as premises. When we have finished the $n$th syllogism, we see we can make still another and this is the $n + 1$th. Thus the number $n$ serves to count a series of successive operations; it is a number obtainable by successive addi-

tions.  This therefore is a number from which we may go back
to unity by *successive subtractions*.  Evidently we could not do
this if we had $n = n - 1$, since then by subtraction we should
always obtain again the same number.  So the way we have been
led to consider this number $n$ implies a definition of the finite
whole number and this definition is the following: A finite whole
number is that which can be obtained by successive additions;
it is such that $n$ is not equal to $n - 1$.

That granted, what do we do?  We show that if there has
been no contradiction up to the $n$th syllogism, no more will there
be up to the $n + 1$th, and we conclude there never will be.  You
say: I have the right to draw this conclusion, since the whole
numbers are by definition those for which a like reasoning is
legitimate.  But that implies another definition of the whole
number, which is as follows: A whole number is that on which we
may reason by recurrence.  In the particular case it is that of
which we may say that, if the absence of contradiction up to the
time of a syllogism of which the number is an integer carries
with it the absence of contradiction up to the time of the syllogism whose number is the following integer, we need fear no
contradiction for any of the syllogisms whose number is an
integer.

The two definitions are not identical; they are doubtless equivalent, but only in virtue of a synthetic judgment *a priori;* we can
not pass from one to the other by a purely logical procedure.
Consequently we have no right to adopt the second, after having
introduced the whole number by a way that presupposes the first.

On the other hand, what happens with regard to the straight
line?  I have already explained this so often that I hesitate to
repeat it again, and shall confine myself to a brief recapitulation
of my thought.  We have not, as in the preceding case, two
equivalent definitions logically irreducible one to the other.  We
have only one expressible in words.  Will it be said there is
another which we feel without being able to word it, since we
have the intuition of the straight line or since we represent to
ourselves the straight line?  First of all, we can not represent it
to ourselves in geometric space, but only in representative space,
and then we can represent to ourselves just as well the objects

which possess the other properties of the straight line, save that of satisfying Euclid's postulate. These objects are 'the non-Euclidean straights,' which from a certain point of view are not entities void of sense, but circles (true circles of true space) orthógonal to a certain sphere. If, among these objects equally capable of representation, it is the first (the Euclidean straights) which we call straights, and not the latter (the non-Euclidean straights), this is properly by definition.

And arriving finally at the third example, the definition of phosphorus, we see the true definition would be: Phosphorus is the bit of matter I see in yonder flask.

## XII

And since I am on this subject, still another word. Of the phosphorus example I said: "This proposition is a real verifiable physical law, because it means that all bodies having all the other properties of phosphorus, save its point of fusion, melt like it at 44°." And it was answered: "No, this law is not verifiable, because if it were shown that two bodies resembling phosphorus melt one at 44° and the other at 50°, it might always be said that doubtless, besides the point of fusion, there is some other unknown property by which they differ."

That was not quite what I meant to say. I should have written, "All bodies possessing such and such properties finite in number (to wit, the properties of phosphorus stated in the books on chemistry, the fusion-point excepted) melt at 44°."

And the better to make evident the difference between the case of the straight and that of phosphorus, one more remark. The straight has in nature many images more or less imperfect, of which the chief are the light rays and the rotation axis of the solid. Suppose we find the ray of light does not satisfy Euclid's postulate (for example by showing that a star has a negative parallax), what shall we do? Shall we conclude that the straight being by definition the trajectory of light does not satisfy the postulate, or, on the other hand, that the straight by definition satisfying the postulate, the ray of light is not straight?

Assuredly we are free to adopt the one or the other definition and consequently the one or the other conclusion; but to adopt the first would be stupid, because the ray of light probably satisfies only imperfectly not merely Euclid's postulate, but the other properties of the straight line, so that if it deviates from the Euclidean straight, it deviates no less from the rotation axis of solids which is another imperfect image of the straight line; while finally it is doubtless subject to change, so that such a line which yesterday was straight will cease to be straight to-morrow if some physical circumstance has changed.

Suppose now we find that phosphorus does not melt at 44°, but at 43.9°. Shall we conclude that phosphorus being by definition that which melts at 44°, this body that we did call phosphorus is not true phosphorus, or, on the other hand, that phosphorous melts at 43.9°? Here again we are free to adopt the one or the other definition and consequently the one or the other conclusion; but to adopt the first would be stupid because we can not be changing the name of a substance every time we determine a new decimal of its fusion-point.

## XIII

To sum up, Russell and Hilbert have each made a vigorous effort; they have each written a work full of original views, profound and often well warranted. These two works give us much to think about and we have much to learn from them. Among their results, some, many even, are solid and destined to live.

But to say that they have finally settled the debate between Kant and Leibnitz and ruined the Kantian theory of mathematics is evidently incorrect. I do not know whether they really believed they had done it, but if they believed so, they deceived themselves.

# CHAPTER V

## I

THE logicians have attempted to answer the preceding considerations. For that, a transformation of logistic was necessary, and Russell in particular has modified on certain points his original views. Without entering into the details of the debate, I should like to return to the two questions to my mind most important: Have the rules of logistic demonstrated their fruitfulness and infallibility? Is it true they afford means of proving the principle of complete induction without any appeal to intuition?

## II

### *The Infallibility of Logistic*

On the question of fertility, it seems M. Couturat has naïve illusions. Logistic, according to him, lends invention 'stilts and wings,' and on the next page: *"Ten years ago,* Peano published the first edition of his *Formulaire."* How is that, ten years of wings and not to have flown!

I have the highest esteem for Peano, who has done very pretty things (for instance his 'space-filling curve,' a phrase now discarded); but after all he has not gone further nor higher nor quicker than the majority of wingless mathematicians, and would have done just as well with his legs.

On the contrary I see in logistic only shackles for the inventor. It is no aid to conciseness—far from it, and if twenty-seven equations were necessary to establish that 1 is a number, how many would be needed to prove a real theorem? If we distinguish, with Whitehead, the individual $x$, the class of which the only member is $x$ and which shall be called $\iota x$, then the class of which the only member is the class of which the only member is $x$ and which shall be called $u x$, do you think these distinctions, useful as they may be, go far to quicken our pace?

Logistic forces us to say all that is ordinarily left to be understood; it makes us advance step by step; this is perhaps surer but not quicker.

It is not wings you logisticians give us, but leading-strings. And then we have the right to require that these leading-strings prevent our falling. This will be their only excuse. When a bond does not bear much interest, it should at least be an investment for a father of a family.

Should your rules be followed blindly? Yes, else only intuition could enable us to distinguish among them; but then they must be infallible; for only in an infallible authority can one have a blind confidence. This, therefore, is for you a necessity. Infallible you shall be, or not at all.

You have no right to say to us: "It is true we make mistakes, but so do you." For us to blunder is a misfortune, a very great misfortune; for you it is death.

Nor may you ask: Does the infallibility of arithmetic prevent errors in addition? The rules of calculation are infallible, and yet we see those blunder *who do not apply these rules;* but in checking their calculation it is at once seen where they went wrong. Here it is not at all the case; the logicians *have applied* their rules, and they have fallen into contradiction; and so true is this, that they are preparing to change these rules and to "sacrifice the notion of class." Why change them if they were infallible?

"We are not obliged," you say, "to solve *hic et nunc* all possible problems." Oh, we do not ask so much of you. If, in face of a problem, you would give *no* solution, we should have nothing to say; but on the contrary you give us *two* of them and those contradictory, and consequently at least one false; this it is which is failure.

Russell seeks to reconcile these contradictions, which can only be done, according to him, "by restricting or even sacrificing the notion of class." And M. Couturat, discovering the success of his attempt, adds: "If the logicians succeed where others have failed, M. Poincaré will remember this phrase, and give the honor of the solution to logistic."

But no! Logistic exists, it has its code which has already had four editions; or rather this code is logistic itself. Is Mr. Russell preparing to show that one at least of the two contradictory reasonings has transgressed the code? Not at all; he is preparing to change these laws and to abrogate a certain number of them. If he succeeds, I shall give the honor of it to Russell's intuition and not to the Peanian logistic which he will have destroyed.

## III

### *The Liberty of Contradiction*

I made two principal objections to the definition of whole number adopted in logistic. What says M. Couturat to the first of these objections?

What does the word *exist* mean in mathematics? It means, I said, to be free from contradiction. This M. Couturat contests. "Logical existence," says he, "is quite another thing from the absence of contradiction. It consists in the fact that a class is not empty." To say: *a*'s exist, is, by definition, to affirm that the class *a* is not null.

And doubtless to affirm that the class *a* is not null, is, by definition, to affirm that *a*'s exist. But one of the two affirmations is as denuded of meaning as the other, if they do not both signify, either that one may see or touch *a*'s which is the meaning physicists or naturalists give them, or that one may conceive an *a* without being drawn into contradictions, which is the meaning given them by logicians and mathematicians.

For M. Couturat, "it is not non-contradiction that proves existence, but it is existence that proves non-contradiction." To establish the existence of a class, it is necessary therefore to establish, by an *example*, that there is an individual belonging to this class: "But, it will be said, how is the existence of this individual proved? Must not this existence be established, in order that the existence of the class of which it is a part may be deduced? Well, no; however paradoxical may appear the assertion, we never demonstrate the existence of an individual. Individuals, just because they are individuals, are always considered as existent. . . . We never have to express that an individual exists, absolutely speaking, but only that it exists in a class." M.

Couturat finds his own assertion paradoxical, and he will certainly not be the only one. Yet it must have a meaning. It doubtless means that the existence of an individual, alone in the world, and of which nothing is affirmed, can not involve contradiction; in so far as it is all alone it evidently will not embarrass any one. Well, so let it be; we shall admit the existence of the individual, 'absolutely speaking,' but nothing more. It remains to prove the existence of the individual 'in a class,' and for that it will always be necessary to prove that the affirmation, ''Such an individual belongs to such a class,'' is neither contradictory in itself, nor to the other postulates adopted.

''It is then,'' continues M. Couturat, ''arbitrary and misleading to maintain that a definition is valid only if we first prove it is not contradictory.'' One could not claim in prouder and more energetic terms the liberty of contradiction. ''In any case, the *onus probandi* rests upon those who believe that these principles are contradictory.'' Postulates are presumed to be compatible until the contrary is proved, just as the accused person is presumed innocent. Needless to add that I do not assent to this claim. But, you say, the demonstration you require of us is impossible, and you can not ask us to jump over the moon. Pardon me; that is impossible for you, but not for us, who admit the principle of induction as a synthetic judgment *a priori*. And that would be necessary for you, as for us.

To demonstrate that a system of postulates implies no contradiction, it is necessary to apply the principle of complete induction; this mode of reasoning not only has nothing 'bizarre' about it, but it is the only correct one. It is not 'unlikely' that it has ever been employed; and it is not hard to find 'examples and precedents' of it. I have cited two such instances borrowed from Hilbert's article. He is not the only one to have used it, and those who have not done so have been wrong. What I have blamed Hilbert for is not his having recourse to it (a born mathematician such as he could not fail to see a demonstration was necessary and this the only one possible), but his having recourse without recognizing the reasoning by recurrence.

## IV

### *The Second Objection*

I pointed out a second error of logistic in Hilbert's article. To-day Hilbert is excommunicated and M. Couturat no longer regards him as of the logistic cult; so he asks if I have found the same fault among the orthodox. No, I have not seen it in the pages I have read; I know not whether I should find it in the three hundred pages they have written which I have no desire to read.

Only, they must commit it the day they wish to make any application of mathematics. This science has not as sole object the eternal contemplation of its own navel; it has to do with nature and some day it will touch it. Then it will be necessary to shake off purely verbal definitions and to stop paying oneself with words.

To go back to the example of Hilbert: always the point at issue is reasoning by recurrence and the question of knowing whether a system of postulates is not contradictory. M. Couturat will doubtless say that then this does not touch him, but it perhaps will interest those who do not claim, as he does, the liberty of contradiction.

We wish to establish, as above, that we shall never encounter contradiction after any number of deductions whatever, provided this number be finite. For that, it is necessary to apply the principle of induction. Should we here understand by finite number every number to which by definition the principle of induction applies? Evidently not, else we should be led to most embarrassing consequences. To have the right to lay down a system of postulates, we must be sure they are not contradictory. This is a truth admitted by *most* scientists; I should have written *by all* before reading M. Couturat's last article. But what does this signify? Does it mean that we must be sure of not meeting contradiction after a *finite* number of propositions, the *finite* number being by definition that which has all properties of recurrent nature, so that if one of these properties fails—if, for instance, we come upon a contradiction—we shall agree to say that the number in question is not finite? In other words, do we mean that we must be sure not to meet contradictions, on condition of agreeing to stop just when we are about to encounter one? To state such a proposition is enough to condemn it.

So, Hilbert's reasoning not only assumes the principle of induction, but it supposes that this principle is given us not as a simple definition, but as a synthetic judgment *a priori*.

To sum up:

A demonstration is necessary.

The only demonstration possible is the proof by recurrence.

This is legitimate only if we admit the principle of induction and if we regard it not as a definition but as a synthetic judgment.

## V

### *The Cantor Antinomies*

Now to examine Russell's new memoir. This memoir was written with the view to conquer the difficulties raised by those Cantor antinomies to which frequent allusion has already been made. Cantor thought he could construct a science of the infinite; others went on in the way he opened, but they soon ran foul of strange contradictions. These antinomies are already numerous, but the most celebrated are:

1. The Burali-Forti antinomy;
2. The Zermelo-König antinomy;
3. The Richard antinomy.

Cantor proved that the ordinal numbers (the question is of transfinite ordinal numbers, a new notion introduced by him) can be ranged in a linear series; that is to say that of two unequal ordinals one is always less than the other. Burali-Forti proves the contrary; and in fact he says in substance that if one could range *all* the ordinals in a linear series, this series would define an ordinal greater than *all* the others; we could afterwards adjoin 1 and would obtain again an ordinal which would be *still greater*, and this is contradictory.

We shall return later to the Zermelo-König antinomy which is of a slightly different nature. The Richard antinomy[1] is as follows: Consider all the decimal numbers definable by a finite

[1] *Revue générale des sciences*, June 30, 1905.

number of words; these decimal numbers form an aggregate $E$, and it is easy to see that this aggregate is countable, that is to say we can *number* the different decimal numbers of this assemblage from 1 to infinity. Suppose the numbering effected, and define a number $N$ as follows: If the $n$th decimal of the $n$th number of the assemblage $E$ is

$$0, 1, 2, 3, 4, 5, 6, 7, 8, 9$$

the $n$th decimal of $N$ shall be:

$$1, 2, 3, 4, 5, 6, 7, 8, 1, 1$$

As we see, $N$ is not equal to the $n$th number of $E$, and as $n$ is arbitrary, $N$ does not appertain to $E$ and yet $N$ should belong to this assemblage since we have defined it with a finite number of words.

We shall later see that M. Richard has himself given with much sagacity the explanation of his paradox and that this extends, *mutatis mutandis,* to the other like paradoxes. Again, Russell cites another quite amusing paradox: *What is the least whole number which can not be defined by a phrase composed of less than a hundred English words?*

This number exists; and in fact the numbers capable of being defined by a like phrase are evidently finite in number since the words of the English language are not infinite in number. Therefore among them will be one less than all the others. And, on the other hand, this number does not exist, because its definition implies contradiction. This number, in fact, is defined by the phrase in italics which is composed of less than a hundred English words; and by definition this number should not be capable of definition by a like phrase.

## VI

### *Zigzag Theory and No-class Theory*

What is Mr. Russell's attitude in presence of these contradictions? After having analyzed those of which we have just spoken, and cited still others, after having given them a form recalling Epimenides, he does not hesitate to conclude: "A propositional function of one variable does not always determine a class.'' A propositional function (that is to say a definition) does not always determine a class. A 'propositional function' or 'norm' may be 'non-predicative.' And this does not mean that these non-predicative propositions determine an empty class, a null class; this does not mean that there is no value of $x$ satisfying the definition and capable of being one of the elements of the class. The elements exist, but they have no right to unite in a syndicate to form a class.

But this is only the beginning and it is needful to know how to recognize whether a definition is or is not predicative. To solve this problem Russell hesitates between three theories which he calls

A. The zigzag theory;

B. The theory of limitation of size;

C. The no-class theory.

According to the zigzag theory ''definitions (propositional functions) determine a class when they are very simple and cease to do so only when they are complicated and obscure.'' Who, now, is to decide whether a definition may be regarded as simple enough to be acceptable? To this question there is no answer, if it be not the loyal avowal of a complete inability: ''The rules which enable us to recognize whether these definitions are predicative would be extremely complicated and can not commend themselves by any plausible reason. This is a fault which might be remedied by greater ingenuity or by using distinctions not yet pointed out. But hitherto in seeking these rules, I have not been able to find any other directing principle than the absence of contradiction.''

This theory therefore remains very obscure; in this night a single light—the word zigzag. What Russell calls the 'zigzaginess' is doubtless the particular characteristic which distinguishes the argument of Epimenides.

According to the theory of limitation of size, a class would cease to have the right to exist if it were too extended. Perhaps it might be infinite, but it should not be too much so. But we always meet again the same difficulty; at what precise moment does it begin to be too much so? Of course this difficulty is not solved and Russell passes on to the third theory.

In the no-classes theory it is forbidden to speak the word 'class' and this word must be replaced by various periphrases. What a change for logistic which talks only of classes and classes of classes! It becomes necessary to remake the whole of logistic. Imagine how a page of logistic would look upon suppressing all the propositions where it is a question of class. There would only be some scattered survivors in the midst of a blank page. *Apparent rari nantes in gurgite vasto.*

Be that as it may, we see how Russell hesitates and the modifications to which he submits the fundamental principles he has hitherto adopted. Criteria are needed to decide whether a definition is too complex or too extended, and these criteria can only be justified by an appeal to intuition.

It is toward the no-classes theory that Russell finally inclines. Be that as it may, logistic is to be remade and it is not clear how much of it can be saved. Needless to add that Cantorism and logistic are alone under consideration; real mathematics, that which is good for something, may continue to develop in accordance with its own principles without bothering about the storms which rage outside it, and go on step by step with its usual conquests which are final and which it never has to abandon.

## VII

### *The True Solution*

What choice ought we to make among these different theories? It seems to me that the solution is contained in a letter of M. Richard of which I have spoken above, to be found in the *Revue générale des sciences* of June 30, 1905. After having set forth the antinomy we have called Richard's antinomy, he gives its explanation. Recall what has already been said of this antinomy. $E$ is the aggregate of *all* the numbers definable by a finite number of words, *without introducing the notion of the aggregate E itself.* Else the definition of $E$ would contain a vicious circle; we must not define $E$ by the aggregate $E$ itself.

Now we have defined $N$ with a finite number of words, it is true, but with the aid of the notion of the aggregate $E$. And this is why $N$ is not part of $E$. In the example selected by M. Richard, the conclusion presents itself with complete evidence and the evidence will appear still stronger on consulting the text of the letter itself. But the same explanation holds good for the other antinomies, as is easily verified. Thus *the definitions which should be regarded as not predicative are those which contain a vicious circle.* And the preceding examples sufficiently show what I mean by that. Is it this which Russell calls the 'zigzaginess'? I put the question without answering it.

## VIII

### *The Demonstrations of the Principle of Induction*

Let us now examine the pretended demonstrations of the principle of induction and in particular those of Whitehead and of Burali-Forti.

We shall speak of Whitehead's first, and take advantage of certain new terms happily introduced by Russell in his recent memoir. Call *recurrent class* every class containing zero, and containing $n + 1$ if it contains $n$. Call *inductive number* every number which is a part of *all* the recurrent classes. Upon what condition will this latter definition, which plays an essential rôle in Whitehead's proof, be 'predicative' and consequently acceptable?

In accordance with what has been said, it is necessary to understand by *all* the recurrent classes, all those in whose definition the notion of inductive number does not enter. Else we fall again upon the vicious circle which has engendered the antinomies.

Now *Whitehead has not taken this precaution.* Whitehead's reasoning is therefore fallacious; it is the same which led to the antinomies. It was illegitimate when it gave false results; it remains illegitimate when by chance it leads to a true result.

A definition containing a vicious circle defines nothing. It is of no use to say, we are sure, whatever meaning we may give to our definition, zero at least belongs to the class of inductive numbers; it is not a question of knowing whether this class is void, but whether it can be rigorously deliminated. A 'non-

32

predicative' class is not an empty class, it is a class whose boundary is undetermined. Needless to add that this particular objection leaves in force the general objections applicable to all the demonstrations.

## IX

Burali-Forti has given another demonstration.[2] But he is obliged to assume two postulates: First, there always exists at least one infinite class. The second is thus expressed:

$$u\epsilon K(K - \iota\Lambda) \cdot \supset \cdot u < v'u.$$

The first postulate is not more evident than the principle to be proved. The second not only is not evident, but it is false, as Whitehead has shown; as moreover any recruit would see at the first glance, if the axiom had been stated in intelligible language, since it means that the number of combinations which can be formed with several objects is less than the number of these objects.

## X

### *Zermelo's Assumption*

A famous demonstration by Zermelo rests upon the following assumption: In any aggregate (or the same in each aggregate of an assemblage of aggregates) we can always choose *at random* an element (even if this assemblage of aggregates should contain an infinity of aggregates). This assumption had been applied a thousand times without being stated, but, once stated, it aroused doubts. Some mathematicians, for instance M. Borel, resolutely reject it; others admire it. Let us see what, according to his last article, Russell thinks of it. He does not speak out, but his reflections are very suggestive.

And first a picturesque example: Suppose we have as many pairs of shoes as there are whole numbers, and so that we can number *the pairs* from one to infinity, how many shoes shall we have? Will the number of shoes be equal to the number of pairs? Yes, if in each pair the right shoe is distinguishable from the left; it will in fact suffice to give the number $2n - 1$ to the right shoe of the $n$th pair, and the number $2n$ to the left

[2] In his article 'Le classi finite,' *Atti di Torino*, Vol. XXXII.

shoe of the $n$th pair. No, if the right shoe is just like the left, because a similar operation would become impossible—unless we admit Zermelo's assumption, since then we could choose *at random* in each pair the shoe to be regarded as the right.

## XI

### *Conclusions*

A demonstration truly founded upon the principles of analytic logic will be composed of a series of propositions. Some, serving as premises, will be identities or definitions; the others will be deduced from the premises step by step. But though the bond between each proposition and the following is immediately evident, it will not at first sight appear how we get from the first to the last, which we may be tempted to regard as a new truth. But if we replace successively the different expressions therein by their definition and if this operation be carried as far as possible, there will finally remain only identities, so that all will reduce to an immense tautology. Logic therefore remains sterile unless made fruitful by intuition.

This I wrote long ago; logistic professes the contrary and thinks it has proved it by actually proving new truths. By what mechanism? Why in applying to their reasonings the procedure just described—namely, replacing the terms defined by their definitions—do we not see them dissolve into identities like ordinary reasonings? It is because this procedure is not applicable to them. And why? Because their definitions are not predicative and present this sort of hidden vicious circle which I have pointed out above; non-predicative definitions can not be substituted for the terms defined. Under these conditions *logistic is not sterile, it engenders antinomies.*

It is the belief in the existence of the actual infinite which has given birth to those non-predicative definitions. Let me explain. In these definitions the word 'all' figures, as is seen in the examples cited above. The word 'all' has a very precise meaning when it is a question of an infinite number of objects; to have another one, when the objects are infinite in number, would require there being an actual (given complete) infinity. Otherwise *all* these objects could not be conceived as postulated anteriorly to their definition, and then if the definition of a notion *N* depends upon *all* the objects *A*, it may be infected with a vicious circle, if among the objects *A* are some indefinable without the intervention of the notion *N* itself.

The rules of formal logic express simply the properties of all possible classifications. But for them to be applicable it is necessary that these classifications be immutable and that we have no need to modify them in the course of the reasoning. If we have to classify only a finite number of objects, it is easy to keep our classifications without change. If the objects are *indefinite* in number, that is to say if one is constantly exposed to seeing new and unforeseen objects arise, it may happen that the appearance of a new object may require the classification to be modified, and thus it is we are exposed to antinomies. *There is no actual (given complete) infinity.* The Cantorians have forgotten this, and they have fallen into contradiction. It is true that Cantorism has been of service, but this was when applied to a real problem whose terms were precisely defined, and then we could advance without fear.

Logistic also forgot it, like the Cantorians, and encountered the same difficulties. But the question is to know whether they went this way by accident or whether it was a necessity for them. For me, the question is not doubtful; belief in an actual infinity is essential in the Russell logic. It is just this which distinguishes it from the Hilbert logic. Hilbert takes the view-point of extension, precisely in order to avoid the Cantorian antinomies. Russell takes the view-point of comprehension. Consequently for him the genus is anterior to the species, and the *summum genus* is anterior to all. That would not be inconvenient if the *summum genus* was finite; but if it is infinite, it is necessary to postulate the infinite, that is to say to regard the infinite as actual (given complete). And we have not only infinite classes; when we pass from the genus to the species in restricting the concept by new conditions, these conditions are still infinite in number. Because they express generally that the envisaged object presents such or such a relation with all the objects of an infinite class.

But that is ancient history. Russell has perceived the peril and takes counsel. He is about to change everything, and, what is easily understood, he is preparing not only to introduce new principles which shall allow of operations formerly forbidden, but he is preparing to forbid operations he formerly thought legitimate. Not content to adore what he burned, he is about to burn what he adored, which is more serious. He does not add a new wing to the building, he saps its foundation.

The old logistic is dead, so much so that already the zigzag theory and the no-classes theory are disputing over the succession. To judge of the new, we shall await its coming.

# BOOK III

---

## THE NEW MECHANICS

### CHAPTER I

#### MECHANICS AND RADIUM

### I

### *Introduction*

THE general principles of Dynamics, which have, since Newton, served as foundation for physical science, and which appeared immovable, are they on the point of being abandoned or
at least profoundly modified? This is what many people have
been asking themselves for some years. According to them, the
discovery of radium has overturned the scientific dogmas we believed the most solid: on the one hand, the impossibility of the
transmutation of metals; on the other hand, the fundamental
postulates of mechanics.

Perhaps one is too hasty in considering these novelties as
finally established, and breaking our idols of yesterday; perhaps
it would be proper, before taking sides, to await experiments
more numerous and more convincing. None the less is it necessary, from to-day, to know the new doctrines and the arguments,
already very weighty, upon which they rest.

In few words let us first recall in what those principles consist:

*A.* The motion of a material point isolated and apart from all
exterior force is straight and uniform; this is the principle of
inertia: without force no acceleration;

*B.* The acceleration of a moving point has the same direction
as the resultant of all the forces to which it is subjected; it is
equal to the quotient of this resultant by a coefficient called
*mass* of the moving point.

The mass of a moving point, so defined, is a constant; it does not depend upon the velocity acquired by this point; it is the same whether the force, being parallel to this velocity, tends only to accelerate or to retard the motion of the point, or whether, on the contrary, being perpendicular to this velocity, it tends to make this motion deviate toward the right, or the left, that is to say to *curve* the trajectory;

*C.* All the forces affecting a material point come from the action of other material points; they depend only upon the *relative* positions and velocities of these different material points.

Combining the two principles *B* and *C*, we reach the *principle of relative motion*, in virtue of which the laws of the motion of a system are the same whether we refer this system to fixed axes, or to moving axes animated by a straight and uniform motion of translation, so that it is impossible to distinguish absolute motion from a relative motion with reference to such moving axes;

*D.* If a material point *A* acts upon another material point *B*, the body *B* reacts upon *A*, and these two actions are two equal and directly opposite forces. This is *the principle of the equality of action and reaction,* or, more briefly, the *principle of reaction.*

Astronomic observations and the most ordinary physical phenomena seem to have given of these principles a confirmation complete, constant and very precise. This is true, it is now said, but it is because we have never operated with any but very small velocities; Mercury, for example, the fastest of the planets, goes scarcely 100 kilometers a second. Would this planet act the same if it went a thousand times faster? We see there is yet no need to worry; whatever may be the progress of automobilism, it will be long before we must give up applying to our machines the classic principles of dynamics.

How then have we come to make actual speeds a thousand times greater than that of Mercury, equal, for instance, to a tenth or a third of the velocity of light, or approaching still more closely to that velocity? It is by aid of the cathode rays and the rays from radium.

We know that radium emits three kinds of rays, designated by the three Greek letters $\alpha$, $\beta$, $\gamma$; in what follows, unless the contrary be expressly stated, it will always be a question of the $\beta$ rays, which are analogous to the cathode rays.

After the discovery of the cathode rays two theories appeared: Crookes attributed the phenomena to a veritable molecular bombardment; Hertz, to special undulations of the ether. This was a renewal of the debate which divided physicists a century ago about light; Crookes took up the emission theory, abandoned for light; Hertz held to the undulatory theory. The facts seem to decide in favor of Crookes.

It has been recognized, in the first place, that the cathode rays carry with them a negative electric charge; they are deviated by a magnetic field and by an electric field; and these deviations are precisely such as these same fields would produce upon projectiles animated by a very high velocity and strongly charged with electricity. These two deviations depend upon two quantities: one the velocity, the other the relation of the electric charge of the projectile to its mass; we cannot know the absolute value of this mass, nor that of the charge, but only their relation; in fact, it is clear that if we double at the same time the charge and the mass, without changing the velocity, we shall double the force which tends to deviate the projectile, but, as its mass is also doubled, the acceleration and deviation observable will not be changed. The observation of the two deviations will give us therefore two equations to determine these two unknowns. We find a velocity of from 10,000 to 30,000 kilometers a second; as to the ratio of the charge to the mass, it is very great. We may compare it to the corresponding ratio in regard to the hydrogen ion in electrolysis; we then find that a cathodic projectile carries about a thousand times more electricity than an equal mass of hydrogen would carry in an electrolyte.

To confirm these views, we need a direct measurement of this velocity to compare with the velocity so calculated. Old experiments of J. J. Thomson had given results more than a hundred times too small; but they were exposed to certain causes of error. The question was taken up again by Wiechert in an arrangement where the Hertzian oscillations were utilized; results were found agreeing with the theory, at least as to order of magnitude; it would be of great interest to repeat these experiments. However that may be, the theory of undulations appears powerless to account for this complex of facts.

The same calculations made with reference to the $\beta$ rays of radium have given velocities still greater: 100,000 or 200,000 kilometers or more yet. These velocities greatly surpass all those we know. It is true that light has long been known to go 300,000 kilometers a second; but it is not a carrying of matter, while, if we adopt the emission theory for the cathode rays, there would be material molecules really impelled at the velocities in question, and it is proper to investigate whether the ordinary laws of mechanics are still applicable to them.

## II

### *Mass Longitudinal and Mass Transversal*

We know that electric currents produce the phenomena of induction, in particular *self-induction*. When a current increases, there develops an electromotive force of self-induction which tends to oppose the current; on the contrary, when the current decreases, the electromotive force of self-induction tends to maintain the current. The self-induction therefore opposes every variation of the intensity of the current, just as in mechanics the inertia of a body opposes every variation of its velocity.

*Self-induction is a veritable inertia.* Everything happens as if the current could not establish itself without putting in motion the surrounding ether and as if the inertia of this ether tended, in consequence, to keep constant the intensity of this current. It would be requisite to overcome this inertia to establish the current, it would be necessary to overcome it again to make the current cease.

A cathode ray, which is a rain of projectiles charged with negative electricity, may be likened to a current; doubtless this current differs, at first sight at least, from the currents of ordinary conduction, where the matter does not move and where the electricity circulates through the matter. This is a *current of convection*, where the electricity, attached to a material vehicle, is carried along by the motion of this vehicle. But Rowland has proved that currents of convection produce the same magnetic effects as currents of conduction; they should produce also the same effects of induction. First, if this were not so, the principle of the conservation of energy would be violated; besides,

Crémieu and Pender have employed a method putting in evidence *directly* these effects of induction.

If the velocity of a cathode corpuscle varies, the intensity of the corresponding current will likewise vary; and there will develop effects of self-induction which will tend to oppose this variation. These corpuscles should therefore possess a double inertia: first their own proper inertia, and then the apparent inertia, due to self-induction, which produces the same effects. They will therefore have a total apparent mass, composed of their real mass and of a fictitious mass of electromagnetic origin. Calculation shows that this fictitious mass varies with the velocity, and that the force of inertia of self-induction is not the same when the velocity of the projectile accelerates or slackens, or when it is deviated; therefore so it is with the force of the total apparent inertia.

The total apparent mass is therefore not the same when the real force applied to the corpuscle is parallel to its velocity and tends to accelerate the motion as when it is perpendicular to this velocity and tends to make the direction vary. It is necessary therefore to distinguish the *total longitudinal mass* from the *total transversal mass*. These two total masses depend, moreover, upon the velocity. This follows from the theoretical work of Abraham.

In the measurements of which we speak in the preceding section, what is it we determine in measuring the two deviations? It is the velocity on the one hand, and on the other hand the ratio of the charge to the *total transversal mass*. How, under these conditions, can we make out in this total mass the part of the real mass and that of the fictitious electromagnetic mass? If we had only the cathode rays properly so called, it could not be dreamed of; but happily we have the rays of radium which, as we have seen, are notably swifter. These rays are not all identical and do not behave in the same way under the action of an electric field and a magnetic field. It is found that the electric deviation is a function of the magnetic deviation, and we are able, by receiving on a sensitive plate radium rays which have been subjected to the action of the two fields, to photograph the curve which represents the relation between these two deviations. This is what Kaufmann has done, deducing from it the relation between the velocity and the ratio of the charge to the total apparent mass, a ratio we shall call $\epsilon$.

One might suppose there are several species of rays, each characterized by a fixed velocity, by a fixed charge and by a fixed mass. But this hypothesis is improbable; why, in fact, would all the corpuscles of the same mass take always the same velocity? It is more natural to suppose that the charge as well as the *real* mass are the same for all the projectiles, and that these differ only by their velocity. If the ratio $\epsilon$ is a function of the velocity, this is not because the real mass varies with this velocity; but, since the fictitious electromagnetic mass depends upon this velocity, the total apparent mass, alone observable, must depend upon it, though the real mass does not depend upon it and may be constant.

The calculations of Abraham let us know the law according to which the *fictitious* mass varies as a function of the velocity; Kaufmann's experiment lets us know the law of variation of the *total* mass.

The comparison of these two laws will enable us therefore to determine the ratio of the *real* mass to the total mass.

Such is the method Kaufmann used to determine this ratio. The result is highly surprising: *the real mass is naught.*

This has led to conceptions wholly unexpected. What had only been proved for cathode corpuscles was extended to all bodies. What we call mass would be only semblance; all inertia would be of electromagnetic origin. But then mass would no longer be constant, it would augment with the velocity; sensibly constant for velocities up to 1,000 kilometers a second, it then would increase and would become infinite for the velocity of light. The transversal mass would no longer be equal to the longitudinal: they would only be nearly equal if the velocity is not too great. The principle $B$ of mechanics would no longer be true.

### III

#### The Canal Rays

At the point where we now are, this conclusion might seem premature. Can one apply to all matter what has been proved only for such light corpuscles, which are a mere emanation of matter and perhaps not true matter? But before entering upon this question, a word must be said of another sort of rays. I refer to the *canal rays*, the *Kanalstrahlen* of Goldstein.

The cathode, together with the cathode rays charged with negative electricity, emits canal rays charged with positive electricity. In general, these canal rays not being repelled by the cathode, are confined to the immediate neighborhood of this cathode, where they constitute the 'chamois cushion,' not very easy to perceive; but, if the cathode is pierced with holes and if it almost completely blocks up the tube, the canal rays spread *back* of the cathode, in the direction opposite to that of the cathode rays, and it becomes possible to study them. It is thus that it has been possible to show their positive charge and to show that the magnetic and electric deviations still exist, as for the cathode rays, but are much feebler.

Radium likewise emits rays analogous to the canal rays, and relatively very absorbable, called $\alpha$ rays.

We can, as for the cathode rays, measure the two deviations and thence deduce the velocity and the ratio $\epsilon$. The results are less constant than for the cathode rays, but the velocity is less, as well as the ratio $\epsilon$; the positive corpuscles are less charged than the negative; or if, which is more natural, we suppose the charges equal and of opposite sign, the positive corpuscles are much the larger. These corpuscles, charged the ones positively, the others negatively, have been called *electrons*.

## IV

### The Theory of Lorentz

But the electrons do not merely show us their existence in these rays where they are endowed with enormous velocities. We shall see them in very different rôles, and it is they that account for the principal phenomena of optics and electricity. The brilliant synthesis about to be noticed is due to Lorentz.

Matter is formed solely of electrons carrying enormous charges, and, if it seems to us neutral, this is because the charges of opposite sign of these electrons compensate each other. We may imagine, for example, a sort of solar system formed of a great positive electron, around which gravitate numerous little planets, the negative electrons, attracted by the electricity of opposite name which charges the central electron. The negative charges of these planets would balance the positive charge of this sun, so that the algebraic sum of all these charges would be naught.

All these electrons swim in the ether. The ether is everywhere identically the same, and perturbations in it are propagated according to the same laws as light or the Hertzian oscillations *in vacuo*. There is nothing but electrons and ether. When a luminous wave enters a part of the ether where electrons are numerous, these electrons are put in motion under the influence of the perturbation of the ether, and they then react upon the ether. So would be explained refraction, dispersion, double refraction and absorption. Just so, if for any cause an electron be put in motion, it would trouble the ether around it and would give rise to luminous waves, and this would explain the emission of light by incandescent bodies.

In certain bodies, the metals for example, we should have fixed electrons, between which would circulate moving electrons enjoying perfect liberty, save that of going out from the metallic body and breaking the surface which separates it from the exterior void or from the air, or from any other non-metallic body.

These movable electrons behave then, within the metallic body, as do, according to the kinetic theory of gases, the molecules of a gas within the vase where this gas is confined. But, under the influence of a difference of potential, the negative movable electrons would tend to go all to one side, and the positive movable electrons to the other. This is what would produce electric currents, and *this is why these bodies would be conductors*. On the other hand, the velocities of our electrons would be the greater the higher the temperature, if we accept the assimilation with the kinetic theory of gases. When one of these movable electrons encounters the surface of the metallic body, whose boundary it can not pass, it is reflected like a billiard ball which has hit the cushion, and its velocity undergoes a sudden change of direction. But when an electron changes direction, as we shall see further on, it becomes the source of a luminous wave, and this is why hot metals are incandescent.

In other bodies, the dielectrics and the transparent bodies, the movable electrons enjoy much less freedom. They remain as if attached to fixed electrons which attract them. The farther they go away from them the greater becomes this attraction and tends to pull them back. They therefore can make only small excursions; they can no longer circulate, but only oscillate about their mean position. This is why these bodies would not be conductors; moreover they would most often be transparent, and they would be refractive, since the luminous vibrations would be communicated to the movable electrons, susceptible of oscillation, and thence a perturbation would result.

I can not here give the details of the calculations; I confine myself to saying that this theory accounts for all the known facts, and has predicted new ones, such as the Zeeman effect.

## V

### *Mechanical Consequences*

We now may face two hypotheses:

1° The positive electrons have a real mass, much greater than their fictitious electromagnetic mass; the negative electrons alone lack real mass. We might even suppose that apart from electrons of the two signs, there are neutral atoms which have only their real mass. In this case, mechanics is not affected; there is no need of touching its laws; the real mass is constant; simply, motions are deranged by the effects of self-induction, as has always been known; moreover, these perturbations are almost negligible, except for the negative electrons which, not having real mass, are not true matter;

2° But there is another point of view; we may suppose there are no neutral atoms, and the positive electrons lack real mass just as the negative electrons. But then, real mass vanishing, either the word *mass* will no longer have any meaning, or else it must designate the fictitious electromagnetic mass; in this case, mass will no longer be constant, the transversal *mass* will no longer be equal to the longitudinal, the principles of mechanics will be overthrown.

First a word of explanation. We have said that, for the same charge, the *total* mass of a positive electron is much greater than that of a negative. And then it is natural to think that this difference is explained by the positive electron having, besides its fictitious mass, a considerable real mass; which takes us back to the first hypothesis. But we may just as well suppose that the real mass is null for these as for the others, but that the fictitious mass of the positive electron is much the greater since this electron is much the smaller. I say advisedly: much the smaller. And, in fact, in this hypothesis inertia is exclusively electromagnetic in origin; it reduces itself to the inertia of the ether; the electrons are no longer anything by themselves; they are solely holes in the ether and around which the ether moves; the smaller these holes are, the more will there be of ether, the greater, consequently, will be the inertia of the ether.

How shall we decide between these two hypotheses? By operating upon the canal rays as Kaufmann did upon the $\beta$ rays? This is impossible; the velocity of these rays is much too slight. Should each therefore decide according to his temperament, the conservatives going to one side and the lovers of the new to the other? Perhaps, but, to fully understand the arguments of the innovators, other considerations must come in.

# CHAPTER II

## MECHANICS AND OPTICS

### I

### *Aberration*

You know in what the phenomenon of aberration, discovered by Bradley, consists. The light issuing from a star takes a certain time to go through a telescope; during this time, the telescope, carried along by the motion of the earth, is displaced. If therefore the telescope were pointed in the *true* direction of the star, the image would be formed at the point occupied by the crossing of the threads of the network when the light has reached the objective; and this crossing would no longer be at this same point when the light reached the plane of the network. We would therefore be led to mis-point the telescope to bring the image upon the crossing of the threads. Thence results that the astronomer will not point the telescope in the direction of the absolute velocity of the light, that is to say toward the true position of the star, but just in the direction of the relative velocity of the light with reference to the earth, that is to say toward what is called the apparent position of the star.

The velocity of light is known; we might therefore suppose that we have the means of calculating the *absolute* velocity of the earth. (I shall soon explain my use here of the word absolute.) Nothing of the sort; we indeed know the apparent position of the star we observe; but we do not know its true position; we know the velocity of the light only in magnitude and not in direction.

If therefore the absolute velocity of the earth were straight and uniform, we should never have suspected the phenomenon of aberration; but it is variable; it is composed of two parts: the velocity of the solar system, which is straight and uniform; the velocity of the earth with reference to the sun, which is variable. If the velocity of the solar system, that is to say if the constant part existed alone, the observed direction would be invariable.

This position that one would thus observe is called the *mean* apparent position of the star.

Taking account now at the same time of the two parts of the velocity of the earth, we shall have the actual apparent position, which describes a little ellipse around the mean apparent position, and it is this ellipse that we observe.

Neglecting very small quantities, we shall see that the dimensions of this ellipse depend only upon the ratio of the velocity of the earth with reference to the sun to the velocity of light, so that the relative velocity of the earth with regard to the sun has alone come in.

But wait! This result is not exact, it is only approximate; let us push the approximation a little farther. The dimensions of the ellipse will depend then upon the absolute velocity of the earth. Let us compare the major axes of the ellipse for the different stars: we shall have, theoretically at least, the means of determining this absolute velocity.

That would be perhaps less shocking than it at first seems; it is a question, in fact, not of the velocity with reference to an absolute void, but of the velocity with regard to the ether, which is taken *by definition* as being absolutely at rest.

Besides, this method is purely theoretical. In fact, the aberration is very small; the possible variations of the ellipse of aberration are much smaller yet, and, if we consider the aberration as of the first order, they should therefore be regarded as of the second order: about a millionth of a second; they are absolutely inappreciable for our instruments. We shall finally see, further on, why the preceding theory should be rejected, and why we could not determine this absolute velocity even if our instruments were ten thousand times more precise!

One might imagine some other means, and in fact, so one has. The velocity of light is not the same in water as in air; could we not compare the two apparent positions of a star seen through a telescope first full of air, then full of water? The results have been negative; the apparent laws of reflection and refraction are not altered by the motion of the earth. This phenomenon is capable of two explanations:

1° It might be supposed that the ether is not at rest, but that

33

it is carried along by the body in motion. It would then not be astonishing that the phenomena of refraction are not altered by the motion of the earth, since all, prisms, telescopes and ether, are carried along together in the same translation. As to the aberration itself, it would be explained by a sort of refraction happening at the surface of separation of the ether at rest in the interstellar spaces and the ether carried along by the motion of the earth. It is upon this hypothesis (bodily carrying along of the ether) that is founded the *theory of Hertz* on the electrodynamics of moving bodies.

2° Fresnel, on the contrary, supposes that the ether is at absolute rest in the void, at rest almost absolute in the air, whatever be the velocity of this air, and that it is partially carried along by refractive media. Lorentz has given to this theory a more satisfactory form. For him, the ether is at rest, only the electrons are in motion; in the void, where it is only a question of the ether, in the air, where this is almost the case, the carrying along is null or almost null; in refractive media, where perturbation is produced at the same time by vibrations of the ether and those of electrons put in swing by the agitation of the ether, the undulations are *partially* carried along.

To decide between the two hypotheses, we have Fizeau's experiment, comparing by measurements of the fringes of interference, the velocity of light in air at rest or in motion. These experiments have confirmed Fresnel's hypothesis of partial carrying along. They have been repeated with the same result by Michelson. *The theory of Hertz must therefore be rejected.*

## II

### *The Principle of Relativity*

But if the ether is not carried along by the motion of the earth, is it possible to show, by means of optical phenomena, the absolute velocity of the earth, or rather its velocity with respect to the unmoving ether? Experiment has answered negatively, and yet the experimental procedures have been varied in all possible ways. Whatever be the means employed there will never be disclosed anything but relative velocities; I mean the velocities of certain material bodies with reference to other material bodies. In fact, if the source of light and the apparatus of observation are on the earth and participate in its motion, the experimental results have always been the same, whatever be the orientation of the apparatus with reference to the orbital motion of the earth. If astronomic aberration happens, it is because the source, a star, is in motion with reference to the observer.

The hypotheses so far made perfectly account for this general result, *if we neglect very small quantities of the order of the square of the aberration.* The explanation rests upon the notion of *local time,* introduced by Lorentz, which I shall try to make clear. Suppose two observers, placed one at $A$, the other at $B$, and wishing to set their watches by means of optical signals. They agree that $B$ shall send a signal to $A$ when his watch marks an hour determined upon, and $A$ is to put his watch to that hour the moment he sees the signal. If this alone were done, there would be a systematic error, because as the light takes a certain time $t$ to go from $B$ to $A$, $A$'s watch would be behind $B$'s the time $t$. This error is easily corrected. It suffices to cross the signals. $A$ in turn must signal $B$, and, after this new adjustment, $B$'s watch will be behind $A$'s the time $t$. Then it will be sufficient to take the arithmetic mean of the two adjustments.

But this way of doing supposes that light takes the same time to go from $A$ to $B$ as to return from $B$ to $A$. That is true if the observers are motionless; it is no longer so if they are carried along in a common translation, since then $A$, for example, will go to meet the light coming from $B$, while $B$ will flee before the light coming from $A$. If therefore the observers are borne along in a common translation and if they do not suspect it, their adjustment will be defective; their watches will not indicate the same time; each will show the *local time* belonging to the point where it is.

The two observers will have no way of perceiving this, if the unmoving ether can transmit to them only luminous signals all of the same velocity, and if the other signals they might send are transmitted by media carried along with them in their translation. The phenomenon each observes will be too soon or too late; it would be seen at the same instant only if the translation did not exist; but as it will be observed with a watch that is wrong, this will not be perceived and the appearances will not be altered.

It results from this that the compensation is easy to explain so long as we neglect the square of the aberration, and for a long time the experiments were not sufficiently precise to warrant taking account of it. But the day came when Michelson imagined a much more delicate procedure: he made rays interfere which had traversed different courses, after being reflected by mirrors; each of the paths approximating a meter and the fringes of interference permitting the recognition of a fraction of a thousandth of a millimeter, the square of the aberration could no longer be neglected, and *yet the results were still negative.* Therefore the theory required to be completed, and it has been by the *Lorentz-Fitzgerald hypothesis.*

These two physicists suppose that all bodies carried along in a translation undergo a contraction in the sense of this translation, while their dimensions perpendicular to this translation remain unchanged. *This contraction is the same for all bodies;* moreover, it is very slight, about one two-hundred-millionth for a velocity such as that of the earth. Furthermore our measuring instruments could not disclose it, even if they were much more precise; our measuring rods in fact undergo the same contraction as the objects to be measured. If the meter exactly fits when applied to a body, if we point the body and consequently the meter in the sense of the motion of the earth, it will not cease to exactly fit in another orientation, and that although the body and the meter have changed in length as well as orientation, and precisely because the change is the same for one as for the other. But it is quite different if we measure a length, not now with a meter, but by the time taken by light to pass along it, and this is just what Michelson has done.

A body, spherical when at rest, will take thus the form of a flattened ellipsoid of revolution when in motion; but the observer will always think it spherical, since he himself has undergone an analogous deformation, as also all the objects serving as points of reference. On the contrary, the surfaces of the waves of light, remaining rigorously spherical, will seem to him elongated ellipsoids.

What happens then? Suppose an observer and a source of light carried along together in the translation: the wave surfaces emanating from the source will be spheres having as centers the successive positions of the source; the distance from this center to the actual position of the source will be proportional to the time elapsed after the emission, that is to say to the radius of the sphere. All these spheres are therefore homothetic one to the other, with relation to the actual position $S$ of the source. But, for our observer, because of the contraction, all these spheres will seem elongated ellipsoids, and all these ellipsoids will moreover be homothetic, with reference to the point $S$; the excentricity of all these ellipsoids is the same and depends solely upon the velocity of the earth. *We shall so select the law of contraction that the point $S$ may be at the focus of the meridian section of the ellipsoid.*

This time the compensation is *rigorous,* and this it is which explains Michelson's experiment.

I have said above that, according to the ordinary theories, observations of the astronomic aberration would give us the absolute velocity of the earth, if our instruments were a thousand times more precise. I must modify this statement. Yes, the observed angles would be modified by the effect of this absolute velocity, but the graduated circles we use to measure the angles would be deformed by the translation: they would become ellipses; thence would result an error in regard to the angle measured, and *this second error would exactly compensate the first.*

This Lorentz-Fitzgerald hypothesis seems at first very extraordinary; all we can say for the moment, in its favor, is that it is only the immediate translation of Michelson's experimental result, if we *define* lengths by the time taken by light to run along them.

However that may be, it is impossible to escape the impression that the principle of relativity is a general law of nature, that one will never be able by any imaginable means to show any but relative velocities, and I mean by that not only the velocities of bodies with reference to the ether, but the velocities of bodies with regard to one another. Too many different experiments have given concordant results for us not to feel tempted to attribute to this principle of relativity a value comparable to that, for example, of the principle of equivalence. In any case, it is proper to see to what consequences this way of looking at things would lead us and then to submit these consequences to the control of experiment.

### III

#### *The Principle of Reaction*

Let us see what the principle of the equality of action and reaction becomes in the theory of Lorentz. Consider an electron *A* which for any cause begins to move; it produces a perturbation in the ether; at the end of a certain time, this perturbation reaches another electron *B*, which will be disturbed from its position of equilibrium. In these conditions there can not be equality between action and reaction, at least if we do not consider the ether, but only the electrons, *which alone are observable,* since our matter is made of electrons.

In fact it is the electron *A* which has disturbed the electron *B*; even in case the electron *B* should react upon *A*, this reaction could be equal to the action, but in no case simultaneous, since the electron *B* can begin to move only after a certain time, necessary for the propagation. Submitting the problem to a more exact calculation, we reach the following result: Suppose a Hertz discharger placed at the focus of a parabolic mirror to which it is mechanically attached; this discharger emits electromagnetic waves, and the mirror reflects all these waves in the same direction; the discharger therefore will radiate energy in a determinate direction. Well, the calculation shows that *the discharger recoils* like a cannon which has shot out a projectile. In the case of the cannon, the recoil is the natural result of the equality of action and reaction. The cannon recoils because the projectile upon which it has acted reacts upon it. But here it is no longer the same. What has been sent out is no longer a material projectile: it is energy, and energy has no mass: it has no counterpart. And, in place of a discharger, we could have considered just simply a lamp with a reflector concentrating its rays in a single direction.

It is true that, if the energy sent out from the discharger or from the lamp meets a material object, this object receives a mechanical push as if it had been hit by a real projectile, and this push will be equal to the recoil of the discharger and of the lamp, if no energy has been lost on the way and if the object absorbs the whole of the energy. Therefore one is tempted to say that there still is compensation between the action and the reaction. But this compensation, even should it be complete, is always belated. It never happens if the light, after leaving its source, wanders through interstellar spaces without ever meeting a material body; it is incomplete, if the body it strikes is not perfectly absorbent.

Are these mechanical actions too small to be measured, or are they accessible to experiment? These actions are nothing other than those due to the *Maxwell-Bartholi* pressures; Maxwell had predicted these pressures from calculations relative to electrostatics and magnetism; Bartholi reached the same result by thermodynamic considerations.

This is how the *tails of comets* are explained. Little particles detach themselves from the nucleus of the comet; they are struck by the light of the sun, which pushes them back as would a rain of projectiles coming from the sun. The mass of these particles is so little that this repulsion sweeps it away against the Newtonian attraction; so in moving away from the sun they form the tails.

The direct experimental verification was not easy to obtain. The first endeavor led to the construction of the *radiometer*. But this instrument *turns backward*, in the sense opposite to the theoretic sense, and the explanation of its rotation, since discovered, is wholly different. At last success came, by making the vacuum more complete, on the one hand, and on the other by not blackening one of the faces of the paddles and directing a pencil of luminous rays upon one of the faces. The radiometric effects and the other disturbing causes are eliminated by a series of painstaking precautions, and one obtains a deviation which is very minute, but which is, it would seem, in conformity with the theory.

The same effects of the Maxwell-Bartholi pressure are forecast likewise by the theory of Hertz of which we have before spoken, and by that of Lorentz. But there is a difference. Suppose that the energy, under the form of light, for example, proceeds from a luminous source to any body through a transparent medium. The Maxwell-Bartholi pressure will act, not alone upon the source at the departure, and on the body lit up at the arrival, but upon the matter of the transparent medium which it traverses. At the moment when the luminous wave reaches a new region of this medium, this pressure will push forward the matter there distributed and will put it back when the wave leaves this region. So that the recoil of the source has for counterpart the forward movement of the transparent matter which is in contact with this source; a little later, the recoil of this same matter has for counterpart the forward movement of the transparent matter which lies a little further on, and so on.

Only, is the compensation perfect? Is the action of the Maxwell-Bartholi pressure upon the matter of the transparent medium equal to its reaction upon the source, and that, whatever be this matter? Or is this action by so much the less as the medium is less refractive and more rarefied, becoming null in the void?

If we admit the theory of Hertz, who regards matter as mechanically bound to the ether, so that the ether may be entirely carried along by matter, it would be necessary to answer yes to the first question and no to the second.

There would then be perfect compensation, as required by the principle of the equality of action and reaction, even in the least refractive media, even in the air, even in the interplanetary void, where it would suffice to suppose a residue of matter, however subtile. If on the contrary we admit the theory of Lorentz, the compensation, always imperfect, is insensible in the air and becomes null in the void.

But we have seen above that Fizeau's experiment does not permit of our retaining the theory of Hertz; it is necessary therefore to adopt the theory of Lorentz, and consequently *to renounce the principle of reaction.*

*Consequences of the Principle of Relativity*

We have seen above the reasons which impel us to regard the principle of relativity as a general law of nature. Let us see to what consequences this principle would lead, should it be regarded as finally demonstrated.

First, it obliges us to generalize the hypothesis of Lorentz and Fitzgerald on the contraction of all bodies in the sense of the translation. In particular, we must extend this hypothesis to the electrons themselves. Abraham considered these electrons as spherical and indeformable; it will be necessary for us to admit that these electrons, spherical when in repose, undergo the Lorentz contraction when in motion and take then the form of flattened ellipsoids.

This deformation of the electrons will influence their mechanical properties. In fact I have said that the displacement of these charged electrons is a veritable current of convection and that their apparent inertia is due to the self-induction of this current: exclusively as concerns the negative electrons; exclusively or not, we do not yet know, for the positive electrons. Well, the deformation of the electrons, a deformation which depends upon their velocity, will modify the distribution of the electricity upon their surface, consequently the intensity of the convection current they produce, consequently the laws according to which the self-induction of this current will vary as a function of the velocity.

At this price, the compensation will be perfect and will conform to the requirements of the principle of relativity, but only upon two conditions:

1° That the positive electrons have no real mass, but only a fictitious electromagnetic mass; or at least that their real mass, if it exists, is not constant and varies with the velocity according to the same laws as their fictitious mass;

2° That all forces are of electromagnetic origin, or at least that they vary with the velocity according to the same laws as the forces of electromagnetic origin.

It still is Lorentz who has made this remarkable synthesis; stop a moment and see what follows therefrom. First, there is no more matter, since the positive electrons no longer have real mass, or at least no constant real mass. The present principles of our mechanics, founded upon the constancy of mass, must therefore be modified. Again, an electomagnetic explanation must be sought of all the known forces, in particular of gravitation, or at least the law of gravitation must be so modified that this force is altered by velocity in the same way as the electromagnetic forces. We shall return to this point.

All that appears, at first sight, a little artificial. In particular, this deformation of electrons seems quite hypothetical. But the thing may be presented otherwise, so as to avoid putting this hypothesis of deformation at the foundation of the reasoning. Consider the electrons as material points and ask how their mass should vary as function of the velocity not to contravene the principle of relativity. Or, still better, ask what should be their acceleration under the influence of an electric or magnetic field, that this principle be not violated and that we come back to the ordinary laws when we suppose the velocity very slight. We shall find that the variations of this mass, or of these accelerations, must be *as if* the electron underwent the Lorentz deformation.

V

*Kaufmann's Experiment*

We have before us, then, two theories: one where the electrons are indeformable, this is that of Abraham; the other where they undergo the Lorentz deformation. In both cases, their mass increases with the velocity, becoming infinite when this velocity becomes equal to that of light; but the law of the variation is not the same. The method employed by Kaufmann to bring to light the law of variation of the mass seems therefore to give us an experimental means of deciding between the two theories.

Unhappily, his first experiments were not sufficiently precise for that; so he decided to repeat them with more precautions, and measuring with great care the intensity of the fields. Under their new form *they are in favor of the theory of Abraham.* Then the principle of relativity would not have the rigorous value we were tempted to attribute to it; there would no longer be reason for believing the positive electrons denuded of real mass like the negative electrons. However, before definitely adopting this conclusion, a little reflection is necessary. The question is of such importance that it is to be wished Kaufmann's experiment were repeated by another experimenter.[1] Unhappily, this experiment is very delicate and could be carried out successfully only by a physicist of the same ability as Kaufmann. All precautions have been properly taken and we hardly see what objection could be made.

There is one point however to which I wish to draw attention: that is to the measurement of the electrostatic field, a measurement upon which all depends. This field was produced between the two armatures of a condenser; and, between these armatures, there was to be made an extremely perfect vacuum, in order to obtain a complete isolation. Then the difference of potential of the two armatures was measured, and the field obtained by dividing this difference by the distance apart of the armatures. That supposes the field uniform; is this certain? Might there not be an abrupt fall of potential in the neighborhood of one of the armatures, of the negative armature, for example? There may be a difference of potential at the meeting of the metal and the vacuum, and it may be that this difference is not the same on the positive side and on the negative side; what would lead me to think so is the electric valve effects between mercury and vacuum. However slight the probability that it is so, it seems that it should be considered.

## VI

### *The Principle of Inertia*

In the new dynamics, the principle of inertia is still true, that is to say that an *isolated* electron will have a straight and uniform motion. At least this is generally assumed; however,

---

[1] At the moment of going to press we learn that M. Bucherer has repeated the experiment, taking new precautions, and that he has obtained, contrary to Kaufmann, results confirming the views of Lorentz.

Lindemann has made objections to this view; I do not wish to take part in this discussion, which I can not here expound because of its too difficult character. In any case, slight modifications to the theory would suffice to shelter it from Lindemann's objections.

We know that a body submerged in a fluid experiences, when in motion, considerable resistance, but this is because our fluids are viscous; in an ideal fluid, perfectly free from viscosity, the body would stir up behind it a liquid hill, a sort of wake; upon departure, a great effort would be necessary to put it in motion, since it would be necessary to move not only the body itself, but the liquid of its wake. But, the motion once acquired, it would perpetuate itself without resistance, since the body, in advancing, would simply carry with it the perturbation of the liquid, without the total vis viva of the liquid augmenting. Everything would happen therefore as if its inertia was augmented. An electron advancing in the ether would behave in the same way: around it, the ether would be stirred up, but this perturbation would accompany the body in its motion; so that, for an observer carried along with the electron, the electric and magnetic fields accompanying this electron would appear invariable, and would change only if the velocity of the electron varied. An effort would therefore be necessary to put the electron in motion, since it would be necessary to create the energy of these fields; on the contrary, once the movement acquired, no effort would be necessary to maintain it, since the created energy would only have to go along behind the electron as a wake. This energy, therefore, could only augment the inertia of the electron, as the agitation of the liquid augments that of the body submerged in a perfect fluid. And anyhow, the negative electrons at least have no other inertia except that.

In the hypothesis of Lorentz, the vis viva, which is only the energy of the ether, is not proportional to $v^2$. Doubtless if $v$ is very slight, the vis viva is sensibly proportional to $v^2$, the quantity of motion sensibly proportional to $v$, the two masses sensibly constant and equal to each other. But *when the velocity tends toward the velocity of light, the vis viva, the quantity of motion and the two masses increase beyond all limit.*

In the hypothesis of Abraham, the expressions are a little more complicated; but what we have just said remains true in essentials.

So the mass, the quantity of motion, the vis viva become infinite when the velocity is equal to that of light.

Thence results that *no body can attain in any way a velocity beyond that of light.* And in fact, in proportion as its velocity increases, its mass increases, so that its inertia opposes to any new increase of velocity a greater and greater obstacle.

A question then suggests itself: let us admit the principle of relativity; an observer in motion would not have any means of perceiving his own motion. If therefore no body in its absolute motion can exceed the velocity of light, but may approach it as nearly as you choose, it should be the same concerning its relative motion with reference to our observer. And then we might be tempted to reason as follows: The observer may attain a velocity of 200,000 kilometers; the body in its relative motion with reference to the observer may attain the same velocity; its absolute velocity will then be 400,000 kilometers, which is impossible, since this is beyond the velocity of light. This is only a seeming, which vanishes when account is taken of how Lorentz evaluates local time.

## VII

### *The Wave of Acceleration*

When an electron is in motion, it produces a perturbation in the ether surrounding it; if its motion is straight and uniform, this perturbation reduces to the wake of which we have spoken in the preceding section. But it is no longer the same, if the motion be curvilinear or varied. The perturbation may then be regarded as the superposition of two others, to which Langevin has given the names *wave of velocity* and *wave of acceleration*. The wave of velocity is only the wave which happens in uniform motion.

As to the wave of acceleration, this is a perturbation altogether analogous to light waves, which starts from the electron at the instant when it undergoes an acceleration, and which is then propagated by successive spherical waves with the velocity of light. Whence follows: in a straight and uniform motion, the energy is wholly conserved; but, when there is an acceleration, there is loss of energy, which is dissipated under the form of luminous waves and goes out to infinity across the ether.

However, the effects of this wave of acceleration, in particular the corresponding loss of energy, are in most cases negligible, that is to say not only in ordinary mechanics and in the motions of the heavenly bodies, but even in the radium rays, where the velocity is very great without the acceleration being so. We may then confine ourselves to applying the laws of mechanics, putting the force equal to the product of acceleration by mass, this mass, however, varying with the velocity according to the laws explained above. We then say the motion is *quasi-stationary.*

It would not be the same in all cases where the acceleration is great, of which the chief are the following:

1° In incandescent gases certain electrons take an oscillatory motion of very high frequency; the displacements are very small, the velocities are finite, and the accelerations very great; energy is then communicated to the ether, and this is why these gases radiate light of the same period as the oscillations of the electron;

2° Inversely, when a gas receives light, these same electrons are put in swing with strong accelerations and they absorb light;

3° In the Hertz discharger, the electrons which circulate in the metallic mass undergo, at the instant of the discharge, an abrupt acceleration and take then an oscillatory motion of high frequency. Thence results that a part of the energy radiates under the form of Hertzian waves;

4° In an incandescent metal, the electrons enclosed in this metal are impelled with great velocity; upon reaching the surface of the metal, which they can not get through, they are reflected and thus undergo a considerable acceleration. This is why the metal emits light. The details of the laws of the emission of light by dark bodies are perfectly explained by this hypothesis;

5° Finally when the cathode rays strike the anticathode, the negative electrons constituting these rays, which are impelled with very great velocity, are abruptly arrested. Because of the acceleration they thus undergo, they produce undulations in the ether. This, according to certain physicists, is the origin of the Röntgen rays, which would only be light rays of very short wave-length.

# CHAPTER III

## THE NEW MECHANICS AND ASTRONOMY

### I

### *Gravitation*

MASS may be defined in two ways:

1° By the quotient of the force by the acceleration; this is the true definition of the mass, which measures the inertia of the body.

2° By the attraction the body exercises upon an exterior body, in virtue of Newton's law. We should therefore distinguish the mass coefficient of inertia and the mass coefficient of attraction. According to Newton's law, there is rigorous proportionality between these two coefficients. But that is demonstrated only for velocities to which the general principles of dynamics are applicable. Now, we have seen that the mass coefficient of inertia increases with the velocity; should we conclude that the mass coefficient of attraction increases likewise with the velocity and remains proportional to the coefficient of inertia, or, on the contrary, that this coefficient of attraction remains constant? This is a question we have no means of deciding.

On the other hand, if the coefficient of attraction depends upon the velocity, since the velocities of two bodies which mutually attract are not in general the same, how will this coefficient depend upon these two velocities?

Upon this subject we can only make hypotheses, but we are naturally led to investigate which of these hypotheses would be compatible with the principle of relativity. There are a great number of them; the only one of which I shall here speak is that of Lorentz, which I shall briefly expound.

Consider first electrons at rest. Two electrons of the same sign repel each other and two electrons of contrary sign attract each other; in the ordinary theory, their mutual actions are proportional to their electric charges; if therefore we have four elec-

512

trons, two positive $A$ and $A'$, and two negative $B$ and $B'$, the charges of these four being the same in absolute value, the repulsion of $A$ for $A'$ will be, at the same distance, equal to the repulsion of $B$ for $B'$ and equal also to the attraction of $A$ for $B'$, or of $A'$ for $B$. If therefore $A$ and $B$ are very near each other, as also $A'$ and $B'$, and we examine the action of the system $A + B$ upon the system $A' + B'$, we shall have two repulsions and two attractions which will exactly compensate each other and the resulting action will be null.

Now, material molecules should just be regarded as species of solar systems where circulate the electrons, some positive, some negative, and *in such a way that the algebraic sum of all the charges is null*. A material molecule is therefore wholly analogous to the system $A + B$ of which we have spoken, so that the total electric action of two molecules one upon the other should be null.

But experiment shows us that these molecules attract each other in consequence of Newtonian gravitation; and then we may make two hypotheses: we may suppose gravitation has no relation to the electrostatic attractions, that it is due to a cause entirely different, and is simply something additional; or else we may suppose the attractions are not proportional to the charges and that the attraction exercised by a charge $+1$ upon a charge $-1$ is greater than the mutual repulsion of two $+1$ charges, or two $-1$ charges.

In other words, the electric field produced by the positive electrons and that which the negative electrons produce might be superposed and yet remain distinct. The positive electrons would be more sensitive to the field produced by the negative electrons than to the field produced by the positive electrons; the contrary would be the case for the negative electrons. It is clear that this hypothesis somewhat complicates electrostatics, but that it brings back into it gravitation. This was, in sum, Franklin's hypothesis.

What happens now if the electrons are in motion? The positive electrons will cause a perturbation in the ether and produce there an electric and a magnetic field. The same will be the case for the negative electrons. The electrons, positive as

34

well as negative, undego then a mechanical impulsion by the
action of these different fields.  In the ordinary theory, the
electromagnetic field, due to the motion of the positive electrons,
exercises, upon two electrons of contrary sign and of the same
absolute charge, equal actions with contrary sign.  We may then
without inconvenience not distinguish the field due to the motion
of the positive electrons and the field due to the motion of the
negative electrons and consider only the algebraic sum of these
two fields, that is to say the resulting field.

In the new theory, on the contrary, the action upon the positive electrons of the electromagnetic field due to the positive
electrons follows the ordinary laws; it is the same with the action
upon the negative electrons of the field due to the negative electrons.  Let us now consider the action of the field due to the
positive electrons upon the negative electrons (or inversely); it
will still follow the same laws, but *with a different coefficient.*
Each electron is more sensitive to the field created by the electrons of contrary name than to the field created by the electrons
of the same name.

Such is the hypothesis of Lorentz, which reduces to Franklin's
hypothesis for slight velocities; it will therefore explain, for
these small velocities, Newton's law.  Moreover, as gravitation
goes back to forces of electrodynamic origin, the general theory
of Lorentz will apply, and consequently the principle of relativity will not be violated.

We see that Newton's law is no longer applicable to great
velocities and that it must be modified, for bodies in motion,
precisely in the same way as the laws of electrostatics for electricity in motion.

We know that electromagnetic perturbations spread with the
velocity of light.  We may therefore be tempted to reject the
preceding theory upon remembering that gravitation spreads,
according to the calculations of Laplace, at least ten million
times more quickly than light, and that consequently it can not
be of electromagnetic origin.  The result of Laplace is well
known, but one is generally ignorant of its signification.  Laplace
supposed that, if the propagation of gravitation is not instantaneous, its velocity of spread combines with that of the body

attracted, as happens for light in the phenomenon of astronomic aberration, so that the effective force is not directed along the straight joining the two bodies, but makes with this straight a small angle. This is a very special hypothesis, not well justified, and, in any case, entirely different from that of Lorentz. Laplace's result proves nothing against the theory of Lorentz.

## II

### *Comparison with Astronomic Observations*

Can the preceding theories be reconciled with astronomic observations?

First of all, if we adopt them, the energy of the planetary motions will be constantly dissipated by the effect of the *wave of acceleration*. From this would result that the mean motions of the stars would constantly accelerate, as if these stars were moving in a resistant medium. But this effect is exceedingly slight, far too much so to be discerned by the most precise observations. The acceleration of the heavenly bodies is relatively slight, so that the effects of the wave of acceleration are negligible and the motion may be regarded as *quasi-stationary*. It is true that the effects of the wave of acceleration constantly accumulate, but this accumulation itself is so slow that thousands of years of observation would be necessary for it to become sensible. Let us therefore make the calculation considering the motion as quasi-stationary, and that under the three following hypotheses:

A. Admit the hypothesis of Abraham (electrons indeformable) and retain Newton's law in its usual form;

B. Admit the hypothesis of Lorentz about the deformation of electrons and retain the usual Newton's law;

C. Admit the hypothesis of Lorentz about electrons and modify Newton's law as we have done in the preceding paragraph, so as to render it compatible with the principle of relativity.

It is in the motion of Mercury that the effect will be most sensible, since this planet has the greatest velocity. Tisserand formerly made an analogous calculation, admitting Weber's law; I recall that Weber had sought to explain at the same time the electrostatic and electrodynamic phenomena in supposing that electrons (whose name was not yet invented) exercise, one upon another, attractions and repulsions directed along the straight joining them, and depending not only upon their distances, but upon the first and second derivatives of these distances, consequently upon their velocities and their accelerations. This law of Weber, different enough from those which to-day tend to prevail, none the less presents a certain analogy with them.

Tisserand found that, if the Newtonian attraction conformed to Weber's law there resulted, for Mercury's perihelion, secular variation of 14", *of the same sense as that which has been observed and could not be explained,* but smaller, since this is 38".

Let us recur to the hypotheses A, B and C, and study first the motion of a planet attracted by a fixed center. The hypotheses B and C are no longer distinguished, since, if the attracting point is fixed, the field it produces is a purely electrostatic field, where the attraction varies inversely as the square of the distance, in conformity with Coulomb's electrostatic law, identical with that of Newton.

The vis viva equation holds good, taking for vis viva the new definition; in the same way, the equation of areas is replaced by another equivalent to it; the moment of the quantity of motion is a constant, but the quantity of motion must be defined as in the new dynamics.

The only sensible effect will be a secular motion of the perihelion. With the theory of Lorentz, we shall find, for this motion, half of what Weber's law would give; with the theory of Abraham, two fifths.

If now we suppose two moving bodies gravitating around their common center of gravity, the effects are very little different, though the calculations may be a little more complicated. The motion of Mercury's perihelion would therefore be 7" in the theory of Lorentz and 5".6 in that of Abraham.

The effect moreover is proportional to $n^3 a^2$, where $n$ is the star's mean motion and $a$ the radius of its orbit. For the planets, in virtue of Kepler's law, the effect varies then inversely as $\sqrt{a^5}$; it is therefore insensible, save for Mercury.

It is likewise insensible for the moon though $n$ is great, because $a$ is extremely small; in sum, it is five times less for Venus, and six hundred times less for the moon than for Mercury. We may add that as to Venus and the earth, the motion of the perihelion (for the same angular velocity of this motion) would be much more difficult to discern by astronomic observations, because the excentricity of their orbits is much less than for Mercury.

To sum up, *the only sensible effect upon astronomic observations would be a motion of Mercury's perihelion, in the same sense as that which has been observed without being explained, but notably slighter.*

That can not be regarded as an argument in favor of the new dynamics, since it will always be necessary to seek another explanation for the greater part of Mercury's anomaly; but still less can it be regarded as an argument against it.

## III

### The Theory of Lesage

It is interesting to compare these considerations with a theory long since proposed to explain universal gravitation.

Suppose that, in the interplanetary spaces, circulate in every direction, with high velocities, very tenuous corpuscles. A body isolated in space will not be affected, apparently, by the impacts of these corpuscles, since these impacts are equally distributed in all directions. But if two bodies $A$ and $B$ are present, the body $B$ will play the rôle of screen and will intercept part of the corpuscles which, without it, would have struck $A$. Then, the impacts received by $A$ in the direction opposite that from $B$ will no longer have a counterpart, or will now be only partially compensated, and this will push $A$ toward $B$.

Such is the theory of Lesage; and we shall discuss it, taking first the view-point of ordinary mechanics.

First, how should the impacts postulated by this theory take place; is it according to the laws of perfectly elastic bodies, or according to those of bodies devoid of elasticity, or according to an intermediate law? The corpuscles of Lesage can not act as prefectly elastic bodies; otherwise the effect would be null, since the corpuscles intercepted by the body $B$ would be replaced by others which would have rebounded from $B$, and calculation proves that the compensation would be perfect. It is necessary then that the impact make the corpuscles lose energy, and this energy should appear under the form of heat. But how much heat would thus be produced? Note that attraction passes through bodies; it is necessary therefore to represent to ourselves the earth, for example, not as a solid screen, but as formed of a very great number of very small spherical molecules, which play individually the rôle of little screens, but between which the corpuscles of Lesage may freely circulate. So, not only the earth is not a solid screen, but it is not even a cullender, since the voids occupy much more space than the plenums. To realize this, recall that Laplace has demonstrated that attraction, in traversing the earth, is weakened at most by one ten-millionth part, and his proof is perfectly satisfactory: in fact, if attraction were absorbed by the body it traverses, it would no longer be proportional to the masses; it would be *relatively* weaker for great bodies than for small, since it would have a greater thickness to traverse. The attraction of the sun for the earth would therefore be *relatively* weaker than that of the sun for the moon, and thence would result, in the motion of the moon, a very sensible inequality. We should therefore conclude, if we adopt the theory of Lesage, that the total surface of the spherical molecules which compose the earth is at most the ten-millionth part of the total surface of the earth.

Darwin has proved that the theory of Lesage only leads exactly to Newton's law when we postulate particles entirely devoid of elasticity. The attraction exerted by the earth on a mass 1 at a distance 1 will then be proportional, at the same time, to the total surface $S$ of the spherical molecules composing it, to the velocity $v$ of the corpuscles, to the square root of the density $\rho$ of the medium formed by the corpuscles. The heat produced will be proportional to $S$, to the density $\rho$, and to the cube of the velocity $v$.

But it is necessary to take account of the resistance experienced by a body moving in such a medium; it can not move, in fact, without going against certain impacts, in fleeing, on the contrary, before those coming in the opposite direction, so that the compensation realized in the state of rest can no longer subsist. The calculated resistance is proportional to $S$, to $\rho$ and to $v$; now, we know that the heavenly bodies move as if they experienced no resistance, and the precision of observations permits us to fix a limit to the resistance of the medium.

This resistance varying as $S\rho v$, while the attraction varies as $S\sqrt{\rho v}$, we see that the ratio of the resistance to the square of the attraction is inversely as the product $Sv$.

We have therefore a lower limit of the product $Sv$. We have already an upper limit of $S$ (by the absorption of attraction by the body it traverses); we have therefore a lower limit of the velocity $v$, which must be at least $24 \cdot 10^{17}$ times that of light.

From this we are able to deduce $\rho$ and the quantity of heat produced; this quantity would suffice to raise the temperature $10^{26}$ degrees a second; the earth would receive in a given time $10^{20}$ times more heat than the sun emits in the same time; I am not speaking of the heat the sun sends to the earth, but of that it radiates in all directions.

It is evident the earth could not long stand such a régime.

We should not be led to results less fantastic if, contrary to Darwin's views, we endowed the corpuscles of Lesage with an elasticity imperfect without being null. In truth, the vis viva of these corpuscles would not be entirely converted into heat, but the attraction produced would likewise be less, so that it would be only the part of this vis viva converted into heat, which would contribute to produce the attraction and that would come to the same thing; a judicious employment of the theorem of the viriel would enable us to account for this.

The theory of Lesage may be transformed; suppress the corpuscles and imagine the ether overrun in all senses by luminous waves coming from all points of space. When a material object receives a luminous wave, this wave exercises upon it a mechanical action due to the Maxwell-Bartholi pressure, just as if it had received the impact of a material projectile. The waves in question could therefore play the rôle of the corpuscles of Lesage. This is what is supposed, for example, by M. Tommasina.

The difficulties are not removed for all that; the velocity of propagation can be only that of light, and we are thus led, for the resistance of the medium, to an inadmissible figure. Besides, if the light is all reflected, the effect is null, just as in the hypothesis of the perfectly elastic corpuscles.

That there should be attraction, it is necessary that the light be partially absorbed; but then there is production of heat. The calculations do not differ essentially from those made in the ordinary theory of Lesage, and the result retains the same fantastic character.

On the other hand, attraction is not absorbed by the body it traverses, or hardly at all; it is not so with the light we know. Light which would produce the Newtonian attraction would have to be considerably different from ordinary light and be, for example, of very short wave length. This does not count that, if our eyes were sensible of this light, the whole heavens should appear to us much more brilliant than the sun, so that the sun would seem to us to stand out in black, otherwise the sun would repel us instead of attracting us. For all these reasons, light which would permit of the explanation of attraction would be much more like Röntgen rays than like ordinary light.

And besides, the X-rays would not suffice; however penetrating they may seem to us, they could not pass through the whole earth; it would be necessary therefore to imagine X'-rays much more penetrating than the ordinary X-rays. Moreover a part of the energy of these X'-rays would have to be destroyed, otherwise there would be no attraction. If you do not wish it transformed into heat, which would lead to an enormous heat production, you must suppose it radiated in every direction under the form of secondary rays, which might be called X'' and which would have to be much more penetrating still than the X'-rays, otherwise they would in their turn derange the phenomena of attraction.

Such are the complicated hypotheses to which we are led when we try to give life to the theory of Lesage.

But all we have said presupposes the ordinary laws of mechanics.

Will things go better if we admit the new dynamics? And first, can we conserve the principles of relativity? Let us give at first to the theory of Lesage its primitive form, and suppose space ploughed by material corpuscles; if these corpuscles were perfectly elastic, the laws of their impact would conform to this principle of relativity, but we know that then their effect would be null. We must therefore suppose these corpuscles are not elastic, and then it is difficult to imagine a law of impact compatible with the principle of relativity. Besides, we should still find a production of considerable heat, and yet a very sensible resistance of the medium.

If we suppress these corpuscles and revert to the hypothesis of the Maxwell-Bartholi pressure, the difficulties will not be less. This is what Lorentz himself has attempted in his Memoir to the Amsterdam Academy of Sciences of April 25, 1900.

Consider a system of electrons immersed in an ether permeated in every sense by luminous waves; one of these electrons, struck by one of these waves, begins to vibrate; its vibration will be synchronous with that of light; but it may have a difference of phase, if the electron absorbs a part of the incident energy. In fact, if it absorbs energy, this is because the vibration of the ether *impels* the electron; the electron must therefore be slower than the ether. An electron in motion is analogous to a convection current; therefore every magnetic field, in particular that due to the luminous perturbation itself, must exert a mechanical action upon this electron. This action is very slight; moreover, it changes sign in the current of the period; nevertheless, the mean action is not null if there is a difference of phase between the vibrations of the electron and those of the ether. The mean action is proportional to this difference, consequently to the energy absorbed by the electron. I can not here enter into the detail of the calculations; suffice it to say only that the final result is an attraction of any two electrons, varying inversely as the square of the distance and proportional to the energy absorbed by the two electrons.

Therefore there can not be attraction without absorption of light and, consequently, without production of heat, and this it is which determined Lorentz to abandon this theory, which, at bottom, does not differ from that of Lesage-Maxwell-Bartholi. He would have been much more dismayed still if he had pushed

the calculation to the end.  He would have found that the temperature of the earth would have to increase $10^{13}$ degrees a second.

## IV

### *Conclusions*

I have striven to give in few words an idea as complete as
possible of these new doctrines; I have sought to explain how
they took birth; otherwise the reader would have had ground
to be frightened by their boldness.  The new theories are not
yet demonstrated; far from it; only they rest upon an aggregate
of probabilities sufficiently weighty for us not to have the right
to treat them with disregard.

New experiments will doubtless teach us what we should
finally think of them.  The knotty point of the question lies in
Kaufmann's experiment and those that may be undertaken to
verify it.

In conclusion, permit me a word of warning.  Suppose that,
after some years, these theories undergo new tests and triumph;
then our secondary education will incur a great danger: certain
professors will doubtless wish to make a place for the new
theories.

Novelties are so attractive, and it is so hard not to seem
highly advanced!  At least there will be the wish to open vistas
to the pupils and, before teaching them the ordinary mechanics,
to let them know it has had its day and was at best good enough
for that old dolt Laplace.  And then they will not form the
habit of the ordinary mechanics.

Is it well to let them know this is only approximative?  Yes;
but later, when it has penetrated to their very marrow, when
they shall have taken the bent of thinking only through it, when
there shall no longer be risk of their unlearning it, then one may,
without inconvenience, show them its limits.

It is with the ordinary mechanics that they must live; this
alone will they ever have to apply.  Whatever be the progress of
automobilism, our vehicles will never attain speeds where it is
not true.  The other is only a luxury, and we should think of
the luxury only when there is no longer any risk of harming
the necessary.

# BOOK IV

## ASTRONOMIC SCIENCE

---

### CHAPTER I

#### THE MILKY WAY AND THE THEORY OF GASES

THE considerations to be here developed have scarcely as yet drawn the attention of astronomers; there is hardly anything to cite except an ingenious idea of Lord Kelvin's, which has opened a new field of research, but still waits to be followed out. Nor have I original results to impart, and all I can do is to give an idea of the problems presented, but which no one hitherto has undertaken to solve. Every one knows how a large number of modern physicists represent the constitution of gases; gases are formed of an innumerable multitude of molecules which, at high speeds, cross and crisscross in every direction. These molecules probably act at a distance one upon another, but this action decreases very rapidly with distance, so that their trajectories remain sensibly straight; they cease to be so only when two molecules happen to pass very near to each other; in this case, their mutual attraction or repulsion makes them deviate to right or left. This is what is sometimes called an impact; but the word *impact* is not to be understood in its usual sense; it is not necessary that the two molecules come into contact, it suffices that they approach sufficiently near each other for their mutual attractions to become sensible. The laws of the deviation they undergo are the same as for a veritable impact.

It seems at first that the disorderly impacts of this innumerable dust can engender only an inextricable chaos before which analysis must recoil. But the law of great numbers, that supreme law of chance, comes to our aid; in presence of a semi-disorder, we must despair, but in extreme disorder, this statistical law

523

reestablishes a sort of mean order where the mind can recover. It is the study of this mean order which constitutes the kinetic theory of gases; it shows us that the velocities of the molecules are equally distributed among all the directions, that the rapidity of these velocities varies from one molecule to another, but that even this variation is subject to a law called Maxwell's law. This law tells us how many of the molecules move with such and such a velocity. As soon as the gas departs from this law, the mutual impacts of the molecules, in modifying the rapidity and direction of their velocities, tend to bring it promptly back. Physicists have striven, not without success, to explain in this way the experimental properties of gases; for example Mariotte's law.

Consider now the milky way; there also we see an innumerable dust; only the grains of this dust are not atoms, they are stars; these grains move also with high velocities; they act at a distance one upon another, but this action is so slight at great distance that their trajectories are straight; and yet, from time to time, two of them may approach near enough to be deviated from their path, like a comet which has passed too near Jupiter. In a word, to the eyes of a giant for whom our suns would be as for us our atoms, the milky way would seem only a bubble of gas.

Such was Lord Kelvin's leading idea. What may be drawn from this comparison? In how far is it exact? This is what we are to investigate together; but before reaching a definite conclusion, and without wishing to prejudge it, we foresee that the kinetic theory of gases will be for the astronomer a model he should not follow blindly, but from which he may advantageously draw inspiration. Up to the present, celestial mechanics has attacked only the solar system or certain systems of double stars. Before the assemblage presented by the milky way, or the agglomeration of stars, or the resolvable nebulae it recoils, because it sees therein only chaos. But the milky way is not more complicated than a gas; the statistical methods founded upon the calculus of probabilities applicable to a gas are also applicable to it. Before all, it is important to grasp the resemblance of the two cases, and their difference.

Lord Kelvin has striven to determine in this manner the dimensions of the milky way; for that we are reduced to counting the stars visible in our telescopes; but we are not sure that behind the stars we see, there are not others we do not see; so that what we should measure in this way would not be the size of the milky way, it would be the range of our instruments.

The new theory comes to offer us other resources. In fact, we know the motions of the stars nearest us, and we can form an idea of the rapidity and direction of their velocities. If the ideas above set forth are exact, these velocities should follow Maxwell's law, and their mean value will tell us, so to speak, that which corresponds to the temperature of our fictitious gas. But this temperature depends itself upon the dimensions of our gas bubble. In fact, how will a gaseous mass let loose in the void act, if its elements attract one another according to Newton's law? It will take a spherical form; moreover, because of gravitation, the density will be greater at the center, the pressure also will increase from the surface to the center because of the weight of the outer parts drawn toward the center; finally, the temperature will increase toward the center: the temperature and the pressure being connected by the law called adiabatic, as happens in the successive layers of our atmosphere. At the surface itself, the pressure will be null, and it will be the same with the absolute temperature, that is to say with the velocity of the molecules.

A question comes here: I have spoken of the adiabatic law, but this law is not the same for all gases, since it depends upon the ratio of their two specific heats; for the air and like gases, this ratio is 1.42; but is it to air that it is proper to liken the milky way? Evidently not; it should be regarded as a monoatomic gas, like mercury vapor, like argon, like helium, that is to say that the ratio of the specific heats should be taken equal to 1.66. And, in fact, one of our molecules would be for example the solar system; but the planets are very small personages, the sun alone counts, so that our molecule is indeed monoatomic. And even if we take a double star, it is probable that the action of a strange star which might approach it would become sufficiently sensible to deviate the motion of general translation of the system much before being able to trouble the relative orbits of the two components; the double star, in a word, would act like an indivisible atom.

However that may be, the pressure, and consequently the temperature, at the center of the gaseous sphere would be by so much the greater as the sphere was larger since the pressure increases by the weight of all the superposed layers. We may suppose that we are nearly at the center of the milky way, and by observing the mean proper velocity of the stars, we shall know that which corresponds to the central temperature of our gaseous sphere and we shall determine its radius.

We may get an idea of the result by the following considerations: make a simpler hypothesis: the milky way is spherical, and in it the masses are distributed in a homogeneous manner; thence results that the stars in it describe ellipses having the same center. If we suppose the velocity becomes nothing at the surface, we may calculate this velocity at the center by the equation of vis viva. Thus we find that this velocity is proportional to the radius of the sphere and to the square root of its density. If the mass of this sphere was that of the sun and its radius that of the terrestrial orbit, this velocity would be (it is easy to see) that of the earth in its orbit. But in the case we have supposed, the mass of the sun should be distributed in a sphere of radius 1,000,000 times greater, this radius being the distance of the nearest stars; the density is therefore $10^{18}$ times less; now, the velocities are of the same order, therefore it is necessary that the radius be $10^9$ times greater, be 1,000 times the distance of the nearest stars, which would give about a thousand millions of stars in the milky way.

But you will say these hypotheses differ greatly from the reality; first, the milky way is not spherical and we shall soon return to this point, and again the kinetic theory of gases is not compatible with the hypothesis of a homogeneous sphere. But in making the exact calculation according to this theory, we should find a different result, doubtless, but of the same order of magnitude; now in such a problem the data are so uncertain that the order of magnitude is the sole end to be aimed at.

And here a first remark presents itself; Lord Kelvin's result, which I have obtained again by an approximative calculation, agrees sensibly with the evaluations the observers have made with their telescopes; so that we must conclude we are very near to piercing through the milky way. But that enables us to answer another question. There are the stars we see because they shine; but may there not be dark stars circulating in the interstellar spaces whose existence might long remain unknown? But then, what Lord Kelvin's method would give us would be the total number of stars, including the dark stars; as his figure is comparable to that the telescope gives, this means there is no dark matter, or at least not so much as of shining matter.

Before going further, we must look at the problem from another angle. Is the milky way thus constituted truly the image of a gas properly so called? You know Crookes has introduced the notion of a fourth state of matter, where gases having become too rarefied are no longer true gases and become what he calls radiant matter. Considering the slight density of the milky way, is it the image of gaseous matter or of radiant matter? The consideration of what is called the *free path* will furnish us the answer.

The trajectory of a gaseous molecule may be regarded as formed of straight segments united by very small arcs corresponding to the successive impacts. The length of each of these segments is what is called the free path; of course this length is not the same for all the segments and for all the molecules; but we may take a mean; this is what is called the *mean path*. This is the greater the less the density of the gas. The matter will be radiant if the mean path is greater than the dimensions of the receptacle wherein the gas is enclosed, so that a molecule has a chance to go across the whole receptacle without undergoing an impact; if the contrary be the case, it is gaseous. From this it follows that the same fluid may be radiant in a little receptacle and gaseous in a big one; this perhaps is why, in a Crookes tube, it is necessary to make the vacuum by so much the more complete as the tube is larger.

How is it then for the milky way? This is a mass of gas of which the density is very slight, but whose dimensions are very great; has a star chances of traversing it without undergoing an impact, that is to say without passing sufficiently near another star to be sensibly deviated from its route? What do we mean by *sufficiently near?* That is perforce a little arbitrary; take it as the distance from the sun to Neptune, which would represent a deviation of a dozen degrees; suppose therefore each of our stars surrounded by a protective sphere of this radius; could a straight pass between these spheres? At the mean distance of the stars of the milky way, the radius of these spheres will be seen under an angle of about a tenth of a second; and we have a thousand millions of stars. Put upon the celestial sphere a thousand million little circles of a tenth of a second radius. Are the chances that these circles will cover a great number of times the celestial sphere? Far from it; they will cover only its sixteen thousandth part. So the milky way is not the image of gaseous matter, but of Crookes' radiant matter. Nevertheless, as our foregoing conclusions are happily not at all precise, we need not sensibly modify them.

But there is another difficulty: the milky way is not spherical, and we have reasoned hitherto as if it were, since this is the form of equilibrium a gas isolated in space would take. To make amends, agglomerations of stars exist whose form is globular and to which would better apply what we have hitherto said. Herschel has already endeavored to explain their remarkable appearances. He supposed the stars of the aggregates uniformly distributed, so that an assemblage is a homogeneous sphere; each star would then describe an ellipse and all these orbits would be passed over in the same time, so that at the end of a period the aggregate would take again its primitive configuration and this configuration would be stable. Unluckily, the aggregates do not appear to be homogeneous; we see a condensation at the center, we should observe it even were the sphere homogeneous, since it is thicker at the center; but it would not be so accentuated. We may therefore rather compare an aggregate to a gas in adiabatic equilibrium, which takes the spherical form because this is the figure of equilibrium of a gaseous mass.

But, you will say, these aggregates are much smaller than the milky way, of which they even in probability make part, and even though they be more dense, they will rather present something analogous to radiant matter; now, gases attain their adiabatic equilibrium only through innumerable impacts of the molecules. That might perhaps be adjusted. Suppose the stars of the aggregate have just enough energy for their velocity to become null when they reach the surface; then they may traverse the aggregate without impact, but arrived at the surface they will go back and will traverse it anew; after a great number of crossings, they will at last be deviated by an impact; under these conditions, we should still have a matter which might be regarded as gaseous; if perchance there had been in the aggregate stars whose velocity was greater, they have long gone away out of it, they have left it never to return. For all these reasons, it would be interesting to examine the known aggregates, to seek to account for the law of the densities, and to see if it is the adiabatic law of gases.

But to return to the milky way; it is not spherical and would rather be represented as a flattened disc. It is clear then that a mass starting without velocity from the surface will reach the center with different velocities, according as it starts from the surface in the neighborhood of the middle of the disc or just on the border of the disc; the velocity would be notably greater in the latter case. Now, up to the present, we have supposed that the proper velocities of the stars, those we observe, must be comparable to those which like masses would attain; this involves a certain difficulty. We have given above a value for the dimensions of the milky way, and we have deduced it from the observed proper velocities which are of the same order of magnitude as that of the earth in its orbit; but which is the dimension we have thus measured? Is it the thickness? Is it the radius of the disc? It is doubtless something intermediate; but what can we say then of the thickness itself, or of the radius of the disc? Data are lacking to make the calculation; I shall confine myself to giving a glimpse of the possibility of basing an evaluation at least approximate upon a deeper discussion of the proper motions.

And then we find ourselves facing two hypotheses: either the stars of the milky way are impelled by velocities for the most part parallel to the galactic plane, but otherwise distributed uniformly in all directions parallel to this plane. If this be so, observation of the proper motions should show a preponderance of components parallel to the milky way; this is to be determined,

35

because I do not know whether a systematic discussion has ever been made from this view-point. On the other hand, such an equilibrium could only be provisory, since because of impacts the molecules, I mean the stars, would in the long run acquire notable velocities in the sense perpendicular to the milky way and would end by swerving from its plane, so that the system would tend toward the spherical form, the only figure of equilibrium of an isolated gaseous mass.

Or else the whole system is impelled by a common rotation, and for that reason is flattened like the earth, like Jupiter, like all bodies that twirl. Only, as the flattening is considerable, the rotation must be rapid; rapid doubtless, but it must be understood in what sense this word is used. The density of the milky way is $10^{25}$ times less than that of the sun; a velocity of rotation $\sqrt{10^{25}}$ times less than that of the sun, for it would, therefore, be the equivalent so far as concerns flattening; a velocity $10^{12}$ times slower than that of the earth, say a thirtieth of a second of arc in a century, would be a very rapid rotation, almost too rapid for stable equilibrium to be possible.

In this hypothesis, the observable proper motions would appear to us uniformly distributed, and there would no longer be a preponderance of components parallel to the galactic plane.

They will tell us nothing about the rotation itself, since we belong to the turning system. If the spiral nebulæ are other milky ways, foreign to ours, they are not borne along in this rotation, and we might study their proper motions. It is true they are very far away; if a nebula has the dimensions of the milky way and if its apparent radius is for example 20″, its distance is 10,000 times the radius of the milky way.

But that makes no difference, since it is not about the translation of our system that we ask information from them, but about its rotation. The fixed stars, by their apparent motion, reveal to us the diurnal rotation of the earth, though their distance is immense. Unluckily, the possible rotation of the milky way, however rapid it may be relatively, is very slow viewed absolutely, and besides the pointings on nebulæ can not be very precise; therefore thousands of years of observations would be necessary to learn anything.

However that may be, in this second hypothesis, the figure of the milky way would be a figure of final equilibrium.

I shall not further discuss the relative value of these two hypotheses since there is a third which is perhaps more probable. We know that among the irresolvable nebulæ, several kinds may be distinguished: the irregular nebulæ like that of Orion, the planetary and annular nebulæ, the spiral nebulæ. The spectra of the first two families have been determined, they are discontinuous; these nebulæ are therefore not formed of stars; besides, their distribution on the heavens seems to depend upon the milky way; whether they have a tendency to go away from it, or on the contrary to approach it, they make therefore a part of the system. On the other hand, the spiral nebulæ are generally considered as independent of the milky way; it is supposed that they, like it, are formed of a multitude of stars, that they are, in a word, other milky ways very far away from ours. The recent investigations of Stratonoff tend to make us regard the milky way itself as a spiral nebula, and this is the third hypothesis of which I wish to speak.

How can we explain the very singular appearances presented by the spiral nebulæ, which are too regular and too constant to be due to chance? First of all, to take a look at one of these representations is enough to see that the mass is in rotation; we may even see what the sense of the rotation is; all the spiral radii are curved in the same sense; it is evident that the *moving wing* lags behind the pivot and that fixes the sense of the rotation. But this is not all; it is evident that these nebulæ can not be likened to a gas at rest, nor even to a gas in relative equilibrium under the sway of a uniform rotation; they are to be compared to a gas in permanent motion in which internal currents prevail.

Suppose, for example, that the rotation of the central nucleus is rapid (you know what I mean by this word), too rapid for stable equilibrium; then at the equator the centrifugal force will drive it away over the attraction, and the stars will tend to break away at the equator and will form divergent currents; but in going away, as their moment of rotation remains constant, while the radius vector augments, their angular velocity will diminish, and this is why the moving wing seems to lag back.

From this point of view, there would not be a real permanent motion, the central nucleus would constantly lose matter which would go out of it never to return, and would drain away progressively. But we may modify the hypothesis. In proportion as it goes away, the star loses its velocity and ends by stopping; at this moment attraction regains possession of it and leads it back toward the nucleus; so there will be centripetal currents. We must suppose the centripetal currents are the first rank and the centrifugal currents the second rank, if we adopt the comparison with a troop in battle executing a change of front; and, in fact, it is necessary that the composite centrifugal force be compensated by the attraction exercised by the central layers of the swarm upon the extreme layers.

Besides, at the end of a certain time a permanent régime establishes itself; the swarm being curved, the attraction exercised upon the pivot by the moving wing tends to slow up the pivot and that of the pivot upon the moving wing tends to accelerate the advance of this wing which no longer augments its lag, so that finally all the radii end by turning with a uniform velocity. We may still suppose that the rotation of the nucleus is quicker than that of the radii.

A question remains; why do these centripetal and centrifugal swarms tend to concentrate themselves in radii instead of disseminating themselves a little everywhere? Why do these rays distribute themselves regularly? If the swarms concentrate themselves, it, is because of the attraction exercised by the already existing swarms upon the stars which go out from the nucleus in their neighborhood. After an inequality is produced, it tends to accentuate itself in this way.

Why do the rays distribute themselves regularly? That is less obvious. Suppose there is no rotation, that all the stars are in two planes at right angles, in such a way that their distribution is symmetric with regard to these two planes.

By symmetry, there would be no reason for their going out of these planes, nor for the symmetry changing. This configuration would give us therefore equilibrium, but *this would be an unstable equilibrium.*

If on the contrary, there is rotation, we shall find an analogous configuration of equilibrium with four curved rays, equal to each other and intersecting at 90°, and if the rotation is sufficiently rapid, this equilibrium is stable.

I am not in position to make this more precise: enough if you see that these spiral forms may perhaps some day be explained by only the law of gravitation and statistical considerations recalling those of the theory of gases.

What has been said of internal currents shows it is of interest to discuss systematically the aggregate of proper motions; this may be done in a hundred years, when the second edition is issued of the chart of the heavens and compared with the first, that we now are making.

But, in conclusion, I wish to call your attention to a question, that of the age of the milky way or the nebulæ. If what we think we see is confirmed, we can get an idea of it. That sort of statistical equilibrium of which gases give us the model is established only in consequence of a great number of impacts. If these impacts are rare, it can come about only after a very long time; if really the milky way (or at least the agglomerations which are contained in it), if the nebulæ have attained this equilibrium, this means they are very old, and we shall have an inferior limit of their age. Likewise we should have of it a superior limit; this equilibrium is not final and can not last always. Our spiral nebulæ would be comparable to gases impelled by permanent motions; but gases in motion are viscous and their velocities end by wearing out. What here corresponds to the viscosity (and which depends upon the chances of impact of the molecules) is excessively slight, so that the present régime may persist during an extremely long time, yet not forever, so that our milky ways can not live eternally nor become infinitely old.

And this is not all. Consider our atmosphere: at the surface must reign a temperature infinitely small and the velocity of the molecules there is near zero. But this is a question only of the mean velocity; as a consequence of impacts, one of these molecules may acquire (rarely, it is true) an enormous velocity, and then it will rush out of the atmosphere, and once out, it will never return; therefore our atmosphere drains off thus with extreme slowness. The milky way also from time to time loses a star by the same mechanism, and that likewise limits its duration.

Well, it is certain that if we compute in this manner the age of the milky way, we shall get enormous figures. But here a difficulty presents itself. Certain physicists, relying upon other considerations, reckon that suns can have only an ephemeral existence, about fifty million years; our minimum would be much greater than that. Must we believe that the evolution of the milky way began when the matter was still dark? But how have the stars composing it reached all at the same time adult age, an age so briefly to endure? Or must they reach there all successively, and are those we see only a feeble minority compared with those extinguished or which shall one day light up? But how reconcile that with what we have said above on the absence of a noteworthy proportion of dark matter? Should we abandon one of the two hypotheses, and which? I confine myself to pointing out the difficulty without pretending to solve it; I shall end therefore with a big interrogation point.

However, it is interesting to set problems, even when their solution seems very far away.

# CHAPTER I1

## FRENCH GEODESY

EVERY one understands our interest in knowing the form and dimensions of our earth; but some persons will perhaps be surprised at the exactitude sought after. Is this a useless luxury? What good are the efforts so expended by the geodesist?

Should this question be put to a congressman, I suppose he would say: "I am led to believe that geodesy is one of the most useful of the sciences; because it is one of those costing us most dear." I shall try to give you an answer a little more precise.

The great works of art, those of peace as well as those of war, are not to be undertaken without long studies which save much groping, miscalculation and useless expense. These studies can only be based upon a good map. But a map will be only a valueless phantasy if constructed without basing it upon a solid framework. As well make stand a human body minus the skeleton.

Now, this framework is given us by geodesic measurements; so, without geodesy, no good map; without a good map, no great public works.

These reasons will doubtless suffice to justify much expense; but these are arguments for practical men. It is not upon these that it is proper to insist here; there are others higher and, everything considered, more important.

So we shall put the question otherwise: can geodesy aid us the better to know nature? Does it make us understand its unity and harmony? In reality an isolated fact is of slight value, and the conquests of science are precious only if they prepare for new conquests.

If therefore a little hump were discovered on the terrestrial ellipsoid, this discovery would be by itself of no great interest. On the other hand, it would become precious if, in seeking the cause of this hump, we hoped to penetrate new secrets.

Well, when, in the eighteenth century, Maupertuis and La Condamine braved such opposite climates, it was not solely to learn the shape of our planet, it was a question of the whole world-system.

If the earth was flattened, Newton triumphed and with him the doctrine of gravitation and the whole modern celestial mechanics.

And to-day, a century and a half after the victory of the Newtonians, think you geodesy has nothing more to teach us?

We know not what is within our globe. The shafts of mines and borings have let us know a layer of 1 or 2 kilometers thickness, that is to say, the millionth part of the total mass; but what is beneath?

Of all the extraordinary journeys dreamed by Jules Verne, perhaps that to the center of the earth took us to regions least explored.

But these deep-lying rocks we can not reach, exercise from afar their attraction which operates upon the pendulum and deforms the terrestrial spheroid. Geodesy can therefore weigh them from afar, so to speak, and tell us of their distribution. Thus will it make us really see those mysterious regions which Jules Verne only showed us in imagination.

This is not an empty illusion. M. Faye, comparing all the measurements, has reached a result well calculated to surprise us. Under the oceans, in the depths, are rocks of very great density; under the continents, on the contrary, are empty spaces.

New observations will modify perhaps the details of these conclusions.

In any case, our venerated dean has shown us where to search and what the geodesist may teach the geologist, desirous of knowing the interior constitution of the earth, and even the thinker wishing to speculate upon the past and the origin of this planet.

And now, why have I entitled this chapter *French Geodesy?* It is because, in each country, this science has taken, more than all others, perhaps, a national character. It is easy to see why.

There must be rivalry. The scientific rivalries are always courteous, or at least almost always; in any case, they are necessary, because they are always fruitful. Well, in those enterprises which require such long efforts and so many collaborators, the individual is effaced, in spite of himself, of course; no one has the right to say: this is my work. Therefore it is not between men, but between nations that rivalries go on.

So we are led to seek what has been the part of France. Her part I believe we are right to be proud of.

At the beginning of the eighteenth century, long discussions arose between the Newtonians who believed the earth flattened, as the theory of gravitation requires, and Cassini, who, deceived by inexact measurements, believed our globe elongated. Only direct observation could settle the question. It was our Academy of Sciences that undertook this task, gigantic for the epoch.

While Maupertuis and Clairaut measured a degree of meridian under the polar circle, Bouguer and La Condamine went toward the Andes Mountains, in regions then under Spain which to-day are the Republic of Ecuador.

Our envoys were exposed to great hardships. Traveling was not as easy as at present.

Truly, the country where Maupertuis operated was not a desert and he even enjoyed, it is said, among the Laplanders those sweet satisfactions of the heart that real arctic voyagers never know. It was almost the region where, in our days, comfortable steamers carry, each summer, hosts of tourists and young English people. But in those days Cook's agency did not exist and Maupertuis really believed he had made a polar expedition.

Perhaps he was not altogether wrong. The Russians and the Swedes carry out to-day analogous measurements at Spitzbergen, in a country where there is real ice-cap. But they have quite other resources, and the difference of time makes up for that of latitude.

The name of Maupertuis has reached us much scratched by the claws of Doctor Akakia; the scientist had the misfortune to displease Voltaire, who was then the king of mind. He was first praised beyond measure; but the flatteries of kings are as much to be dreaded as their displeasure, because the days after are terrible. Voltaire himself knew something of this.

Voltaire called Maupertuis, my amiable master in thinking, marquis of the polar circle, dear flattener out of the world and Cassini, and even, flattery supreme, Sir Isaac Maupertuis; he wrote him: "Only the king of Prussia do I put on a level with you; he only lacks being a geometer." But soon the scene changes, he no longer speaks of deifying him, as in days of yore the Argonauts, or of calling down from Olympus the council of the gods to contemplate his works, but of chaining him up in a madhouse.  He speaks no longer of his sublime mind, but of his despotic pride, plated with very little science and much absurdity.

I care not to relate these comico-heroic combats; but permit me some reflections on two of Voltaire's verses.  In his 'Discourse on Moderation' (no question of moderation in praise and criticism), the poet has written:

> You have confirmed in regions drear
> What Newton discerned without going abroad.

These two verses (which replace the hyperbolic praises of the first period) are very unjust, and doubtless Voltaire was too enlightened not to know it.

Then, only those discoveries were esteemed which could be made without leaving one's house.

To-day, it would rather be theory that one would make light of.

This is to misunderstand the aim of science.

Is nature governed by caprice, or does harmony rule there? That is the question.  It is when it discloses to us this harmony that science is beautiful and so worthy to be cultivated.  But whence can come to us this revelation, if not from the accord of a theory with experiment?  To seek whether this accord exists or if it fails, this therefore is our aim.  Consequently these two terms, which we must compare, are as indispensable the one as the other.  To neglect one for the other would be nonsense.  Isolated, theory would be empty, experiment would be blind; each would be useless and without interest.

Maupertuis therefore deserves his share of glory.  Truly, it will not equal that of Newton, who had received the spark divine; nor even that of his collaborator Clairaut.  Yet it is not to be despised, because his work was necessary, and if France, outstripped by England in the seventeenth century, has so well taken her revenge in the century following, it is not alone to the genius of Clairauts, d'Alemberts, Laplaces that she owes it; it is also to the long patience of the Maupertuis and the La Condamines.

We reach what may be called the second heroic period of geodesy. France is torn within. All Europe is armed against her; it would seem that these gigantic combats might absorb all her forces. Far from it; she still has them for the service of science. The men of that time recoiled before no enterprise, they were men of faith.

Delambre and Méchain were commissioned to measure an arc going from Dunkerque to Barcelona. This time there was no going to Lapland or to Peru; the hostile squadrons had closed to us the ways thither. But, though the expeditions are less distant, the epoch is so troubled that the obstacles, the perils even, are just as great.

In France, Delambre had to fight against the ill-will of suspicious municipalities. One knows that the steeples, which are visible from so far, and can be aimed at with precision, often serve as signal points to geodesists. But in the region Delambre traversed there were no longer any steeples. A certain proconsul had passed there, and boasted of knocking down all the steeples rising proudly above the humble abode of the sansculottes. Pyramids then were built of planks and covered with white cloth to make them more visible. That was quite another thing: with white cloth! What was this rash person who, upon our heights so recently set free, dared to raise the hateful standard of the counter-revolution? It was necessary to border the white cloth with blue and red bands.

Méchain operated in Spain; the difficulties were other; but they were not less. The Spanish peasants were hostile. There steeples were not lacking: but to install oneself in them with mysterious and perhaps diabolic instruments, was it not sacrilege? The revolutionists were allies of Spain, but allies smelling a little of the stake.

"Without cease," writes Méchain, "they threaten to butcher us." Fortunately, thanks to the exhortations of the priests, to the pastoral letters of the bishops, these ferocious Spaniards contented themselves with threatening.

Some years after Méchain made a second expedition into Spain: he proposed to prolong the meridian from Barcelona to the Balearics. This was the first time it had been attempted to make the triangulations overpass a large arm of the sea by observing signals installed upon some high mountain of a far-away isle. The enterprise was well conceived and well prepared; it failed however.

The French scientist encountered all sorts of difficulties of which he complains bitterly in his correspondence. "Hell," he writes, perhaps with some exaggeration—"hell and all the scourges it vomits upon the earth, tempests, war, the plague and black intrigues are therefore unchained against me!"

The fact is that he encountered among his collaborators more of proud obstinacy than of good will and that a thousand accidents retarded his work. The plague was nothing, the fear of the plague was much more redoubtable; all these isles were on their guard against the neighboring isles and feared lest they should receive the scourge from them. Méchain obtained permission to disembark only after long weeks upon the condition of covering all his papers with vinegar; this was the antisepsis of that time.

Disgusted and sick, he had just asked to be recalled, when he died.

Arago and Biot it was who had the honor of taking up the unfinished work and carrying it on to completion.

Thanks to the support of the Spanish government, to the protection of several bishops and, above all, to that of a famous brigand chief, the operations went rapidly forward. They were successfully completed, and Biot had returned to France when the storm burst.

It was the moment when all Spain took up arms to defend her independence against France. Why did this stranger climb the mountains to make signals? It was evidently to call the French army. Arago was able to escape the populace only by becoming a prisoner. In his prison, his only distraction was reading in the Spanish papers the account of his own execution. The papers of that time sometimes gave out news prematurely. He had at least the consolation of learning that he died with courage and like a Christian.

Even the prison was no longer safe; he had to escape and reach Algiers. There, he embarked for Marseilles on an Algerian vessel. This ship was captured by a Spanish corsair, and behold Arago carried back to Spain and dragged from dungeon to dungeon, in the midst of vermin and in the most shocking wretchedness.

If it had only been a question of his subjects and his guests, the dey would have said nothing. But there were on board two lions, a present from the African sovereign to Napoleon. The dey threatened war.

The vessel and the prisoners were released. The port should have been properly reached, since they had on board an astronomer; but the astronomer was seasick, and the Algerian seamen, who wished to make Marseilles, came out at Bougie. Thence Arago went to Algiers, traversing Kabylia on foot in the midst of a thousand perils. He was long detained in Africa and threatened with the convict prison. Finally he was able to get back to France; his observations, which he had preserved and safeguarded under his shirt, and, what is still more remarkable, his instruments had traversed unhurt these terrible adventures.

Up to this point, not only did France hold the foremost place, but she occupied the stage almost alone.

In the years which follow, she has not been inactive and our staff-office map is a model. However, the new methods of observation and calculation have come to us above all from Germany and England. It is only in the last forty years that France has regained her rank. She owes it to a scientific officer, General Perrier, who has successfully executed an enterprise truly audacious, the junction of Spain and Africa. Stations were installed on four peaks upon the two sides of the Mediterranean. For long months they awaited a calm and limpid atmosphere. At last was seen the little thread of light which had traversed 300 kilometers over the sea. The undertaking had succeeded.

To-day have been conceived projects still more bold. From a mountain near Nice will be sent signals to Corsica, not now for geodesic determinations, but to measure the velocity of light. The distance is only 200 kilometers; but the ray of light is to make the journey there and return, after reflection by a mirror installed in Corsica. And it should not wander on the way, for it must return exactly to the point of departure.

Ever since, the activity of French geodesy has never slackened. We have no more such astonishing adventures to tell; but the scientific work accomplished is immense. The territory of France beyond the sea, like that of the mother country, is covered by triangles measured with precision.

We have become more and more exacting and what our fathers admired does not satisfy us to-day. But in proportion as we seek more exactitude, the difficulties greatly increase; we are surrounded by snares and must be on our guard against a thousand unsuspected causes of error. It is needful, therefore, to create instruments more and more faultless.

Here again France has not let herself be distanced. Our appliances for the measurement of bases and angles leave nothing to desire, and I may also mention the pendulum of Colonel Defforges, which enables us to determine gravity with a precision hitherto unknown.

The future of French geodesy is at present in the hands of the Geographic Service of the army, successively directed by General Bassot and General Berthaut. We can not sufficiently congratulate ourselves upon it. For success in geodesy, scientific aptitudes are not enough; it is necessary to be capable of standing long fatigues in all sorts of climates; the chief must be able to win obedience from his collaborators and to make obedient his native auxiliaries. These are military qualities. Besides, one knows that, in our army, science has always marched shoulder to shoulder with courage.

I add that a military organization assures the indispensable unity of action. It would be more difficult to reconcile the rival pretensions of scientists jealous of their independence, solicitous of what they call their fame, and who yet must work in concert, though separated by great distances. Among the geodesists of former times there were often discussions, of which some aroused long echoes. The Academy long resounded with the quarrel of Bouguer and La Condamine. I do not mean to say that soldiers are exempt from passion, but discipline imposes silence upon a too sensitive self-esteem.

Several foreign governments have called upon our officers to organize their geodesic service: this is proof that the scientific influence of France abroad has not declined.

Our hydrographic engineers contribute also to the common achievement a glorious contingent. The survey of our coasts, of our colonies, the study of the tides, offer them a vast domain of research. Finally I may mention the general leveling of France which is carried out by the ingenious and precise methods of M. Lallamand.

With such men we are sure of the future. Moreover, work for them will not be lacking; our colonial empire opens for them immense expanses illy explored. That is not all: the International Geodetic Association has recognized the necessity of a new measurement of the arc of Quito, determined in days of yore by La Condamine. It is France that has been charged with this operation; she had every right to it, since our ancestors had made, so to speak, the scientific conquest of the Cordilleras. Besides, these rights have not been contested and our government has undertaken to exercise them.

Captains Maurain and Lacombe completed a first reconnaissance, and the rapidity with which they accomplished their mission, crossing the roughest regions and climbing the most precipitous summits, is worthy of all praise. It won the admiration of General Alfaro, President of the Republic of Ecuador, who called them 'los hombres de hierro,' the men of iron.

The final commission then set out under the command of Lieutenant-Colonel (then Major) Bourgeois. The results obtained have justified the hopes entertained. But our officers have encountered unforeseen difficulties due to the climate. More than once, one of them has been forced to remain several months at an altitude of 4,000 meters, in the clouds and the snow, without seeing anything of the signals he had to aim at and which refused to unmask themselves. But thanks to their perseverance and courage, there resulted from this only a delay and an increase of expense, without the exactitude of the measurements suffering therefrom.

# GENERAL CONCLUSIONS

WHAT I have sought to explain in the preceding pages is how the scientist should guide himself in choosing among the innumerable facts offered to his curiosity, since indeed the natural limitations of his mind compel him to make a choice, even though a choice be always a sacrifice. I have expounded it first by general considerations, recalling on the one hand the nature of the problem to be solved and on the other hand seeking to better comprehend that of the human mind, which is the principal instrument of the solution. I then have explained it by examples; I have not multiplied them indefinitely; I also have had to make a choice, and I have chosen naturally the questions I had studied most. Others would doubtless have made a different choice; but what difference, because I believe they would have reached the same conclusions.

There is a hierarchy of facts; some have no reach; they teach us nothing but themselves. The scientist who has ascertained them has learned nothing but a fact, and has not become more capable of foreseeing new facts. Such facts, it seems, come once, but are not destined to reappear.

There are, on the other hand, facts of great yield; each of them teaches us a new law. And since a choice must be made, it is to these that the scientist should devote himself.

Doubtless this classification is relative and depends upon the weakness of our mind. The facts of slight outcome are the complex facts, upon which various circumstances may exercise a sensible influence, circumstances too numerous and too diverse for us to discern them all. But I should rather say that these are the facts we think complex, since the intricacy of these circumstances surpasses the range of our mind. Doubtless a mind vaster and finer than ours would think differently of them. But what matter; we can not use that superior mind, but only our own.

The facts of great outcome are those we think simple; may be they really are so, because they are influenced only by a small

544

number of well-defined circumstances, may be they take on an appearance of simplicity because the various circumstances upon which they depend obey the laws of chance and so come to mutually compensate. And this is what happens most often. And so we have been obliged to examine somewhat more closely what chance is.

Facts where the laws of chance apply become easy of access to the scientist who would be discouraged before the extraordinary complication of the problems where these laws are not applicable. We have seen that these considerations apply not only to the physical sciences, but to the mathematical sciences. The method of demonstration is not the same for the physicist and the mathematician. But the methods of invention are very much alike. In both cases they consist in passing up from the fact to the law, and in finding the facts capable of leading to a law.

To bring out this point, I have shown the mind of the mathematician at work, and under three forms: the mind of the mathematical inventor and creator; that of the unconscious geometer who among our far distant ancestors, or in the misty years of our infancy, has constructed for us our instinctive notion of space; that of the adolescent to whom the teachers of secondary education unveil the first principles of the science, seeking to give understanding of the fundamental definitions. Everywhere we have seen the rôle of intuition and of the spirit of generalization without which these three stages of mathematicians, if I may so express myself, would be reduced to an equal impotence.

And in the demonstration itself, the logic is not all; the true mathematical reasoning is a veritable induction, different in many regards from the induction of physics, but proceeding like it from the particular to the general. All the efforts that have been made to reverse this order and to carry back mathematical induction to the rules of logic have eventuated only in failures, illy concealed by the employment of a language inaccessible to the uninitiated. The examples I have taken from the physical sciences have shown us very different cases of facts of great outcome. An experiment of Kaufmann on radium rays revolutionizes at the same time mechanics, optics and astronomy. Why? Because in proportion as these sciences have developed,

36

we have the better recognized the bonds uniting them, and then we have perceived a species of general design of the chart of universal science.   There are facts common to several sciences, which seem the common source of streams diverging in all directions and which are comparable to that knoll of Saint Gothard whence spring waters which fertilize four different valleys.

And then we can make choice of facts with more discernment than our predecessors who regarded these valleys as distinct and separated by impassable barriers.

It is always simple facts which must be chosen, but among these simple facts we must prefer those which are situated upon these sorts of knolls of Saint Gothard of which I have just spoken.

And when sciences have no direct bond, they still mutually throw light upon one another by analogy.   When we studied the laws obeyed by gases we knew we had attacked a fact of great outcome; and yet this outcome was still estimated beneath its value, since gases are, from a certain point of view, the image of the milky way, and those facts which seemed of interest only for the physicist, ere long opened new vistas to astronomy quite unexpected.

And finally when the geodesist sees it is necessary to move his telescope some seconds to see a signal he has set up with great pains, this is a very small fact; but this is a fact of great outcome, not only because this reveals to him the existence of a small protuberance upon the terrestrial globe, that little hump would be by itself of no great interest, but because this protuberance gives him information about the distribution of matter in the interior of the globe and through that about the past of our planet, about its future, about the laws of its development.

# INDEX